Electromagnetic Wave Absorbing Materials

Wiley Series in Materials for Electronic & Optoelectronic Applications

www.wiley.com/go/meoa

Series Editors

Professor Richard Curry, University of Manchester, Manchester, UK
Professor Harry Ruda, University of Toronto, Toronto, Canada
Professor Jun Luo, Chinese Academy of Sciences, Beijing, China

Honorary Series Editors

Professor Arthur Willoughby, University of Southampton, Southampton, UK
Dr Peter Capper, Ex-Leonardo MW Ltd, Southampton, UK
Professor Safa Kasap, University of Saskatchewan, Saskatoon, Canada

Published Titles

Bulk Crystal Growth of Electronic, Optical and Optoelectronic Materials, Edited by P. Capper
Properties of Group-IV, III—V and II—VI Semiconductors, S. Adachi
Charge Transport in Disordered Solids with Applications in Electronics, Edited by S. Baranovski
Optical Properties of Condensed Matter and Applications, Edited by J. Singh
Thin Film Solar Cells: Fabrication, Characterization, and Applications, Edited by J. Poortmans and V. Arkhipov
Dielectric Films for Advanced Microelectronics, Edited by M. R. Baklanov, M. Green, and K. Maex
Liquid Phase Epitaxy of Electronic, Optical and Optoelectronic Materials, Edited by P. Capper and M. Mauk
Molecular Electronics: From Principles to Practice, M. Petty
Luminescent Materials and Applications, A. Kitai
CVD Diamond for Electronic Devices and Sensors, Edited by R. S. Sussmann
Properties of Semiconductor Alloys: Group-IV, III—V and II—VI Semiconductors, S. Adachi
Mercury Cadmium Telluride, Edited by P. Capper and J. Garland
Zinc Oxide Materials for Electronic and Optoelectronic Device Applications, Edited by C. Litton, D. C. Reynolds, and T. C. Collins
Lead-Free Solders: Materials Reliability for Electronics, Edited by K. N. Subramunian
Silicon Photonics: Fundamentals and Devices, M. Jamal Deen and P. K. Basu
Nanostructured and Subwavelength Waveguides: Fundamentals and Applications, M. Skorobogatiy
Photovoltaic Materials: From Crystalline Silicon to Third-Generation Approaches, Edited by G. Conibeer and A. Willoughby
Glancing Angle Deposition of Thin Films: Engineering the Nanoscale, M. M. Hawkeye, M. T. Taschuk and M. J. Brett
Physical Properties of High-Temperature Superconductors, R. Wesche
Spintronics for Next Generation Innovative Devices, Edited by K. Sato and E. Saitoh
Inorganic Glasses for Photonics: Fundamentals, Engineering and Applications, A. Jha
Amorphous Semiconductors: Structural, Optical and Electronic Properties, K. Morigaki, S. Kugler and K. Shimakawa
Microwave Materials and Applications, Two-volume set, Edited by M. T. Sebastian, R. Ubic and H. Jantunen
Molecular Beam Epitaxy: Materials and Applications for Electronics and Optoelectronics, Edited by H. Asahi and Y. Korikoshi
Metalorganic Vapor Phase Epitaxy (MOVPE): Growth, Materials Properties, and Applications, Edited by S. Irvine and P. Capper
Optical Properties of Materials and Their Applications, Second Edition, Edited by J. Singh
Oxide Electronics, Edited by A. Ray
Organic Semiconductors for Optoelectronics, Edited by H. Naito
Phosphors for Radiation Detection, Edited by T. Yanagida and M. Koshimizu
Spintronics: Materials, Devices and Applications, Edited by Kaiyou Wang, Meiyin Yang and Jun Luo
Photoconductivity and Photoconductive Materials: Fundamentals, Techniques and Applications, Edited by Safa O. Kasap
Electromagnetic Wave Absorbing Materials: Fundamentals and Applications, First Edition. Edited by Hongjing Wu, Jun Luo, and Meiyin Yang.

Electromagnetic Wave Absorbing Materials

Fundamentals and Applications

Edited by

Hongjing Wu
Northwestern Polytechnical University
Shaanxi, China

Jun Luo and Meiyin Yang
Chinese Academy of Sciences
Beijing, China

WILEY

This edition first published 2024.

Registered Offices
John Wiley & Sons, Inc., 111 River Street, Hoboken, NJ 07030, USA
John Wiley & Sons Ltd, The Atrium, Southern Gate, Chichester, West Sussex, PO19 8SQ, UK
For details of our global editorial offices, customer services, and more information about Wiley products visit us at www.wiley.com.

Wiley also publishes its books in a variety of electronic formats and by print-on-demand. Some content that appears in standard print versions of this book may not be available in other formats.

Library of Congress Cataloging-in-Publication Data:

Names: Wu, Hongjing (Professor), author. | Luo, Jun (Professor of microelectronics), author. | Yang, Meiyin, author.
Title: Electromagnetic wave absorbing materials : fundamentals and applications / Hongjing Wu, Jun Luo and Meiyin Yang.
Description: Hoboken, NJ : Wiley, 2024. | Includes index.
Identifiers: LCCN 2024011190 (print) | LCCN 2024011191 (ebook) | ISBN 9781119699347 (hardback) | ISBN 9781119699323 (adobe pdf) | ISBN 9781119699224 (epub)
Subjects: LCSH: Absorption. | Electromagnetic waves--Transmission.
Classification: LCC QC182 .W8 2024 (print) | LCC QC182 (ebook) | DDC 621.38028/4--dc23/eng/20240409
LC record available at https://lccn.loc.gov/2024011190
LC ebook record available at https://lccn.loc.gov/2024011191

Cover Design: Wiley
Cover Images: © bubaone/Getty Images, © oxygen/Getty Images

Set in 10/12pt Times LT Std by Straive, Chennai, India

Contents

List of Contributors

Qing Chang MOE Key Laboratory of Material Physics and Chemistry Under Extraordinary Conditions, School of Physical Science and Technology, Northwestern Polytechnical University, Xi'an, China

Geng Chen MOE Key Laboratory of Material Physics and Chemistry Under Extraordinary Conditions, School of Physical Science and Technology, Northwestern Polytechnical University, Xi'an, China

Weibin Deng Shaanxi Engineering Laboratory for Graphene New Carbon Materials and Applications, School of Materials Science and Engineering, Northwestern Polytechnical University, Xi'an, China

Shengchong Hui MOE Key Laboratory of Material Physics and Chemistry Under Extraordinary Conditions, School of Physical Science and Technology, Northwestern Polytechnical University, Xi'an, China

Zijing Li MOE Key Laboratory of Material Physics and Chemistry Under Extraordinary Conditions, School of Physical Science and Technology, Northwestern Polytechnical University, Xi'an, China

Hongsheng Liang MOE Key Laboratory of Material Physics and Chemistry Under Extraordinary Conditions, School of Physical Science and Technology, Northwestern Polytechnical University, Xi'an, China

Bin Shi MOE Key Laboratory of Material Physics and Chemistry Under Extraordinary Conditions, School of Physical Science and Technology, Northwestern Polytechnical University, Xi'an, China

Yuntong Wang MOE Key Laboratory of Material Physics and Chemistry Under Extraordinary Conditions, School of Physical Science and Technology, Northwestern Polytechnical University, Xi'an, China

Jiaming Wen MOE Key Laboratory of Material Physics and Chemistry Under Extraordinary Conditions, School of Physical Science and Technology, Northwestern Polytechnical University, Xi'an, China

Tao Zhang MOE Key Laboratory of Material Physics and Chemistry Under Extraordinary Conditions, School of Physical Science and Technology, Northwestern Polytechnical University, Xi'an, China

Overview of the Work

Description

Electromagnetic Wave-Absorbing Materials: Fundamentals and Applications introduces the fundamentals and basic concepts of the mainstream electromagnetic wave absorption materials, including MOF-derived materials, MXene-derived materials, and high-entropy materials. It covers the design and preparation of these wave-absorbing materials, including their characterization, properties, manipulation, and related research. The book provides a comprehensive description of the various microscopic electromagnetic wave loss mechanisms (dielectric loss, magnetic loss) that are currently available, their causes of formation, and effects on the electromagnetic response properties of the materials. It also describes and summarizes the design, preparation, and application of the mainstream electromagnetic wave absorption devices, such as foam, gel, and film-based materials.

The book details the latest methods, technologies, and applications currently available in the field of electromagnetic wave absorption. It will support the studies of academic researchers and graduate students, as well as the many industrial research and development efforts of engineers and materials scientists. In addition, it will be of great value to those directly involved in industrial sectors related to electromagnetic wave absorption materials, as well as researchers in materials science, materials chemistry, nanomaterials, and other electromagnetic materials fields.

Key Features

Electromagnetic Wave-Absorbing Materials: Fundamentals and Applications mainly describes how to design and prepare high-performance electromagnetic wave absorption materials.

The book includes the synthesis of new wave absorption materials and the new technology of regulating the electromagnetic wave-absorbing properties of materials. The latest progress in the field of electromagnetic wave-absorbing materials is reviewed, which provides valuable insights for the future development direction of electromagnetic wave-absorbing materials.

Acknowledgments

We gratefully acknowledge the financial support provided by the National Science Foundation of China (grants nos. 51872238, 52074227, and 21806129), the Fundamental Research Funds for the Central Universities (grant nos. 3102018zy045 and 3102019AX11), the Natural Science Basic Research Plan in Shaanxi Province of China (grant nos. 2017JQ5116 and 2020JM-118), and the Key Laboratory of Icing and Anti/De-icing of CARDC (IADL20220401).

1

Metal–Organic Framework-Based Electromagnetic Wave Absorption Materials

Zijing Li

MOE Key Laboratory of Material Physics and Chemistry Under Extraordinary Conditions, School of Physical Science and Technology, Northwestern Polytechnical University, Xi'an, China

CHAPTER MENU

1.1 Brief Introduction to Metal–Organic Frameworks

Metal–organic frameworks (MOFs), also known as porous coordination polymers (PCPs), first proposed by Yaghi and Li et al. in the 1990s, are a class of crystalline porous materials consisting of metal ions/clusters and organic ligands assembled in an orderly manner [1, 2]. MOFs are exquisite porous materials composed of metal ions or clusters and organic ligands. These components are meticulously arranged in a crystalline structure, held together by coordination bonds formed between the metal centers and the organic ligands. Additionally, depending on the specific composition and structure, other forces such as hydrogen bonding, van der Waals forces, and π-π stacking also play a significant role. The composition of MOFs is diverse, covering almost all kinds of metal ions, such as main group, transition group, lanthanides, rare earth metals, and so on, while their organic ligands are mostly polyamines, carboxyl groups, pyridines, porphyrins, cyano groups, crown ethers, and phosphoric acid [3–7]. In recent years, MOFs have gained significant attention due to their remarkable characteristics, including the ability to design and control their components, achieve a regular and highly adjustable pore structure, exhibit high crystallinity, possess a

Electromagnetic Wave Absorbing Materials: Fundamentals and Applications, First Edition.
Edited by Hongjing Wu, Jun Luo, and Meiyin Yang.

large specific surface area, exhibit unique liquid and glassy properties, and showcase diverse topologies. Moreover, their environmentally friendly preparation aligns perfectly with sustainable development strategies [8–11]. Therefore, MOFs and their derived materials have received extensive attention in the fields of catalysis [12, 13], clean energy [14, 15], gas storage and separation [16, 17], sensors [18, 19], and medicine [20, 21]. Up to now, more than 20,000 MOF materials have been synthesized through the selection of various metal centers and organic ligands. Common MOF materials [5, 8, 22] include isotropic metal–organic frameworks (IRMOFs), coordination columnar layers (CCL), zeolitic imidazolate frameworks (ZIFs), porous coordination networks (PhCNs) [23–25], and so on. With the development of science and technology and the rising demand of the times, the types and quantities of MOFs and their derived materials will further increase, and their application scenarios and fields will further expand.

Due to their exceptional flexibility and diverse structural components, MOFs exhibit significant research value and hold great potential for development in the field of electromagnetic wave (EMW) absorption materials. The unique combination of metal and organic components, along with the highly adjustable microstructure, makes MOFs an ideal candidate for EMW absorption materials. Through careful control of the pyrolysis process, organic linkers can be transformed into a porous carbon network while preserving the original porous skeleton. Simultaneously, metal ions/clusters can be reduced to metal composites such as metals, metal oxides, and metal carbides via the carbothermal reduction process. These metal composites are homogeneously embedded within the network structure of the porous carbon, resulting in carbon-based composites with abundant interfaces and defects, a continuous conductive network, and exceptional magnetic response properties [26–29]. By precisely controlling the components, structural form, and composition, MOFs can be combined with other materials to optimize conductive loss, enhance dipole polarization, construct interface polarization, and improve magnetic response. This strategic approach ultimately leads to the development of MOF-based composite wave-absorbing materials that leverage a diverse range of microphysical loss mechanisms. Notably, when compared to other absorbing materials, MOFs and their derivatives offer exceptional versatility in modulating magnetic loss, dielectric loss, and impedance match properties [30–34]. By incorporating diverse component structures and facilitating the in situ generation of metal/metal oxide nanoparticles (NPs) or clusters, MOF-based materials demonstrate an extraordinary ability to attenuate incident EMWs on a substantial scale. This achievement underscores the immense potential of MOFs in the field of EMW absorption [35–37]. In addition, the highly porous structure inherited from the pristine MOFs facilitates the multiple scattering of incident microwaves and the lengthening of the transmission path, thus promoting their absorption of EMWs. Therefore, MOF-based EMW absorption materials are prepared and studied in large quantities at present and are also moving toward more novel and microscopic directions, such as multidimensional design, quantum dot modification, supramolecular cross-linking, and so on.

1.2 Preparation Method of MOF Materials

MOFs and MOF-derived materials have been synthesized by a variety of methods. Theoretically, the conditions for the synthesis of MOFs should be chosen so that metal–ligand bonds can not only be formed but also be broken and reorganized to prompt the structure propagation.

Dynamic bonds can induce changes in the morphology of the materials by breakage and reorganization of the structure under specific conditions. Therefore, dynamic bonds play an important role in the formation of crystals and ordered structures, where any faulty bonding that may lead to disorder or premature structural termination can be corrected [38–43].

1.2.1 Solvothermal Method

Solvothermal synthesis is a widely employed and straightforward method for the preparation of MOFs. Typically, this method involves the combination of metal salts with multipoint organic ligands in high-boiling solvents such as dimethylformamide (DMF), dimethyl sulfoxide (DMSO), and dimethylformamide (DEF). The resulting mixture is heated in a PTFE-lined stainless steel reactor or immersed in a nonflammable silicone-based oil bath on a hot plate within an oven or fume hood. Generally, the reaction is allowed to proceed for 12–48 hours. When selecting a stainless steel reactor, it is important to consider factors such as the reaction size, solvent volume, and target temperature to ensure sufficient space for potential pressure buildup. The hydrothermal reaction method offers the ability to modulate the properties of synthesized MOF materials by systematically varying parameters including reaction temperature, time, solvent, reagent concentration, and pH. These parameters not only impact the material's topology but also influence crystal size and phase purity. It is worth noting that the synthesis of MOFs is a dynamic process and is highly sensitive to even small changes in the reaction mixture. This sensitivity allows for an exceptionally high degree of tunability in MOFs prepared via the hydrothermal method. For instance, the use of a mixture of metal chloride (MCl) and a multipoint carboxylic acid linker can result in the formation of a lesser stoichiometric amount of HCl. This, in turn, can dissolve the forming MOFs and slow down crystal growth to a certain extent. Conversely, if metal acetylacetone (M(acac)) is employed as the metal salt, the resulting acetylacetone by-product does not exert a constraining effect on crystal formation. The hydrothermal method has been extensively utilized to synthesize numerous MOFs and MOF-derived materials, which have found applications in diverse fields. Wang and coworkers [38] synthesized carbon quantum dots (CQD)/Ni-MOF composites by a one-pot hydrothermal method using $NiCl_3$-$6H_2O$ as the metal salt, p-phthalic acid (PTA) as the ligand, and DMF, ethanol, CQD solution, and deionized water as the solvents. By controlling the amount of CQDs, the conductivity and micro-morphology of Ni-MOF can be effectively regulated, and the number of active sites can be controlled, which in turn improves its electrochemical properties. Yuan and coworkers [44] used 2-amino terephthalic acid, tetrabutyl titanate ($C_{16}H_{36}O_4Ti$), 1,4-benzene dicarboxylic acid (BDC), DMF, methanol, sodium chloride ($C_{16}H_{36}O_4Ti$), and methanol (CH_3OH) as raw materials. Two different MOF materials, MIL-125(Ti) and amino-functionalized NH_2-MIL-125(Ti), were successfully synthesized by a solvent-thermal method, and they relied on their valence-electron-transferring properties and exhibited effective photocatalytic activity in visible light for the reduction of Cr(VI) in aqueous solution. Lang and coworkers [45] employed a large-scale bottom-up solvent-based solvatochromic approach to prepare Ni, Fe, Al, Co, Mn, Zn, Cd sulfates, *N,N*-dimethylacetamide (DMAC), *N,N*-dimethylformamide (DMF), formamide, N-2-methylpyrrolidinone (NMP), *N,N*-dimethylformamide (DEF), and Ni-NiMOFs using a large-scale bottom-up solvent-based solvatochromic approach. Ni-M-M-MOF (M = Fe, Al, Co, Mn, Zn, and Cd) nanosheets (NSs) prepared by a large-scale bottom-up solvent-thermal method, with a thickness of only a few atomic layers, are capable of directly acting as highly efficient electrocatalysts for oxygen precipitation reactions (as shown in Figure 1.1).

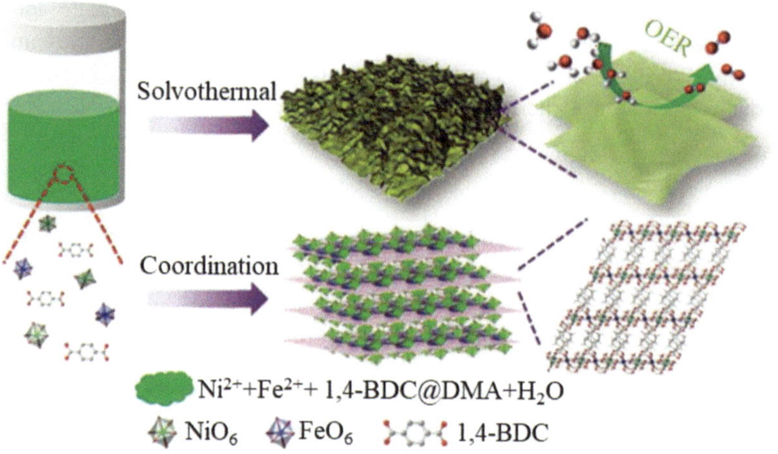

Figure 1.1 Synthetic procedure for the production of ultrathin metal–organic framework nanosheets and their utilization for the oxygen evolution reaction. Source: Li et al. [45]/with permission of John Wiley & Sons.

1.2.2 Microwave-Assisted Synthesis Method

The microwave-assisted method can effectively transfer the energy generated by EMWs to the reactants for more efficient and faster heating. Microwave irradiation, as a more efficient method of energy input than conventional heating methods, is expected to significantly reduce the reaction time without affecting the product yield or the quality of the finished product, and this means is suitable for industrial production. This effect has already been demonstrated by several researchers in laboratory-scale MOF synthesis. Unlike conventional heating methods, during microwave-assisted synthesis, crystallization of the synthesis takes place on hot spots formed by direct heating of the solvent, enabling the formation of small-sized particles in a short period. The theory of efficient thermal conversion by microwaves is favorable for the high-yield synthesis of MOF, but the limited penetration depth of microwaves into the absorbing medium limits the size of the reactor, which then hinders the large-scale industrial production of MOF.

Piao and coworkers [46] synthesized FeMo-MIL-88B materials by an ultrafast microwave-assisted method using $FeCl_3$-$6H_2O$, urea, DMF, terephthalic acid, and $Na_2MoO_4{\cdot}2H_2O$ as raw materials, which facilitated the hydrogen-extraction reaction (HER), oxygen-extraction reaction (OER), and overall hydrolysis reaction. Huo and coworkers [47] synthesized nickel-iron-based trimetallic MOF NSs through a simple microwave-assisted method using $NiCl_2{\cdot}6H_2O$, $CoCl_2{\cdot}6H_2O$, $FeCl_3{\cdot}6H_2O$, terephthalic acid, and DMF as raw materials, which enabled simultaneous and rapid synthesis and activation of MOFs used for oxygen precipitation reactions. Cai and coworkers [48] prepared a Zn-doped nickel-based metal–organic framework (Ni-MOF) material with a honeycomb-layered spherical structure through the microwave-assisted method using $NiCl_2{\cdot}6H_2O$, PTA, $Zn(Ac)_2{\cdot}2H_2O$, and DMF as the raw materials (as shown in Figure 1.2), and excellent electrochemical properties were obtained by regulating the doping amount of Zn.

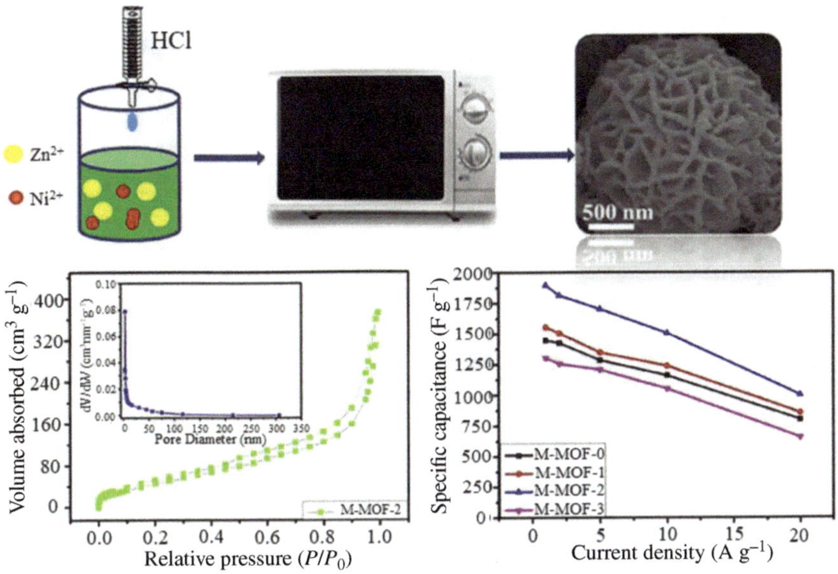

Figure 1.2 Schematic illustration of the overall formation process of the Zn-doped nickel-based metal–organic framework (Ni-MOF). Source: Chen et al. [48]/with permission of Elsevier.

1.2.3 Electrochemical Synthesis Method

Cathodic deposition and anodic dissolution are the two main methods for the electrochemical synthesis of MOFs. In the cathodic deposition method, a solution consisting of metal ions, organic linkers, and metal salts is in contact with the cathode surface. In the anodic dissolution method, a metal electrode rather than a metal salt is used as the metal cation source to avoid corrosion by anions such as nitrates and acetates. This method provides the metal ions continuously required for MOF formation, while organic linkers and electrolytes can be added as needed. Due to the increase in pH near the cathode surface, an electrochemical reduction reaction is triggered, which in turn leads to the deposition of the MOF film. In electrochemical synthesis, the yield and structure of the material are influenced by the electrolyte, solvent, voltage, density, and temperature. Compared to conventional methods, this method has the advantages of a fast reaction rate and low temperature, and the metal source is not salt. In addition, it is a discontinuous process that enables higher synthesis efficiency.

Easun and coworkers [49] prepared Mn-MOF by electrochemical synthesis using 2,5-diamino terephthalic acid and DMF as raw materials and obtained MOF materials with high CO_2 uptake by regulating the crystal synthesis and yield through the parameters of electrochemical synthesis, such as current density, electrolyte dosage, and reaction time (as shown in Figure 1.3a). Limin and coworkers [50] prepared ZIF-8 thin films by electrochemical deposition, which possessed a three-dimensional ordered macroporous metal–organic skeleton with a thickness of 4 cm and excellent optical, catalytic, and biosensing properties (as shown in Figure 1.3b). Liu et al. [51] proposed an electrochemical synthesis method for the preparation of large-area $Cu_3(HHTP)_2$ MOF films on single-crystal Cu anodes by charge-induced molecular assembly to achieve a surface reaction, and the synthesized MOF films possessed high electrical conductivity (as shown in Figure 1.3c).

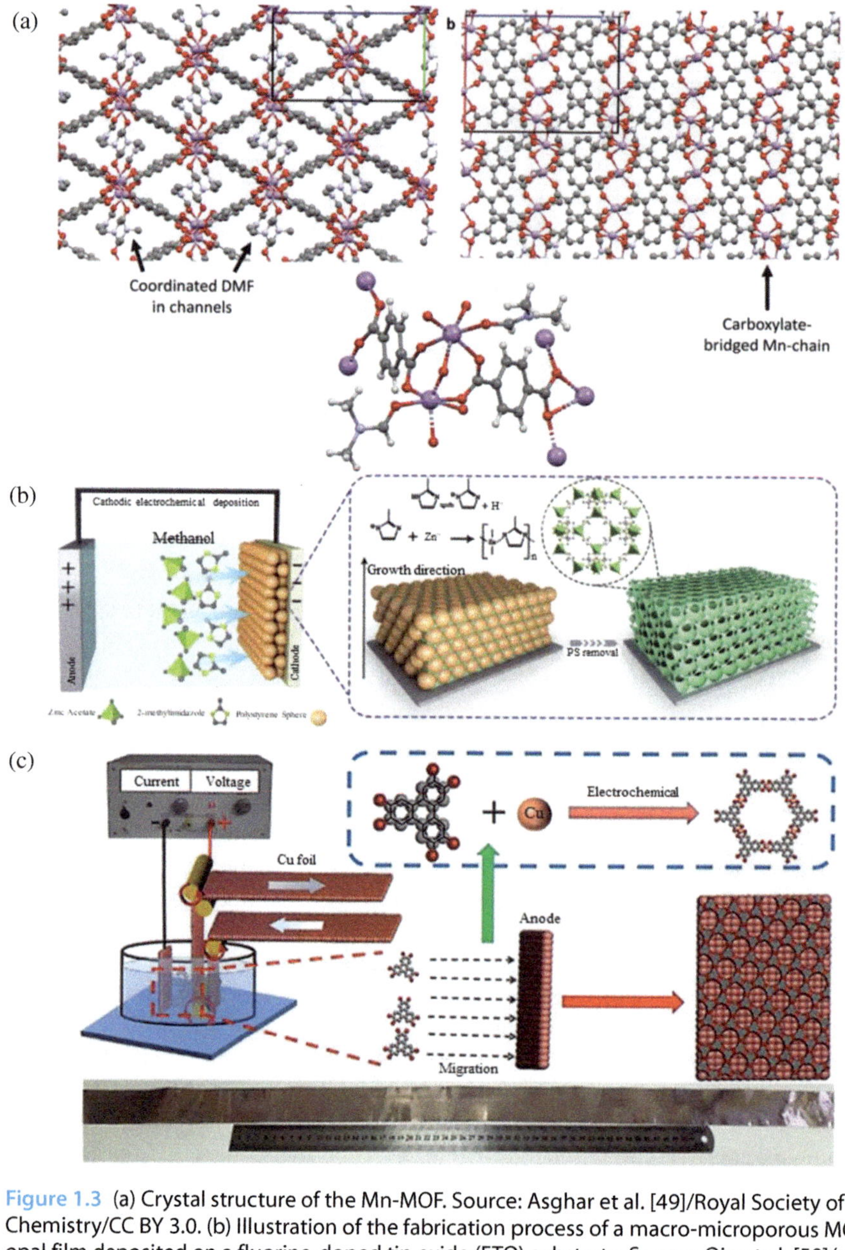

Figure 1.3 (a) Crystal structure of the Mn-MOF. Source: Asghar et al. [49]/Royal Society of Chemistry/CC BY 3.0. (b) Illustration of the fabrication process of a macro-microporous MOF inverse opal film deposited on a fluorine-doped tin oxide (FTO) substrate. Source: Qin et al. [50]/with permission of Royal Society of Chemistry. (c) Electrochemical reaction cell for the preparation of a $Cu_3(HHTP)_2$ film on Cu foil and the schematic diagram of coordination reaction between Cu^{2+} and the HHTP ion. Source: Liu et al. [51]/with permission of John Wiley & Sons.

1.2.4 Ultrasonic Method

The ultrasound (US)-assisted method is a simple, economical, and effective method for the preparation of MOFs. It has been found that US with a frequency of 20 KHz–1 MHz can cause chemical reactions of molecules in the solvent. The process of bubble formation, growth, and rupture during ultrasonication leads to an increase in the local temperature of the solution, which can result in homogeneous nucleation of the material, reduction of the crystallization time, and formation of smaller crystal sizes. Dissolving metal ions and organic ligands in organic solvents and then synergistically preparing MOFs in the presence of US is also an extremely effective approach.

Heydari and coworkers [52] prepared a cerium-organic skeleton (Ce-UIO-66MOF) rapidly and efficiently with 1,4-benzene dicarboxylic acid, DMF, and $(NH_4)_2Ce(NO_3)_6$ as raw materials under high-density US-assisted radiation at 305 W. It showed good catalytic activity for aerobic oxidation reactions. Haque and Jhung [53] compared US-assisted, microwave (MW) irradiation and conventional electrical heating methods for the synthesis of isomeric CPO-27S. The ultrasonic method resulted in faster nucleation and crystallization than the other heating methods due to the strong hot spot of US irradiation. The US-assisted synthesis of CPO-27-Co crystals with the smallest size and the highest porosity demonstrated that the US-assisted method can obtain small-sized MOF crystals with low energy consumption (i.e. shorter reaction time and/or lower temperature). Ding and coworkers [54] investigated the US-assisted rapid synthesis of caffeic acid (CA)-loaded and antimicrobial-use cyclodextrin metal–organic skeletons (U-CD-MOF), which reduced the preparation time to a few minutes (as shown in Figure 1.4c). It was found that ultrasonic power, reaction time, and temperature affected the morphology and size of the resulting crystals and the antimicrobial properties of the materials.

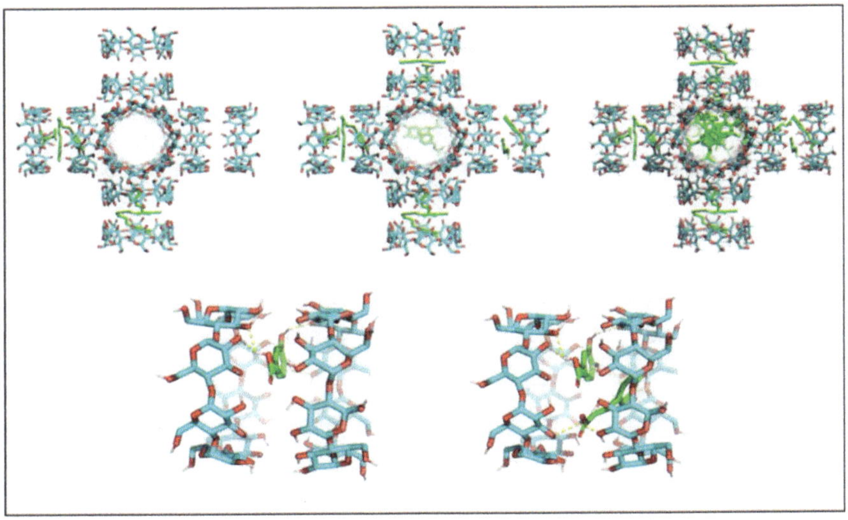

Figure 1.4 Molecular docking simulations of caffeic acid molecules distributed in CD-MOF. Source: Shen et al. [54]/with permission of Elsevier.

1.2.5 Mechanochemistry Method

Mechanochemical synthesis refers to the process of grinding a solid to facilitate a quantitative chemical reaction using grinding; this process usually does not use any solvent or uses only one solvent. This method enables faster and more efficient synthesis of MOFs than traditional hydrothermal methods. Impressively, MOF precursors with low solubility can also be adapted to this method. Typical mechanochemical methods include ball milling, pressure application, and extrusion. The ball milling method is the most common mechanochemical synthesis method of MOFs, which usually adopts a high-energy ball milling process to make solid or solid–liquid mixtures chemically react in a short time. The synthesis process of this method is simple, environmentally friendly, and low cost, which has good application prospects. However, this method still has some drawbacks, such as small yield, long equipment downtime, and difficult product cleaning. The pressurized synthesis method usually uses static/dynamic high pressure to pressurize the reactants to promote interface contact and improve reaction efficiency. This technique simplifies the synthesis process and reduces waste generation. However, this method suffers from low capacity in the large-scale preparation of MOFs because the feedstock is difficult to react completely. Extrusion is also an efficient and continuous mechanochemical method for the preparation of MOFs. The method requires little to no solvent; then the MOF synthesis can be achieved continuously. This mechanochemical technique offers the possibility of large-scale production of MOFs. It should be noted that to make the synthesized MOF have higher structural properties, the chemical ratio of the precursors should be strictly controlled so that the precursors can be as fully reacted as possible, as well as to avoid clogging of the pores by unreacted precursors. In addition, for the mechanochemical synthesis method, to fully carry out the reaction or reduce the generation of by-products, it is necessary to select appropriate precursors to avoid the formation of complex by-products, thus reducing the impact on the pore structure.

Tao and coworkers [55] proposed a simple and rapid mechanochemical synthesis of a series of HKUST-1, which can be used for the efficient adsorption of SF_6. Experiments were carried out using trimeric hexanoic acid (H3BTC) and copper nitrate doped with three different hybrid ligands, namely imidazole (Im), 1,2,4-1*H*-triazole (Trtz), and tetrazole (Tetz), respectively, and three different hetero ligands were successfully prepared. X@HKUST-1 (X = Im, Trtz, or Tetz) was successfully prepared. Notably, the water stability of the doped imidazole Im@HKUST-1 was significantly improved, compared to HUST-1. Xie and coworkers [56] prepared Pt-doped Zn-MOF-74 (PtZn-MOF-74) from Pt-doped ZnO (Pt-ZnO) by mechanochemical transformation. The lack of a large number of solvents limited the solvation and diffusion of the growing substance during the mechanochemical transformation process, thus preventing the agglomeration of the Pt dopant in PtZn-MOF-74. Liu and coworkers [57] formed MOFs by rapid encapsulation of a series of substrates, such as imidazole, phosphonic acid, urea, and amino sulfonic acid, into the pores of MOF NENU-3 via a one-pot mechanochemical synthesis (as shown in Figure 1.5). The synthesis of MOFs loaded with functional guests, which used to take days and require multiple steps, can now be done in one step within minutes. The proton conductivity of the resulting composite is improved by two to three times of magnitude over NENU-3. Compared with other methods of synthesizing MOFs, the mechanochemical technique has an extremely high synthesis efficiency, can significantly reduce the reaction time, and possesses great potential for large-scale and industrial production.

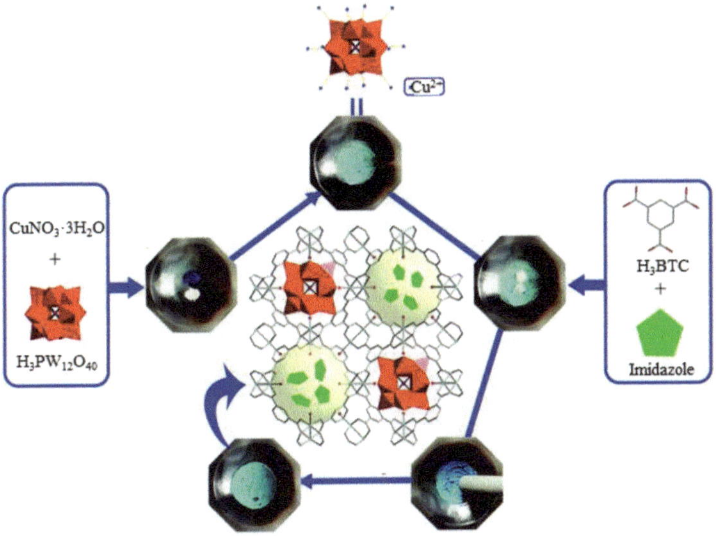

Figure 1.5 The synthesis process of Im@NENU-3 by a one-pot mechanochemical method. Source: Wang et al. [57]/with permission of Royal Society of Chemistry.

1.2.6 Steam-Assisted Conversion Method

Steam-assisted reforming is a method of MOF synthesis that uses small, submicron droplets generated by heating to facilitate a chemical reaction. Steam-assisted reforming typically uses smaller reactors and produces less waste than conventional hydrothermal synthesis. In this method, only a thin layer of mixed reaction precursors is spread in the reaction apparatus to facilitate the completion of the reaction. In addition, the use of a combination of spray drying and constant steam-assisted conversion is effective enough to achieve large-scale preparation of MOF. This combination of spray drying and steam-assisted conversion is expected to be an efficient, simple, low-cost, and low-waste method for MOF production, potentially leading to industrial-scale MOF production.

Dong and coworkers [58] successfully prepared ZIF-8 and ZIF-67 in place of organic DMF in water at 120 °C and 1.9848 bar with high solids concentration and significant solid–liquid separation. Notably, this is the first time that water has been used as the sole solvent for the preparation of ZIF materials. Jhung and coworkers [59] achieved rapid production of nano-porous MIL via steam-assisted conversion of −100(Fe) for rapid production. Here, the use of zero-valent iron, instead of Fe metal salts, avoided the associated negative effects. Furthermore, Kim and coworkers [60] prepared MIL-101(Cr) by steam-assisted conversion in an autoclave containing porous polytetrafluoroethylene plates at 220 °C, achieving a 90% product yield with an SSA of up to 4100 $m^2\ g^{-1}$ and good reproducibility (as shown in Figure 1.6).

1.2.7 Fluid Chemistry Method

Flow chemistry, in which chemical reactions are made to take place in a continuous flowing fluid rather than in a reaction vessel, shows advantages such as high surface area and volume

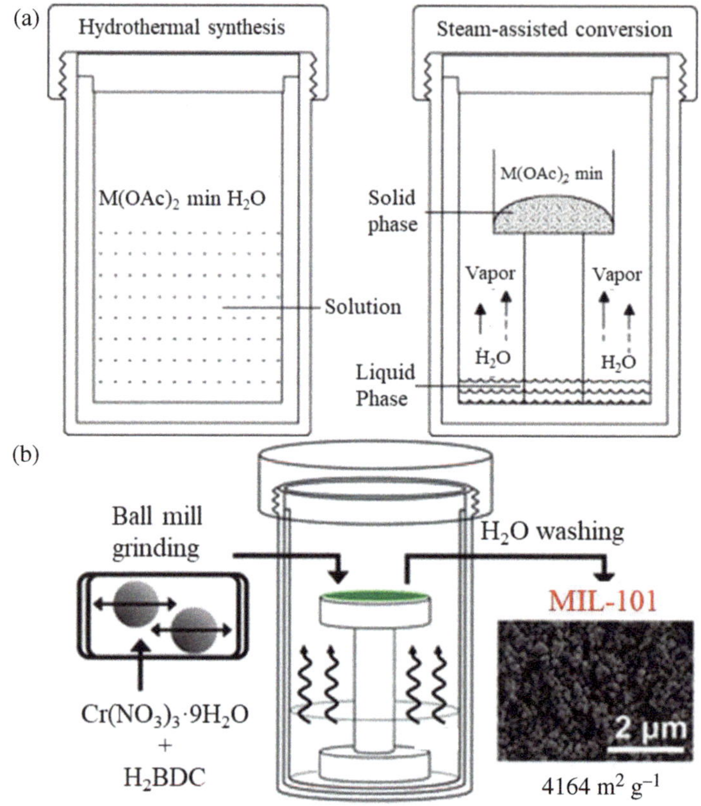

Figure 1.6 (a) Schematic diagram of the reaction vessels for ZIF synthesis. Left: hydrothermal synthesis (HTS), and right: the steam-assisted conversion (SAC) method; M-Zn, Co. Source: Shi et al. [58]/with permission of John Wiley & Sons. (b) Dry-gel conversion synthesis of Cr-MIL-101 aided by grinding. Source: Kim et al. [60]/with permission of Royal Society of Chemistry.

ratios of the reaction mixture, rapid mass, and heat transfer. Flow chemistry is considered to be a cost-effective method to meet the requirements of industrial production. Studies have been carried out to synthesize MOFs using flow chemistry techniques. Microfluidic synthesis is a typical fluid chemistry method. Microfluidic synthesis provides another effective way to synthesize MOFs on a large scale due to its advantages of automatic control and online quality monitoring. Fluid chemistry is globally recognized as a green and safe preparation technology that overturns the traditional synthesis methods. It provides greater flexibility and selectivity in the design of MOF synthesis routes, which can improve the synthesis efficiency and reduce pollution from the source. The fluid chemistry method is an important development trend for future MOF synthesis and is expected to achieve commercial mass production of MOFs for industrial applications in the future.

Herman and coworkers [61] designed a segmented reactor combining continuous flow and microwave radiation for the synthesis of MOF-74(Ni) at a pressure of 2.5 bar with a reagent conversion of 95.5%. Segmented flow improves the mixing state of the feedstock, reduces particle contamination, and allows the reaction to be carried out for a long period, with the

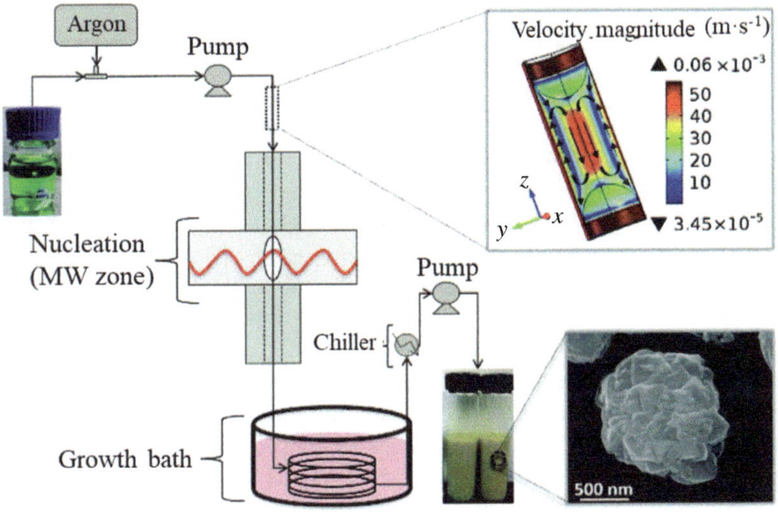

Figure 1.7 Schematic of the segmented, continuous-flow, microwave-assisted synthetic reactor for MOF-74(Ni). The inset on the top right, obtained using Comsol, shows turbulence in-between slugs in segmented flow. Source: Albuquerque et al. [61]/with permission of American Chemical Society.

results possessing a high degree of reproducibility. Thus, this strategy is time-efficient and allows for the continuous preparation of high-quality MOFs and composites, which is a favorable factor for large-scale production (as shown in Figure 1.7).

1.3 MOF-Derived EMW Absorption Materials

With the vigorous development of electronic information technology, the intelligent era is promoting the fourth industrial revolution, and the electronic information physical system based on the internet and intelligent entity system has comprehensively improved the types and application fields of electromagnetic equipment. However, the resulting electromagnetic pollution is also becoming more serious. Therefore, it is of great significance to use EMW absorption materials to absorb useless EMW pollution and eliminate harmful radiation. So far, a variety of nanomaterials with different topological, chemical, and physical properties have been applied to EMW absorption. In particular, MOF-derived materials with a unique composition and structure, owing to their unique micro-topological structure and diversified metal coordination, not only exhibit high EMW absorption performance and a flexible absorption frequency band but also possess abundant absorption mechanisms, thus attracting increasing attention. The design concept and research direction of MOF-derived EMW absorption materials can be summarized in the following stages: (1) selecting organic ligands and metal ions, (2) controlling the topology through self-assembly behavior, (3) obtaining different kinds of derivatives through post-processing, and (4) testing the absorbability of materials to EMWs and analyzing their microscopic electromagnetic response mechanisms.

1.3.1 Monometallic MOF-Derived Absorption Materials

Metals, alloys, ferrites, carbon nanofibers, carbon nanotubes (CNTs), graphene, and conductive polymers all possess certain EMW absorption capabilities. However, metals and alloys have limited applications due to their narrow absorption bands, while pure carbon materials exhibit a single loss mechanism that hinders further improvement in their EMW absorption capabilities. As a result, research has focused on modifying materials by combining carbon materials with single metal materials, leading to the development of materials with advantages such as high absorption strength, lightweight, and broad absorption frequency. Among these materials, MOFs synthesized using transition metal elements such as Fe, Co, Ni, Cu, and Zn serve as excellent precursors. These MOFs can be transformed into magnetic or nonmagnetic metal derivatives/carbon matrix composites at different temperatures. MOF-derived materials with nonmagnetic metals as coordination centers primarily concentrate on constructing microscopic conductive networks and modulating dielectric loss. On the other hand, MOF-derived materials with magnetic metals as coordination centers not only modulate polarization and conductivity loss but also enhance the magnetic loss capability by manipulating magnetic eddy currents and magnetic resonance effects.

Gao and coworkers [62] synthesized rod-like Cu-MOF-74 precursors using 2,5-dihydroxyterephthalic acid as a ligand. The precursor was treated with a KOH alkaline solution for 180 min to transform its morphology into a circular dodecahedron. Subsequently, calcined at 300 °C, the calcined samples were placed in an aqueous Na_2S solution for treatment, resulting in hollow dodecahedrons with Cu_2S shells. After mixing the sample with paraffin wax at 20 wt%, the optimum effective absorption bandwidth of the sample was achieved at 2.3 mm, which reached a value of 6.2 GHz, covering 11.8–18.0 GHz. The material has a large number of defects and nonhomogeneous interfaces, which can effectively generate polarization relaxation. The structure of the hollow core can enhance the multi-scattering ability of the material to a certain degree, further optimizing the impedance-matching properties and improving the EMW absorption performance of the material.

Che and coworkers [63] synthesized octahedral ZIF-67 precursors using 2-methylimidazole as a ligand by controlling the solvent and cobalt/linker molar ratio. After calcination at 500 °C, Co@N-doped carbon composites (Co@NC) were obtained, with the initial MOF precursor determining the distribution of the carbon backbone and magnetic cobalt NPs. The carbonization yielded Co@NC composites with controllable size and shape. Additionally, the magnetic carbon composites have the advantages of rich interfacial polarization and a strong magnetic coupling network, and the MOF-derived dielectric carbon skeleton provides electron transport paths and enhances the conductive dissipation. The optimized octa-decahedral Co@NC sample exhibits the best microwave absorption of −53.0 dB at a thickness of 1.8 mm and an effective absorption bandwidth of 6.2 GHz (as shown in Figure 1.8).

1.3.2 Multi-metal MOF-Derived Absorption Materials

Compared with monometallic MOFs, multi-metal MOFs possess a richer variety of active sites, which are conducive to the improvement of the stability of the material backbone structure, chemical activity, alteration of the state of charge distribution, and magnetic properties of the materials. Numerous parameters of the polymetallic MOFs can be designed and regulated according to the ratio of the adjusting metals so that MOF materials with desired properties can be obtained. Therefore, it is easier to regulate the EMW absorption

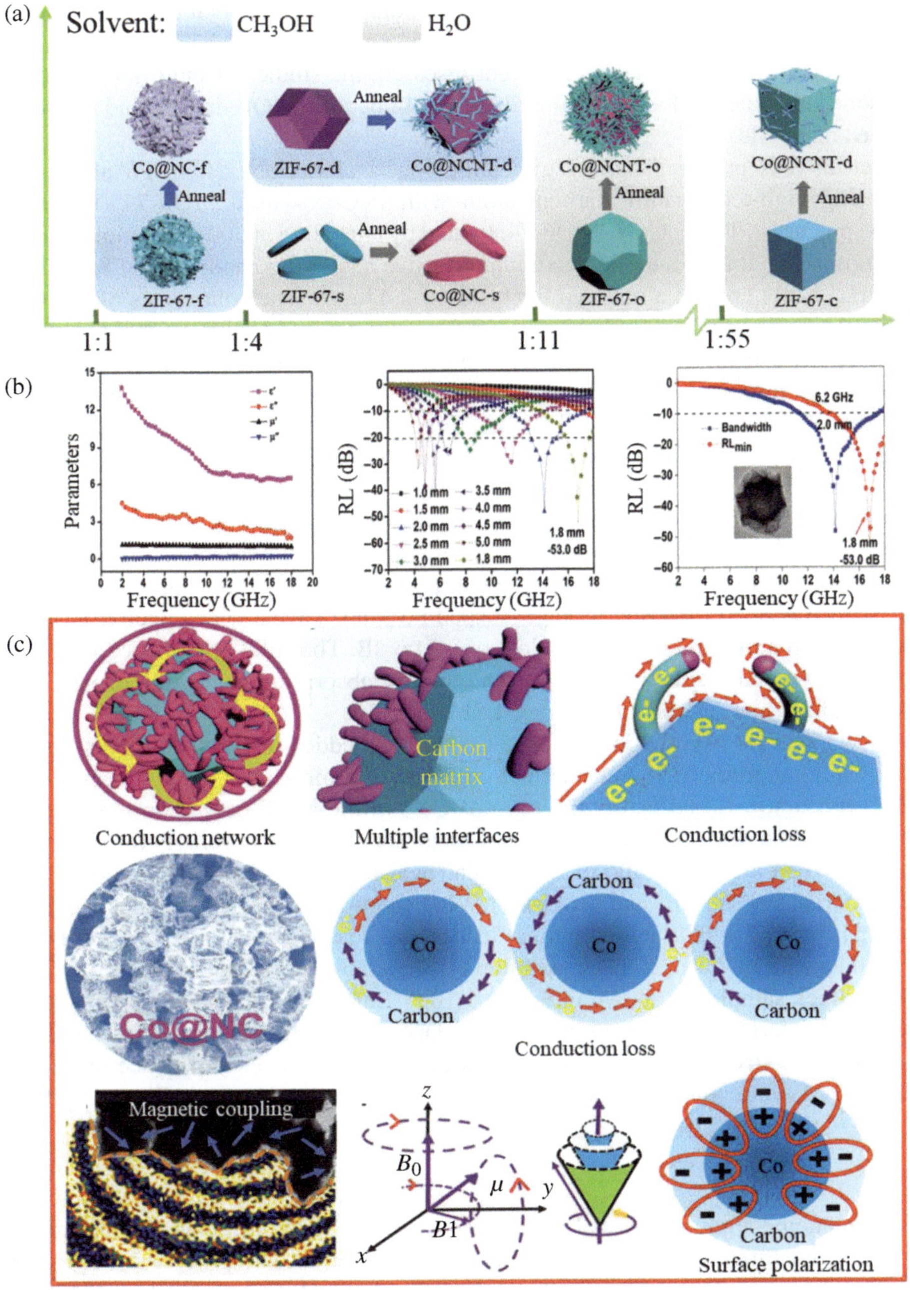

Figure 1.8 (a) Schematic preparation process of Co@NC composites. (b) Frequency dependencies of electromagnetic parameters, calculated reflection loss curves, maximum microwave RL curves, and EAB of Co@NC composites. (c) Schematic illustration of the EMW absorption mechanism. Source: Huang et al. [63]/with permission of John Wiley & Sons.

properties of polymetallic MOFs with different morphologies in the metal centers with different electronic configurations. Their electromagnetic parameters can be easily optimized by reasonable control of temperature. In recent research, the studies of multi-metal MOF-derived absorption materials focused primarily on the bimetallic MOF-derived and tri-metal MOF-derived materials.

Ouyang et al. [64] prepared hollow trimetallic spherical FeCoNi@C EMW-absorbing materials using high-temperature carbonization with FeCoNi-based MOF-74 (FeCoNi-MOF) as the precursor. The FeCoNi-MOF-74, treated by annealing at 700 °C, exhibits excellent EMW-absorbing ability, and the maximum effective absorbing bandwidth of 8.08 GHz (9.92–18 GHz) is obtained at a material thickness of 2.47 mm. The composites of FeCoNi ternary alloy and carbon matrix polymorphs are not only equipped with the magnetic loss brought by magnetic metals but also made the interfacial polarization and dipole polarization of the material increase, which greatly enhances the dielectric loss capability and impedance matching characteristics of the material. In addition, the electronic leaps generated in the alternating electromagnetic field between the conductive FeCoNi NPs and the amorphous carbon will jointly form a highly conductive network, thus enhancing the conductive loss.

Kong and coworkers [65] successfully prepared hollow-cage or solid-box magnetic/dielectric Fe/C/C and dielectric Fe/Mn/C EMW-absorbing nanomaterials by pyrolyzing Prussian blue analogs with controllable topology and phase composition (as shown in Figure 1.9a). The solid box–shaped Fe/Co/C and hollow cage–shaped Fe/Mn/C have a wide effective absorption bandwidth and a minimum reflection loss of −54.6 dB. The solid box–shaped Fe/Co/C nano-compound prepared at 900 °C exhibits an effective absorption bandwidth of 8.8 GHz at a thickness of 2.5 mm (as shown in Figure 1.9b).

Hu and coworkers [66] synthesized tunable Co NPs embedded in porous nitrogen-doped carbon (Co/NC) nanorods by direct, thermal cleavage of a bimetallic organic framework (CoZn-ZIF) precursor. By adjusting the ratio of Co^{2+} in the MOF precursor, the content and distribution of the Co NPs in the composite absorber change accordingly (as shown in Figure 1.10a). When the molar ratio of Co^{2+} to Zn^{2+} was 3:1, the carbonized composite had the largest external surface area and the best EMW absorption performance. When the filler mass loading is only 15 wt%, the minimum reflection loss reaches −52.3 dB at 10.1 GHz for a thin layer thickness of 2.5 mm. The maximum effective absorption bandwidth is 5.0 GHz for a 2.0 mm thick sample (as shown in Figure 1.10b). Enhanced conductive and magnetic losses, abundant interfacial polarization, dipole polarization, defect-induced dipole polarization, layered pore structure, and geometrical effects result in better impedance matching and enhanced EMW attenuation of Co/NC absorbing materials (as shown in Figure 1.10c).

1.3.3 MOF-Carbon Composite Absorption Materials

Most of the monometallic/multi-metal MOF-derived EMW-absorbing materials are made through high-temperature carbonization of the MOF precursor. The processed MOF can form a composite material of metal or metal oxide and carbon, which has excellent dielectric and magnetic losses. However, its inherent disadvantage is the lack of a good conductive network, which also restricts the ion transmission rate and prevents the further enhancement of the wave-absorbing properties. Carbon-based materials themselves have a good attenuation ability for EMWs and lower density, which possesses strong stability. The composite with MOF is expected to address the lack of a conductive network [67]. Common carbon

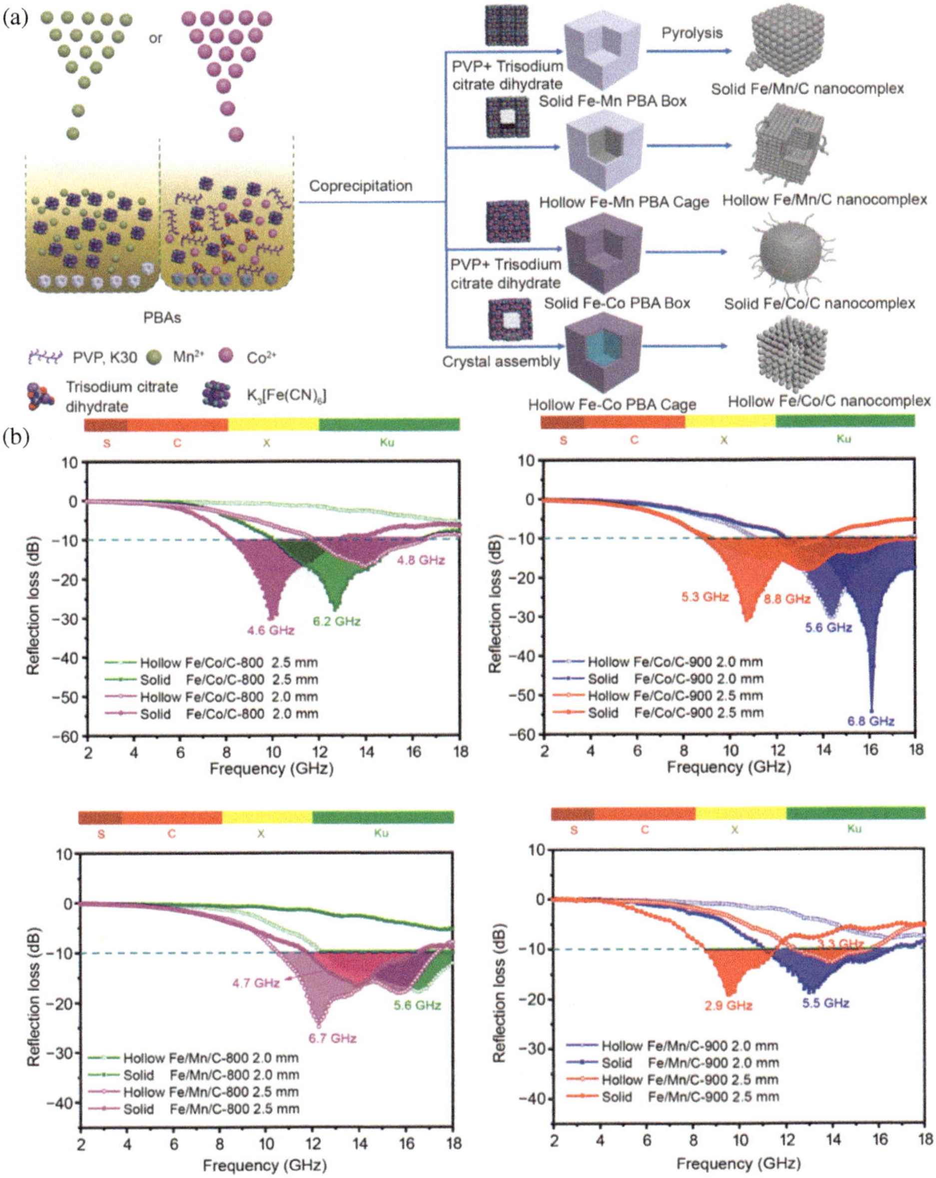

Figure 1.9 (a) Schematic illustration of the fabrication route toward hollow Prussian blue analogs (PBA) cages and solid PBA boxes and their derived nanocomplexes. (b) The comparison of electromagnetic wave absorption performance for pyrolyzed Fe/Co/C and Fe/Mn/C nanocomplexes. Source: Miao et al. [65]/with permission of Springer Nature.

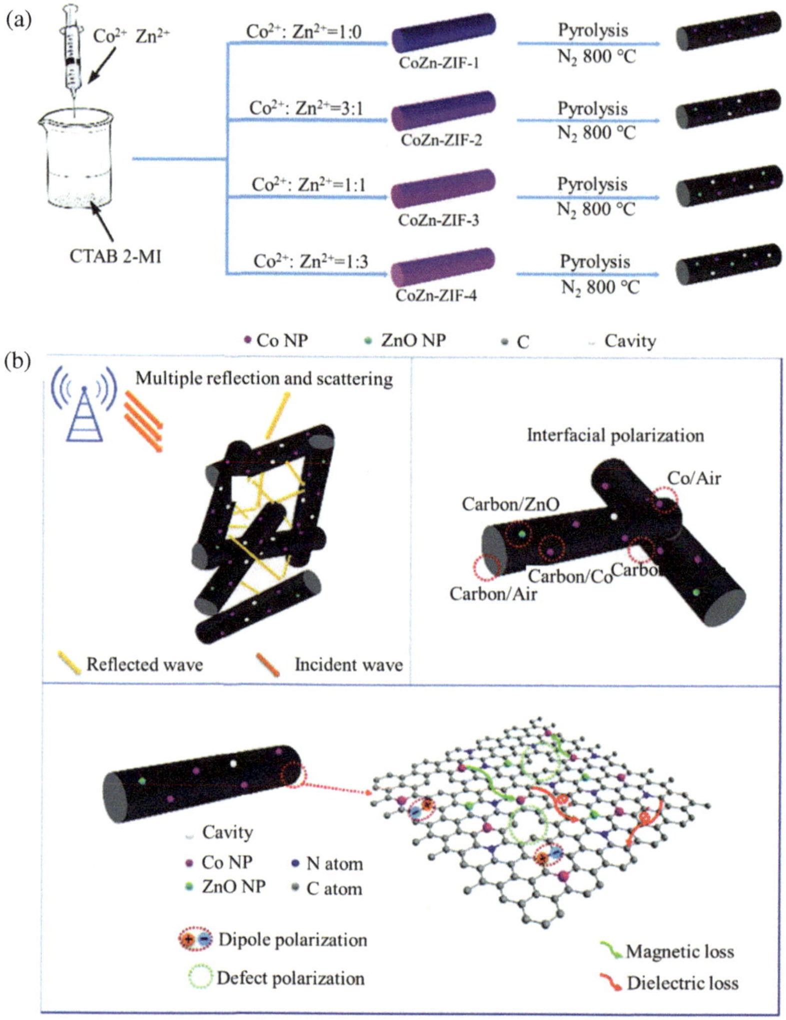

Figure 1.10 (a) Schematic illustration of the synthesis process of Co/NC nanorods. (b) Calculated RL curves and 3D reflection loss map presentations of Co/NC-1, Co/NC-2, Co/NC-3, and Co/NC-4 nanorods, respectively. (c) Schematic diagram of microwave attenuation in the Co/NC composites.
Source: Pan et al. [66]/with permission of Royal Society of Chemistry.

materials include carbon black [68], carbon fiber [69], CNTs [70], and graphene [71], and their derivatives. The introduction of carbon materials is not only conducive to the construction of a wide range of conductive networks to improve charge transport efficiency, but also the structural defects and non-uniform interfaces of carbon materials in MOFs can enhance dielectric loss, which can effectively improve the overall wave absorption performance of the materials. Carbon nanotubes are characterized by a large range of π-bonds formed by the P electrons on the carbon atoms, which give them special electrical properties due to the remarkable properties of conjugation effects. Carbon nanotubes have a one-dimensional hollow tubular structure, the wall of which is surrounded by a single layer or multiple layers of graphene sheets, and their properties change depending on the way the graphene sheets are curled, reflecting metallic or semiconducting properties. Their electrical conductivity has a close relationship with their diameters and chirality, which reflect highly tunable properties. Therefore, they are widely used in EMW-absorbing materials. In recent years, a large amount of the MOF-CNT absorption materials has been successfully prepared.

Gu and coworkers [72] successfully prepared three-dimensional hybrid nanostructures (CNT/FeCoNi@C) consisting of MOF-derived magnetic nanospheres and Fe-filled CNT sponges. The minimum reflection loss and effective absorption bandwidth of the prepared CNT/FeCoNi@C sponges reached −51.7 dB and 6.0 GHz, respectively, which is better than most reported MOF-based EMW-absorbing materials. One-dimensional CNTs have excellent conductive channels to dissipate current, resulting in large conductive losses. CNTs, FeCoNi, and carbon can construct a large number of heterogeneous interfaces with each other, resulting in interface polarization loss. At the same time, there are a large number of defects in the FeCoNi@C multiphase alloy formed after carbonization, which can form abundant defect-induced dipole polarization, resulting in polarization relaxation loss. In addition, the porous cavities of CNT sponges and FeCoNi@C microspheres can enhance the scattering ability of the materials against EMWs and optimize the impedance-matching characteristics.

Carbon fiber, with its more than 90% carbon content and high-strength fiber characteristics, makes it an excellent material for manufacturing aerospace and other equipment. It has the characteristics such as low specific gravity, high strength, stable physical and chemical properties, excellent electrical conductivity, excellent elastic modulus, and a thermal expansion coefficient close to zero or even negative. Therefore, it is widely used in strategies such as mechanical strength enhancement and electrical conductivity optimization of materials.

Liu and coworkers [73] successfully synthesized a layered composite hollow carbon fiber@nitrogen-doped carbon/Co (HCF@NC/Co) by in situ growing two-dimensional Co metal–organic skeleton (ZIF-67) NSs on the surface of hollow luffa fiber (HLF) and calcination at a high temperature. The aggregation of carbonized MOFs is effectively avoided, and a uniform hierarchical one-dimensional structure is constructed. When the mass ratio of HCF@NC/Co to paraffin wax is 14 wt% and the material thickness is 2.25 mm, the minimum reflection loss of HCF@NC/Co is −50.14 dB, and the maximum effective absorption bandwidth reaches 7.36 GHz. The electrons excited by alternating electromagnetic fields migrate and jump in the one-dimensional conductive network constructed by HCF@NC/Co, converting electromagnetic energy into heat energy and causing conduction loss. Secondly, a large number of heterogeneous interfaces can induce strong interfacial polarization and improve dielectric loss capacity. The doped N and Co particles in graphitic carbon and the defect sites in the lattice also lead to a large number of dipole polarization losses. The magnetic Co particles also provide magnetic loss through eddy current loss, exchange resonance, and natural resonance.

Hu and coworkers [74] prepared a series of NiCo@nanoporous carbon@carbon fiber nanocomposites (NiCo@NPC@CF) consisting of micro- and nanoscale porous carbon structures and flower-like NiCo@C particles using Ni/Co-MOF particles and biomass cotton as precursors (as shown in Figure 1.11a). The hierarchically structured NiCo@NPC@CF at 3.5 mm thickness possesses a minimum reflection loss of −46.5 dB and an effective absorption bandwidth of 10.0 GHz (as shown in Figure 1.11b). The three-dimensional carbon fiber

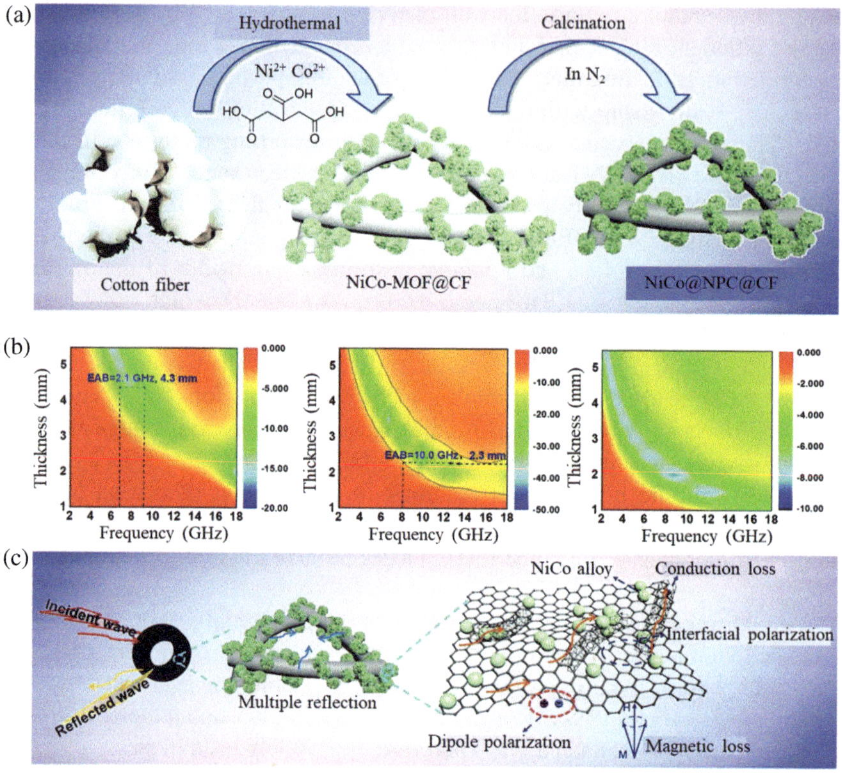

Figure 1.11 (a) Schematic of the synthetic procedure for NiCo@NPC@CF composites. (b) The corresponding 2D curves of NiCo@NPC@CF. (c) Microwave absorption mechanisms of the NiCo@NPC@CF composite. Source: Jin et al. [74]/with permission of Royal Society of Chemistry.

interconnecting network derived from biomass cotton can increase the electrical conductivity and thus enhance the conductive loss. Multiple interfaces between NiCo alloys, CNTs, and carbon fibers are conducive to enhancing the interfacial polarization effect and the loss capability. In addition, the magnetic NiCo alloy can further enhance the electromagnetic loss capability of NiCo@NPC@CF through natural resonance and exchange resonance (as shown in Figure 1.11c).

Graphene, as a two-dimensional atomic crystal, has excellent mechanical stiffness, strength, and elasticity, as well as very high electrical and thermal conductivity, which suggests that graphene can replace other materials in existing applications. Many properties of graphene exceed those obtained in any other material, and its electron migration rate, Young's modulus, and thermal conductivity all reach the limits predicted by theory. Moreover, due to its easy chemical functionalization, it has been widely modified and applied in various fields. In recent years, many graphene composites have been frequently applied to EMW absorption.

Zhao and coworkers [75] successfully synthesized Fe–Co/N-doped carbon/reduced graphene oxide (Fe–Co/NC/rGO) composites with a layered porous structure by in situ growth of Fe–Co organic framework on porous cocoon-like rGO sheets and then calcination at high temperature (as shown in Figure 1.12a). When the sample thickness was 2.5 mm, the

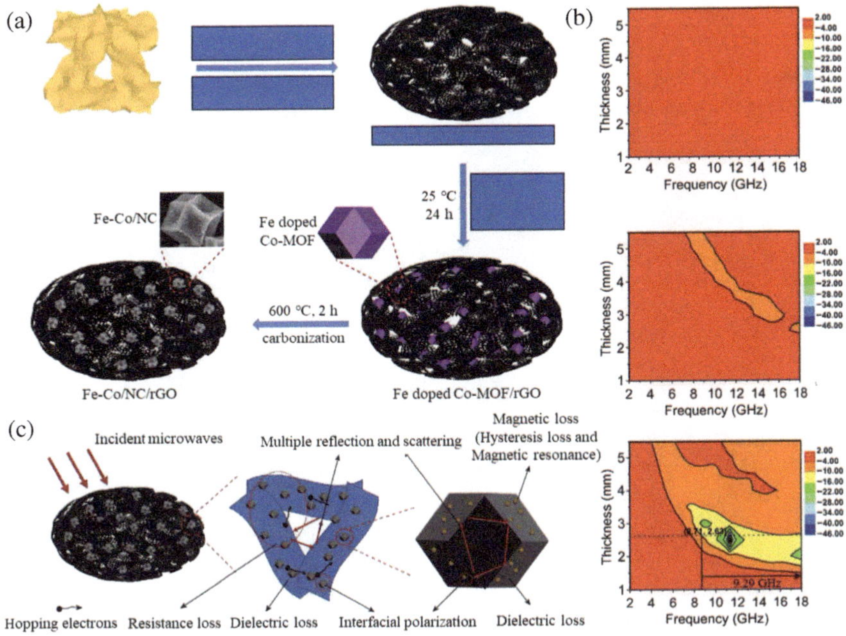

Figure 1.12 (a) Schematic drawings illustrating the fabrication process of hierarchically porous Fe–Co/NC/rGO. (b) Calculated 2D reflection loss of Co/NC, Fe–Co/NC, and Fe–Co/NC/rGO with a mass filling ratio of 25 wt%. (c) Schematic diagram of the microwave absorption mechanism of Fe–Co/NC/rGO. Source: Wang et al. [75]/Springer Nature/CC BY 4.0.

effective absorption bandwidth is as high as 9.12 GHz (8.88–18.0 GHz), and the minimum reflection loss value reaches −48.26 dB at 11.28 GHz (as shown in Figure 1.12b). The magnetic hysteresis loss and magnetic resonance of Fe and Co magnetic metals are the causes of the magnetic loss of composite materials. FeCo alloy embedded in N-doped C can form many interfaces, which greatly enhance the interfacial polarization of the material. After the carbonization of FeCo-MOF, N doping introduces a large number of defect sites into the C lattice and the rGO, thereby enhancing defect-induced dipole polarization. In addition, the conductive network formed by overlapping graphene oxide (GO) sheets can further enhance the conductivity loss (as shown in Figure 1.12c).

Che and coworkers [76] successfully synthesized accordion-like CeO_{2-x}/reduced graphene oxide (CeO_{2-x}/rGO) hybrid materials by converting Ce-MOF into NPs and combining with GO, inducing two-dimensional GO assembly into three-dimensional accordion-like composite materials (as shown in Figure 1.13a). The prepared lightweight, 3D layered CeO_{2-x}/RGO composite has excellent EMW absorption ability; its effective absorption bandwidth reaches 5.84 GHz, and the minimum reflection loss reaches −50.6 dB (as shown in Figure 1.13b). The charge accumulation at the interface caused by the inherent conductivity difference between CeO_{2-x}, NPs, and RGO enhances the interface polarization effect of the material. The residual oxygen functional groups and structural defects in RGO produce abundant polarization sites, which further enhance the dipole polarization. In addition, the highly conductive GO promotes the conductivity loss within the GO sheet (as shown in Figure 1.13c).

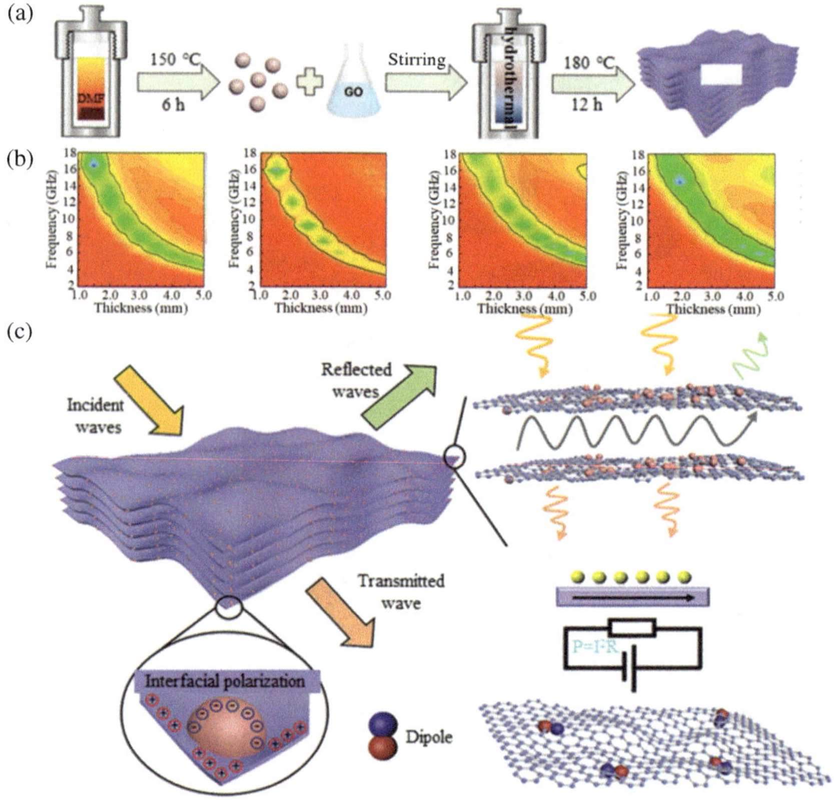

Figure 1.13 (a) Preparation scheme of the accordion-like CeO_{2-x}/RGO composite. (b) Contour map of absolute values of the accordion-like CeO_{2-x}/RGO composite. (c) The schematic of microwave mechanisms of accordion-like CeO_{2-x}/RGO absorbers. Source: Li et al. [76]/with permission of John Wiley & Sons.

1.3.4 MOF-MXene Composite Absorption Materials

Among two-dimensional nanomaterials, transition metal carbides, nitrides, or carbon-nitrides (MXenes) have demonstrated excellent performance in a wide range of applications spanning energy storage, sensing, optoelectronics, catalysis, and biomedicine due to their unique metal conductivity, solution processability, high mechanical properties, and broadly tunable properties. In recent years, MXene and its derivative materials have been widely used for EMW absorption.

Zhang and coworkers [77] constructed an interwoven one-dimensional heterostructure using the assembly of MXene and MOF and controlled the growth of CNTs by adjusting the load of CoNi-MOF on MXene, thus achieving the regulation of electromagnetic parameters. When the matching thickness of the prepared sample is 1.6 mm, the minimum reflection loss of the absorber reaches −51.6 dB, and the effective absorption bandwidth covers 4.5 GHz (13.2–17.7 GHz). The synergistic effect of the double one-dimensional heterostructure and the construction of the three-dimensional conductive network give excellent impedance match characteristics and conductance loss capability. A large number of one-dimensional heterostructural heterogeneous interfaces between CoNi alloy, CNTs, MXene, and TiO_2 greatly improved the interfacial polarization effect. Defects introduced after carbonization and dipole polarization induced by CoNi NPs further enhance the dielectric loss capacity of the material. In addition, the natural resonance and eddy current loss of CoNi alloy can further increase the magnetic loss capacity of the material.

Lu and coworkers [78] designed and successfully prepared a novel sandwich-like two-dimensional layered Fe&TiO_2 nanoparticles@C nanocomposite based on MXene-MOF hybrid (as shown in Figure 1.14a). The composite obtained a wide effective absorption bandwidth of 6.5 GHz at a thickness of only 1.6 mm and a minimum reflection loss of −51.8 dB at 6.6 GHz (as shown in Figure 1.14b). The conductive network composed of conductive carbon sheets formed by MXene can enhance the conductivity loss ability of the material.

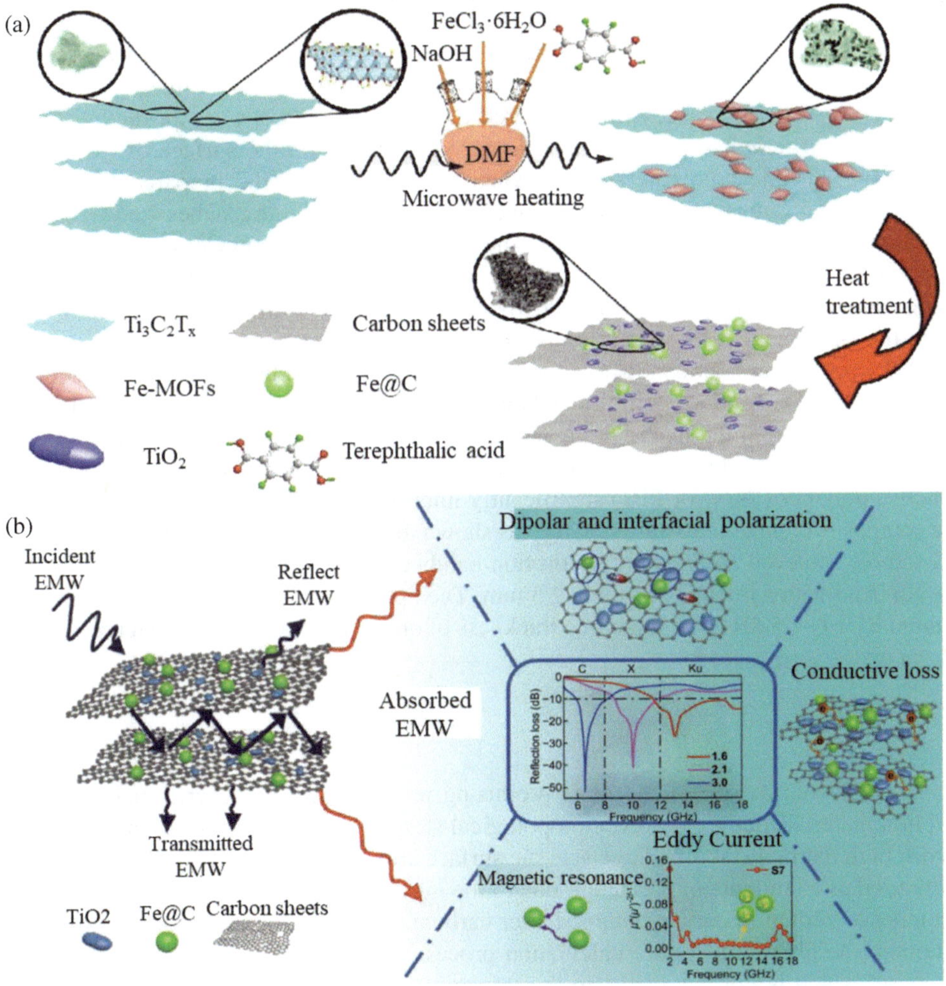

Figure 1.14 (a) Schematic representation of the facile synthesis route of the Fe&TiO_2@C by the heating reaction followed by heat treatment. (b) Illustration of electromagnetic microwave absorption mechanisms for Fe&TiO_2@C nanocomposites. Source: Deng et al. [78]/Springer Nature/CC BY 4.0.

The dipoles in Fe, TiO_2, and carbon can provide a large amount of dipole polarization loss. In addition, the dielectric loss generated by the capacitance-like interface between them will make a significant contribution to the conversion of EM energy into heat. Moreover, Fe, TiO_2 NPs, and carbon sheets with different electrical conductivity can form a capacitance-like effect, which can cause the accumulation and rearrangement of space charges, resulting in a large amount of interface polarization.

1.3.5 MOF-Conductive Polymer Composite Absorption Materials

Conductive polymers are also often combined with MOFs as excellent absorbing materials due to their high dielectric properties, low density, flexibility, corrosion resistance, low cost, and tunable absorption frequency properties [79, 80]. Commonly used conductive polymers are polyaniline (PANI) [81] and polypyrrole (PPy) [82]. When a conductive polymer with adjustable conductivity is combined with the MOF, the electromagnetic parameters of the material can be adjusted so that the absorbent can move toward a lower filling rate and a thinner thickness, while still ensuring a good effective absorption bandwidth.

Ma and coworkers [83] successfully prepared CoZn/C@MoS_2@PPy composite by applying the MOF self-template method and coating polypyridine on the surface of CoZn/C@MoS_2 through chemical polymerization. When the thickness of the prepared CoZn/C@MoS_2@PPy composite is 2.0 mm, the effective absorption bandwidth reaches 4.56 GHz, and the minimum reflection loss is −49.18 dB. The introduction of Zn and Co particles into the carbon matrix, as well as MoS_2 and PPy, provides rich interfacial polarization. The presence of a large number of metal particles in the carbon matrix can cause dipole polarization, which leads to the enhancement of dielectric loss. Graphitized carbon substrates and highly conductive PPy provide greater conductive losses. In addition, the exchange resonance induced by magnetic metal Co and Zn particles and the natural ferromagnetic resonance enhance the magnetic loss ability of the material. Ye and coworkers [84] synthesized $LaCrO_3$/PANI and $LaCrO_3$/PPy composites by in situ chemical oxidation polymerization. The introduction of a conductive polymer (PANI or PPy) significantly improved the EMW absorption properties of the composite compared to pure $LaCrO_3$ (as shown in Figure 1.15a). The optimal $LaCrO_3$/PANI sample achieved an effective absorption bandwidth of 5.8 GHz, covering almost the entire Ku band with a thickness of only 2.5 mm. The optimal $LaCrO_3$/PPy sample showed a minimum RL of −45 dB and a matching thickness of only 3 mm (as shown in Figure 1.15b).

1.4 Summarize and Prospect

MOF is considered as one of the most promising precursors of high-performance EMW absorption materials due to its flexible topological structure, various preparation processes, low cost, high thermal stability, large specific surface area, and high porosity. When MOF is directly used as a template or precursor, due to the inherent characteristics of MOF, most of the composite carbon materials obtained after carbonization show a microporous structure. In addition, the high-temperature calcination process may also cause part of the structure to collapse, which is not conducive to the transport and transfer of electrons/ions. It is a common way to improve the dielectric properties of MOF by constructing or combining a conductive network with high conductivity efficiency. Furthermore, the structure of MOF

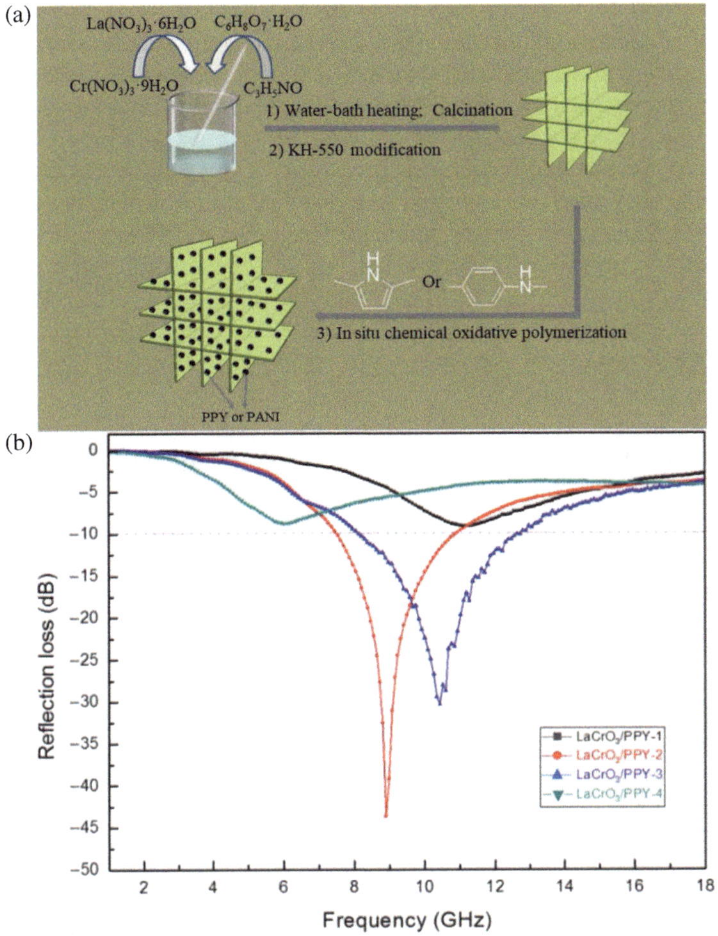

Figure 1.15 (a) Schematic representation of the facile synthesis route of the $LaCrO_3$/PANI and $LaCrO_3$/PPy composites. (b) Frequency dependence of RL values at different thicknesses for $LaCrO_3$/PANI and $LaCrO_3$/PPy composites. Source: Yu et al. [84]/Springer Nature.

derivatives also affects their electromagnetic response characteristics, and the different topological structures of MOF have significant effects on the dielectric and magnetic properties of materials. The regulation of structures and the design and construction of new structures show great potential in the development of new MOF-derived wave-absorbing materials. Defect engineering and interface engineering also improve the polarization loss (interface polarization and dipole polarization) ability of MOF-derived materials, but the quantitative estimation of the influence of various mechanisms still lacks sufficient theoretical support, so both classification and quantification need to be further explored. Hybrid engineering is a very promising research direction because it can create many active sites that promote the absorption of EMWs. However, at present, many researchers pursue the reflection loss value and effective absorption bandwidth of absorbing materials too much, which makes the development of MOF-derived absorbing materials seriously disconnected from practical applications. Most of the current MOF-derived materials exist in powder form. To cope with the development trend of lightweight, miniaturization, and integration in the current electronic information technology and communication industry and to realize the timely absorption of materials, it is inevitable to prepare MOF-derived absorbing materials with high mechanical properties such as films, fabrics, fibers, gels, and foams. In addition, to adapt to diverse and demanding practical application scenarios such as high altitude or deep sea, these materials should also have high durability, corrosion resistance, low density, sustainability, wear resistance, high- and low-temperature resistance, transparency, and other characteristics. Therefore, in this era of rapid development, the development of multifunctional intelligent devices based on MOF-derived MAMs will be a promising field. Over the past decade, MOF's novel conditioning technology has made encouraging progress in structural optimization and performance regulation of EMW absorption materials. Despite the promising prospects, it is clear that much work remains to be done. It is hoped that this chapter can promote the innovative development of MOF-derived EMW absorption materials and overcome the bottleneck of MOF design in the field of EMW absorption materials.

References

1 Yaghi, O.M. and Li, H. (1995). Hydrothermal synthesis of a metal-organic framework containing large rectangular channels. *J. Am. Chem. Soc.* 117: 10401–10402.

2 Li, H., Eddaoudi, M., O'Keeffe, M., and Yaghi, O.M. (1999). Design and synthesis of an exceptionally stable and highly porous metal-organic framework. *Nature* 402: 276–279.

3 Férey, G. (2008). Hybrid porous solids: past, present, future. *Chem. Soc. Rev.* 37: 191–214.

4 Kitagawa, S., Kitaura, R., and Noro, S.-i. (2004). Functional porous coordination polymers. *Angew. Chem. Int. Ed.* 43: 2334–2375.

5 Yu, J., Cui, Y., Wu, C. et al. (2012). Second-order nonlinear optical activity induced by ordered dipolar chromophores confined in the pores of an anionic metal-organic framework. *Angew. Chem. Int. Ed.* 51: 10542–10545.

6 Talin, A.A., Centrone, A., Ford, A.C. et al. (2014). Tunable electrical conductivity in metal-organic framework thin-film devices. *Science* 2014 (343): 66–69.

7 Kitaura, R., Kitagawa, S., Kubota, Y. et al. (2002). Formation of a one-dimensional array of oxygen in a microporous metal-organic solid. *Science* 298: 2358–2361.

8 Gao, Z., Yang, K., Zhao, Z. et al. (2023). Design principles in MOF-derived electromagnetic wave absorption materials: review and perspective. *Int. J. Miner. Metall. Mater.* 30 (3): 405–427.

9 Ashworth, C. (2017). Metal-organic frameworks: molten MOFs. *Nat. Rev. Mater.* 2: 17074.

10 Chatterjee, A., Seyfferth, J., Lucci, J. et al. (2016). MOF acetyltransferase regulates transcription and respiration in mitochondria. *Cell* 167: 722–738.

11 Li, Y., Wen, G., Li, J. et al. (2022). Synthesis and shaping of metal-organic frameworks: a review. *Chem. Commun.* 58: 11488.

12 Wang, Q. and Astruc, D. (2020). State of the art and prospects in metal-organic framework (MOF)-based and MOF-derived nanocatalysis. *Chem. Rev.* 120: 1438–1511.

13 Wang, K., Zhang, J., Hsu, Y. et al. (2023). Bioinspired framework catalysts: from enzyme immobilization to biomimetic catalysis. *Chem. Rev.* 123: 5347–5420.

14 Lia, S. and Qiang, X. (2013). Metal-organic frameworks as platforms for clean energy. *Energy Environ. Sci.* 6: 1656.

15 Chen, Z., Li, P., Anderson, R. et al. (2020). Balancing volumetric and gravimetric uptake in highly porous materials for clean energy. *Science* 368: 297–303.

16 Erdosy, D.P., Wenny, M.B., Cho, J. et al. (2022). Microporous water with high gas solubilities. *Nature* 608: 712–718.

17 Amooghin, A.E., Garcia, H., Sanaeepur, H. et al. (2022). Fluorinated metal-organic frameworks for gas separation. *Chem. Soc. Rev.* 51: 7427–7508.

18 Li, H., Zhao, S., Zang, S., and Li, J. (2020). Functional metal-organic frameworks as effective sensors of gases and volatile compounds. *Chem. Soc. Rev.* 49: 6364.

19 Dolgopolova, E.A., Rice, A.M., Martin, C.R., and Shustova, N.B. (2018). Photochemistry and photophysics of MOFs: steps towards MOF-based sensing enhancements. *Chem. Soc. Rev.* 47: 4710–4728.

20 Peng, X., Tang, S., Tang, D. et al. (2023). Autonomous metal-organic framework nanorobots for active mitochondria-targeted cancer therapy. *Sci. Adv.* 9: 1736.

21 Zou, J., Wei, G., Xiong, C. et al. (2022). Efficient oral insulin delivery enabled by transferrin-coated acid-resistant metal-organic framework nanoparticles. *Sci. Adv.* 8: 4677.

22 Chamanehpour, E., Sayadi, M.H., and Hajiani, M. (2022). A hierarchical graphitic carbon nitride supported by metal-organic framework and copper nanocomposite as a novel bifunctional catalyst with long-term stability for enhanced carbon dioxide photoreduction under solar light irradiation. *Adv. Compos. Hybrid Mater.* 5: 2461–2477.

23 Yan Guo, H., Liu, D.W., El-Bahy, Z.M. et al. (2022). Engineering hierarchical heterostructure material based on metal-organic frameworks and cotton fiber for high-efficient microwave absorber. *Nano Res.* 15: 6841–6850.

24 Sun, Z., Wang, M., Fan, J. et al. (2021). TiO_2@MIL-101(Cr) nanocomposites as an efficient photocatalyst for degradation of toluene. *Adv. Compos. Hybrid. Mater.* 4: 1322–1329.

25 Cheng, W., Wang, Y., Ge, S. et al. (2021). One-step microwave hydrothermal preparation of Cd/Zr-bimetallic metal-organic frameworks for enhanced photochemical properties. *Adv. Compos. Hybrid. Mater.* 4: 150–161.

26 Zhang, P., Zhang, X., Li, B. et al. (2021). Enhanced microwave absorption performance in an ultralight porous single-atom Co-N-C absorber. *Adv. Compos. Hybrid. Mater.* 4: 1292–1301.

27 Zhang, Z., Cai, Z., Wang, Z. et al. (2021). A review on metal-organic framework-derived porous carbon-based novel microwave absorption materials. *Nano-Micro Lett.* 13: 56.

28 Zhang, X., Qiao, J., Jiang, Y. et al. (2021). Carbon-based MOF derivatives: emerging efficient electromagnetic wave absorption agents. *Nano-Micro Lett.* 13: 135.

29 Vasanth Kumar, K., Charalambopoulou, G., Kainourgiakis, M. et al. (2015). Insights on the physical adsorption of hydrogen and methane in UiO series of MOFs using molecular simulations. *Comput. Theor. Chem.* 1061: 36–45.

30 Jung, K., Kim, J., and Choi, J. (2020). Synthesis of magnetic porous carbon composite derived from metal-organic framework using recovered terephthalic acid from polyethylene terephthalate (PET) waste bottles as organic ligand and its potential as adsorbent for antibiotic tetracycline hydrochloride. *Compos. Part. B: Eng.* 187: 107867.

31 Guan, B.Y., Yan, L., Wang, Y. et al. (2018). Porous iron-cobalt alloy/nitrogen-doped carbon cages synthesized via pyrolysis of complex metal-organic framework hybrids for oxygen reduction. *Adv. Funct. Mater.* 28: 1706738.

32 Gao, S., Zhang, G., Wang, Y. et al. (2021). MOFs derived magnetic porous carbon microspheres constructed by core-shell Ni@C with high-performance microwave absorption. *J. Mater. Sci. Technol.* 88: 56–65.

33 Zhang, C., Li, H., Zhang, Q. et al. (2021). Core-shell structured porous carbon nanofibers integrated with ultra-small SnO_2 nanocrystals for fast and stable lithium storage. *Chem. Eng. J.* 420: 127705.

34 Zeng, Q., Wang, L., Li, X. et al. (2021). Double ligand MOF-derived pomegranate-like Ni@C microspheres as high-performance microwave absorber. *Appl. Surf. Sci.* 538: 148051.

35 Hailong, X., Yin, X., Zhu, M. et al. (2019). Constructing hollow graphene nano-spheres confined in porous amorphous carbon particles for achieving full X band microwave absorption. *Carbon* 142: 346–353.

36 Wang, L., Li, X., Shi, X. et al. (2021). Recent progress of microwave absorption microspheres by magnetic-dielectric synergy. *Nanoscale* 13: 2136–2156.

37 Shu, J., Yang, X., Zhang, X. et al. (2020). Tailoring MOF-based materials to tune electromagnetic property for great microwave absorbers and devices. *Carbon* 162: 157–171.

38 Pan, Y., Yan, S., Liu, Y. et al. (2022). Significantly enhanced electrochemical performance of 2D Ni-MOF by carbon quantum dot for high-performance supercapacitors. *Electrochim. Acta* 422: 140560.

39 Ren, J., Dyosiba, X., Musyoka, N.M. et al. (2017). Review on the current practices and efforts towards pilot-scale production of metal-organic frameworks (MOFs). *Coord. Chem. Rev.* 352: 187–219.

40 Rubio-Martinez, M., Avci-Camur, C., Thornton, A.W. et al. (2017). New synthetic routes towards MOF production at scale. *Chem. Soc. Rev.* 46: 3453–3480.

41 Wang, Z., Liu, L., Li, Z. et al. (2022). Shaping of metal-organic frameworks: a review. *Energy Fuels* 36: 2927–2944.

42 Liu, X.M., Xie, L.H., and Wu, Y. (2020). Recent advances in the shaping of metal–organic frameworks. *Inorg. Chem. Front.* 7: 2840–2866.

43 Valizadeh, B., Nguyen, T.N., and Stylianou, K.C. (2018). Shape engineering of metal-organic frameworks. *Polyhedron* 145: 1–15.

44 Wang, H., Yuan, X., Yan, W. et al. (2015). Facile synthesis of amino-functionalized titanium metal-organic frameworks and their superior visible-light photocatalytic activity for Cr (VI) reduction. *J. Hazard. Mater.* 286: 187–194.

45 Li, F., Wang, P., Huang, X. et al. (2019). Large-scale, bottom-up synthesis of binary metal-organic framework nanosheets for efficient water oxidation. *Angew. Chem. Int. Ed.* 58: 7051–7056.

46 Choi, Y., Chen, T., Kim, D. et al. (2021). Transformation of microwave synthesized highly uniform FeMo-MIL-88B nanorod to oxynitride derivate for overall water splitting reaction. *Appl. Mater. Today* 24: 101093.

47 Li, Q., Liu, Y., Niu, S. et al. (2021). Microwave-assisted rapid synthesis and activation of ultrathin trimetal-organic framework nanosheets for efficient electrocatalytic oxygen evolution. *J. Colloid Interf. Sci.* 603: 148–156.

48 Chen, Y., Ni, D., Yang, X. et al. (2018). Microwave-assisted synthesis of honeycomblike hierarchical spherical Zn-doped Ni-MOF as a high-performance battery-type supercapacitor electrode material. *Electrochim. Acta* 278: 114–123.

49 Asghar, A., Iqbal, N., Noor, T. et al. (2021). Efficient electrochemical synthesis of a manganese-based metal-organic framework for H_2 and CO_2 uptake. *Green Chem.* 23: 1220–1227.

50 Qin, S., Zhang, S., Chen, M., and Limin, W. (2022). Electrochemical fabrication of long-range ordered macro-microporous metal-organic framework films. *J. Mater. Chem. A* 10: 9497–9505.

51 Liu, Y., Wei, Y., Liu, M. et al. (2021). Electrochemical synthesis of large area two-dimensional metal-organic framework films on copper anodes. *Angew. Chem. Int. Ed.* 60: 2887–2891.

52 Karimi, M., Mohebali, H., Sadeghi, S. et al. (2021). Additive-free aerobic C-H oxidation through a defect-engineered Ce-MOF catalytic system. *Micropor. Mesopor. Mat.* 322: 111054.

53 Haque, E. and Jhung, S.H. (2011). Synthesis of isostructural metal-organic frameworks, CPO-27s, with ultrasound, microwave, and conventional heating: effect of synthesis methods and metal ions. *Chem. Eng. J.* 173: 866–872.

54 Shen, M., Zhou, J., Elhadidy, M. et al. (2022). Cyclodextrin metal-organic framework by ultrasound-assisted rapid synthesis for caffeic acid loading and antibacterial application. *Ultrason. Sonochem.* 86: 106003.

55 Liu, P., Zhao, T., Cai, K. et al. (2022). Rapid mechanochemical construction of HKUST-1 with enhancing water stability by hybrid ligands assembly strategy for efficient adsorption of SF6. *Chem. Eng. J.* 437: 135364.

56 Yi, B., Zhao, H., Cao, L. et al. (2022). A direct mechanochemical conversion of Pt-doped metal-organic framework-74 from doped metal oxides for CO oxidation. *Materials Today Nano* 17: 100158.

57 Wang, Y., Ying, L., Li, Z. et al. (2021). One-pot mechanochemical synthesis to encapsulate functional guests into a metal-organic framework for proton conduction. *Chem. Commun.* 57: 8933–8936.

58 Shi, Q., Chen, Z., Song, Z. et al. (2011). Synthesis of ZIF-8 and ZIF-67 by steam-assisted conversion and an investigation of their tribological behaviors. *Angew. Chem. Int. Ed.* 50: 672–675.

59 Ahmed, I., Jeon, J., Khan, N.A., and Jhung, S.H. (2012). Synthesis of a metal-organic framework, iron-benezenetricarboxylate, from dry gels in the absence of acid and salt. *Cryst. Growth Des.* 12: 5878–5881.

60 Kim, J., Lee, Y.-R., and Ahn, W.-S. (2013). Dry-gel conversion synthesis of Cr-MIL-101 aided by grinding: high surface area and high yield synthesis with minimum purification. *Chem. Commun.* 49: 7647–7649.

61 Albuquerque, G.H., Fitzmorris, R.C., Ahmadi, M. et al. (2015). Gas-liquid segmented flow microwave-assisted synthesis of MOF-74(Ni) under moderate pressures. *CrystEngComm* 17: 5502–5510.

62 Li, W., Chen, J., and Gao, P. (2022). MOFs-derived hollow Copper-based sulfides for optimized electromagnetic behaviors. *J. Colloid Interf. Sci.* 606: 719–727.

63 Huang, M., Wang, L., Pei, K. et al. (2020). Multidimension-controllable synthesis of MOF-derived Co@N-doped carbon composite with magnetic-dielectric synergy toward strong microwave absorption. *Small* 16: 2000158.

64 Ouyang, J., He, Z., Zhang, Y. et al. (2019). Trimetallic FeCoNi@C nanocomposite hollow spheres derived from metal-organic frameworks with superior electromagnetic wave absorption ability. *ACS Appl. Mater. Interfaces* 11: 39304–39314.

65 Miao, P., Chen, J., Tang, Y. et al. (2020). Highly efficient and broad electromagnetic wave absorbers tuned via topology-controllable metalorganic frameworks. *Sci. China Mater.* 63: 2050–2061.

66 Pan, J., Huadong Yang, Q., Hong, H.W. et al. (2021). Bimetal-organic frameworks derived tuneable Co nanoparticles embedded in porous nitrogen-doped carbon nanorods as high-performance electromagnetic wave absorption materials. *J. Mater. Chem. C* 9: 7302–7309.

67 Liang, C., He, J., Zhang, Y. et al. (2022). MOF-derived CoNi@C-silver nanowires/cellulose nanofiber composite papers with excellent thermal management capability for outstanding electromagnetic interference shielding. *Compos. Sci. Technol.* 224: 109445.

68 Wang, W., Liu, D., Cheng, H. et al. (2022). Structural design and broadband radar absorbing performance of multi-layer patch using carbon black. *Adv. Compos. Hybrid Mater.* 5: 3137–3145.

69 Li, Z., Xie, W., Yao, F. et al. (2022). Comprehensive electrocatalytic degradation of tetracycline in wastewater by electrospun perovskite manganite nanoparticles supported on carbon nanofibers. *Adv. Compos. Hybrid Mater.* 5: 2092–2105.

70 Wu, N., Qiao, J., Liu, J. et al. (2018). Strengthened electromagnetic absorption performance derived from synergistic effect of carbon nanotube hybrid with Co@ C beads. *Adv. Compos. Hybrid Mater.* 1: 149–159.

71 Ruan, K. and Gu, J. (2022). Ordered alignment of liquid crystalline graphene fluoride for significantly enhancing thermal conductivities of liquid crystalline polyimide composite films. *Macromolecules* 55: 4134–4145.

72 Qingmei, H., Yang, R., Yang, S. et al. (2022). Metal-organic framework-derived core-shell nanospheres anchored on Fe-filled carbon nanotube sponge for Strong wideband microwave absorption. *ACS Appl. Mater. Interfaces* 14: 10577–10587.

73 Guo, Y., Wang, D., Jingwen Wang, Y. et al. (2022). Hierarchical HCF@NC/Co derived from hollow loofah fiber anchored with metal-organic frameworks for highly efficient microwave absorption. *ACS Appl. Mater. Interfaces* 14: 2038–2050.

74 Jin, H., Wen, H., Hong, Q. et al. (2022). NiCo@NPC@CF nanocomposites derived from NiCo-MOF/cotton for high-performance electromagnetic wave absorption. *J. Mater. Chem. C* 10: 8310–8320.

75 Wang, S., Yingchun, X., Ruru, F. et al. (2019). Rational construction of hierarchically porous Fe-Co/N-doped carbon/rGO composites for broadband microwave absorption. *Nano-Micro Lett.* 11: 1–16.

76 Li, Q., Zhao, Y., Li, X. et al. (2020). MOF induces 2D GO to assemble into 3D accordion-like composites for tunable and optimized microwave absorption performance. *Small* 16: 2003905.

77 Fei, W., Liu, Z., Wang, J. et al. (2021). Template-free self-assembly of MXene and CoNi-bimetal MOF into intertwined one-dimensional heterostructure and its microwave absorbing properties. *Chem. Eng. J.* 422: 130591.

78 Deng, B., Xiang, Z., Xiong, J. et al. (2020). Sandwich-like Fe&TiO_2@C nanocomposites derived from MXene/Fe-MOFs hybrids for electromagnetic absorption. *Nano-Micro Lett.* 12: 1–16.

79 Guo, J., Chen, Z., El-Bahy, Z.M. et al. (2022). Tunable negative dielectric properties of magnetic $CoFe_2O_4$/graphite-polypyrrole metacomposites. *Adv. Compos. Hybrid Mater.* 5: 899–906.

80 Liang, C., Gu, Z., Zhang, Y. et al. (2021). Structural design strategies of polymer matrix composites for electromagnetic interference shielding: a review. *Nano-Micro Lett.* 13: 181.

81 Luo, W., Wei, Y., Zhuang, Z. et al. (2022). Fabrication of Ti_3C_2Tx MXene/polyaniline composite films with adjustable thickness for high performance flexible all-solid-state symmetric supercapacitors. *Electrochim. Acta* 406: 139871.

82 Wang, W., Ren, J., Wang, C. et al. (2022). Magnetic Fe_3O_4/polypyrrole-salicylaldehyde composite for efficient removal of Mn (VII) from aqueous solution by double-layer adsorption. *J. Appl. Polym. Sci.* 139: e52515.

83 Bi, Y., Ma, M., Liu, Y. et al. (2021). Microwave absorption enhancement of 2-dimensional CoZn/C@MoS_2@PPy composites derived from metal-organic framework. *J. Colloid Interf. Sci.* 600: 209–218.

84 Yu, G., Ye, M., Han, A. et al. (2021). Conductive polymer-decorated lanthanum chromate with remarkable electromagnetic wave absorption properties. *J. Mater. Sci: Mater. Electron.* 32: 24681–24697.

2

2D MXenes for Electromagnetic Wave Absorption

Weibin Deng

Shaanxi Engineering Laboratory for Graphene New Carbon Materials and Applications, School of Materials Science and Engineering, Northwestern Polytechnical University, Xi'an, China

CHAPTER MENU

Electromagnetic Wave Absorbing Materials: Fundamentals and Applications, First Edition.
Edited by Hongjing Wu, Jun Luo, and Meiyin Yang.

2.1 Introduction to MXenes

Since the identification of MXenes by Prof. Yury Gogotsi in 2011 [1], MXenes have garnered significant attention from the scientific community. In recent years, the synthesis of MXenes has witnessed remarkable progress. Due to their unique structure and exceptional properties, MXenes have emerged as one of the most promising materials to date. Their distinctive 2D-layered structure, combined with outstanding mechanical strength, metallic conductivity, favorable hydrophilicity, superior biocompatibility, and versatile tunable surface terminations, has positioned MXenes as versatile materials with wide-ranging applications.

In particular, MXenes have unique prospects in microwave absorption (MA) and electromagnetic interference shielding. The high electrical conductivity makes MXenes easily construct conductive networks and brings about conduction loss. Interestingly, abundant functional groups (OH, O, and/or F groups) lead to a dipole, contributing to polarization loss. Moreover, various structures endow MXenes with tunable dielectric properties, which could optimize MA performance. Nevertheless, the pristine MXenes suffer from a few problems, such as self-agglomeration, easy oxidation, a single-loss mechanism, and impedance mismatch, which limit MA properties and practical application. To improve the MA performance and environmental adaptability of MXenes, strategies of composition and structure design have been extensively developed. Surface modification and combination with lossy materials lead to a richer synergetic loss mechanism, a stronger polarization loss ability, and a good impedance balance. Likewise, micro-/macrostructure manipulation, such as 3D microspheres, fibers, and aerogels, also gains interest from researchers. The coupling effect of multidimensional units enhance the attenuation capacity and endows the MXene-based MA materials with an ultra-low filling ratio. In addition, environmental adaptability can be improved to some extent.

This chapter focuses on the preparation and properties of MXenes, as well as the MA performance of MXene-based materials designed through composition optimization and structure regulation. First, the preparation method of MXenes is presented. Second, the properties of MXenes are briefly introduced. Subsequently, the recent advances in MXene-based MA materials are summarized. In detail, strategies based on components (such as dielectric, magnetism, and multiple loss materials) and micro-/macrostructure manipulation (such as 3D microspheres, fibers, sandwiches, and hierarchical structures) are categorized, while the highlights and MA properties of related works are detailed to facilitate the innovative material design and explore MA mechanism in depth. Finally, the development potential of absorbers based on MXenes has been projected, and the remaining problems are analyzed.

2.2 Preparation Method of MXenes

MXenes are composed of few-atom-thick layers of transition metal nitrides, carbides, or carbonitriles. MXenes have a general chemical formula of $M_{n+1}X_nT_x$ (n = 1–3) and are synthesized by the delamination of these compounds called MAX phases. In the chemical formula, M signifies a transition metal, X is C and/or N, and T_x represents the oxygen, fluorine, or hydroxyl terminals resulting from the synthetic precursors (Figure 2.1a).

As the precursor to MXenes, the MAX phases are generally a ternary layered compound, in which M is a transition metal, A is an element of group IIIA or IVA in the periodic

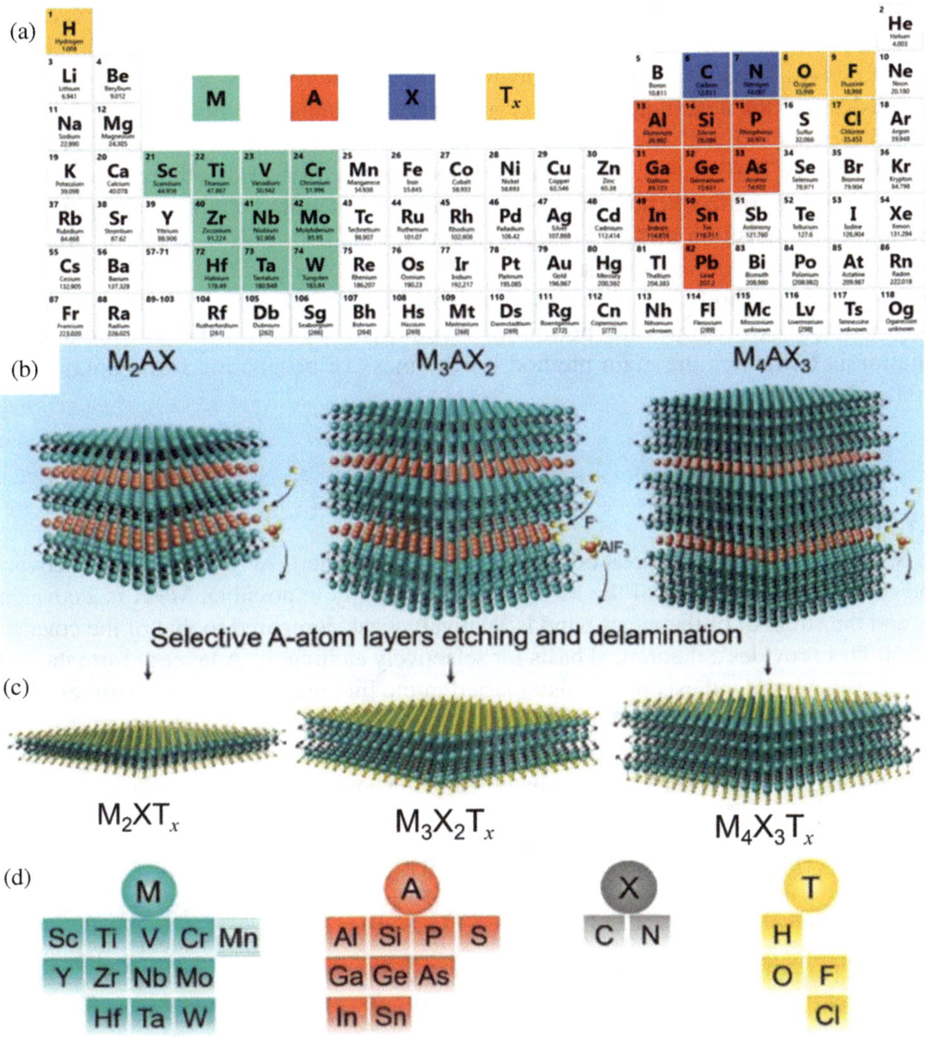

Figure 2.1 (a) Composition of MAX and MXene phases by the elements of the periodic table. Source: Zhang et al. [2]/with permission of John Wiley & Sons. (b) Three different types of precursors, M_2AX, M_3AX_2, and M_4AX_3, and selective etching of the A group layers (represented in red color). (c) MXenes are formed after selective etching with surface terminals (presented in yellow). (d) Possible elements of MAX and MXenes. Source: Verma et al. [3]/with permission of John Wiley & Sons.

table, and X is carbon or nitrogen (Figure 2.1b). Graphite possesses a weak van der Waals force among the layers, allowing for micromechanical exfoliation to obtain graphene flakes. Unlike graphite, the M—X covalent bonds and M—A metallic bonds in the MAX phase are quite strong; one can hardly break these bonds to make MXenes through a mechanical exfoliation method. Moreover, MAX is not susceptible to chemical reactions in common acids (HCl, H_2SO_4, and HNO_3), bases (NaOH, KOH), or salts (Na_2SO_4, K_2SO_4), regardless of their concentration. Fortunately, M—A bonds are weaker than the M—X bonds; therefore, $M_{n+1}X_n$ layers are quickly delaminated. The reactive M—A bonds can be broken, and the A element layers can be selectively removed, resulting in multilayered m-MXenes by employing strong etchants such as hydrofluoric acid (HF), lithium fluoride-hydrochloric acid mixtures (LiF-HCl), or ammonium hydrogen bifluoride (NH_4HF_2).

Obviously, after the delamination of A group layers, the transition metals (13 or 16 group elements) are exposed, and M-elements end with T_x surface functional groups to form corresponding $M_{n+1}X_nT_x$ (Figure 2.1c). The arrangement of exterior terminations is determined using the chemistry of the etching environment and post-treatment methods, along with the nature of the M (transition metal). Figure 2.1d denotes the list of M, T, and X, as well as a few A group elements of MAX phases.

To date, many methods have been used to fabricate 2D MXenes, including top-down etching-assisted exfoliation and bottom-up synthesis. Note that the etching-assisted exfoliation is nowadays the main method for the massive production of monolayer and multilayer MXenes.

2.2.1 Top-Down Strategy

2.2.1.1 HF Etching

In the MAX phase, M—A is a metal bond. Due to the strong interlayer chemical bond between M and A, mechanical peeling of the MAX phase is virtually impossible. M—X is a covalent bond, and the strength of the metal bond is relatively weak compared to that of the covalent bond [4]. This provides a theoretical basis for selectively etching the A layer to form the MX 2D nanolayer. Most synthesis protocols reported in the literature use HF. An early example is $M_{n+1}X_nT_x$ ($Ti_3C_2F_2$ and $Ti_3C_2(OH)_2$), created by the removal of the weakly bound Al-atom layers in the Ti_3AlC_2. The synthetic reactions occur at room temperature with aqueous HF as the etchant (Figure 2.2a). The reactions involved in the aqueous HF are as follows:

$$Ti_3AlC_2(s) + 3HF(aq) = Ti_3C_2(s) + AlF_3(aq) + \frac{3}{2}H_2(g) \quad (2.1)$$

$$Ti_3C_2(s) + 2HF(aq) = Ti_3C_2F_2(s) + H_2(g) \quad (2.2)$$

$$Ti_3C_2(s) + 2H_2O(aq) = Ti_3C_2(OH)_2(s) + H_2(g) \quad (2.3)$$

Reactions (2.2) and (2.3) lead to different surface terminations with hydroxyl or fluorine groups. The as-prepared MXenes consist of multilayer flakes that stack together with weak interlayer interactions. Ultrasonication of the resulting products leads to the exfoliation of MXene nanosheets.

2.2.1.2 In Situ HF Etching

Given that HF is hazardous, an alternative milder etchant was developed for the in situ production of HF for selective etching. LiF-HCl serves as one of the most widely used etchants currently for the synthesis of high-quality MXenes. Li^+ acts as a pre-intercalation layer, weakening the interaction between the plates and making ultrasonic dispersion easier. Moreover, this mild fabrication method leads to a larger planar size with fewer atomic defects than pure HF etching (Figure 2.2b).

$$LiF(aq) + 2HCl(aq) = HF(aq) + LiCl(aq) \quad (2.4)$$

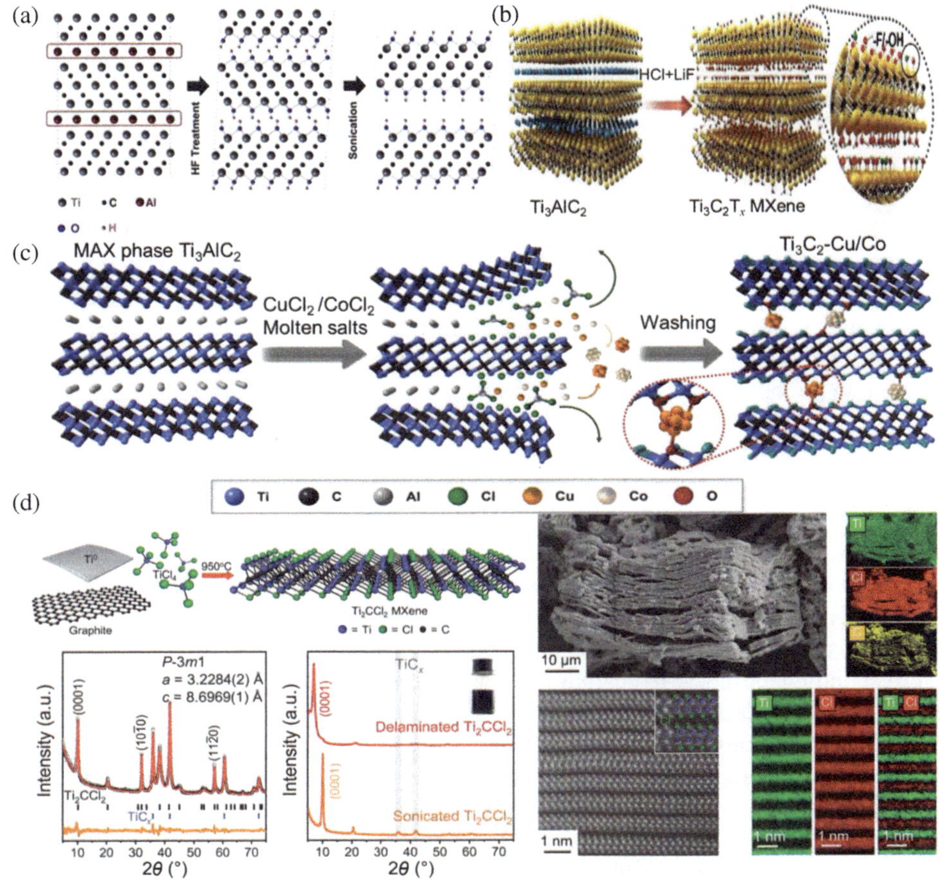

Figure 2.2 (a) Schematic diagram of the process for HF etching and delamination of MXenes. Source: Naguib et al. [1]/John Wiley & Sons. (b) Schematic illustration of the preparation of $Ti_3C_2T_x$ via in situ HF etching. Source: Li et al. [5]/with permission of John Wiley & Sons. (c) Ti_3AlC_2 MAX phase is immersed in $CuCl_2$ Lewis molten salt. The reaction between Ti_3AlC_2 and $CuCl_2$ results in the formation of $Ti_3C_2T_x$. Source: Bai et al. [6]/with permission of Springer Nature. (d) Direct synthesis and characterization of Ti_2CCl_2 MXene via chemical vapor deposition (CVD). Source: Wang et al. [7]/with permission of American Association for the Advancement of Science.

2.2.1.3 Alkaline Solution Chemical Etching

Alkalis are expected to achieve selective etching of the MAX phase. Because of the extremely strong combination ability of alkaline and amphoteric element Al, it is theoretically feasible to utilize the alkaline solution to etch the Al element of the "A" layer in MAX. High temperature and concentrated sodium hydroxide can easily dissolve aluminum (oxide) hydroxide to form fluorine-free MXenes. The fluorine-free nature provides more OH and O terminals, which contribute to the improvement of electrochemical performance. In addition, the substitution of aluminum atoms by OH groups can form 2D $Ti_3C_2(OH)_2$ nanosheets with larger lateral dimensions. In this method, the concentration of alkali can reach 87.5%. In conclusion, etching the MAX phase with concentrated alkali is effective and can yield highly hydrophilic products without F at the ends.

2.2.1.4 Electrochemical Etching

Ti_2CT_x can be fabricated from Ti_2AlC in aqueous HCl using an electrochemical strategy. The electrochemical etching strategy can ensure that no fluoride ions appear in the etching process, and there will only be chloride (–Cl) and hydroxyl (–OH) group components on the surface of the as-prepared MXenes. The electrochemical etching process for fabricating $Ti_3C_2T_x$ is depicted, in which two identical Ti_3AlC_2 blocks are utilized as symmetrical electrodes [8]. The Ti_3AlC_2 cathode remains intact while the Ti_3AlC_2 anode is partially intact. Al was corroded by Cl^- containing electrolytes because of the vigorous reaction between Cl^- and Al. In addition, alkaline solutions can also corrode Ti_3AlC_2 because of the violent reaction of OH^- ions with Al. The bonding strength of Ti—C bonds is stronger than that of Ti—Al bonds, resulting in the preferential removal of Al. Moreover, Li ions enter the electrolyte and can be inserted into the $Ti_3C_2T_x$ sheet, which not only weakens the interaction between adjacent layers and widens the interlayer spacing but also promotes corrosion progress and accelerates m-$Ti_3C_2T_x$ delamination.

2.2.1.5 Molten Salt Etching

The aqueous etching approaches are suitable for most Al-containing MAX phases under low operating temperatures, producing hydrophilic MXenes. However, the aqueous system fails with the non-Al MAX or nitride MAX phases. The theoretical estimation confirmed that the transformation of $Ti_{n+1}AlN_n$ to $Ti_{n+1}N_n$ had a higher energy barrier than that of the $Ti_{n+1}AlC_n$ to $Ti_{n+1}C_n$, reflecting the thermodynamic limitation to etch the $Ti_{n+1}AlN_n$ due to strong bonding between Ti and Al atoms. In addition, the cohesive energy of $Ti_{n+1}N_n$ was lower than that of its corresponding carbides, indicating its poor structural stability, which would allow its dissolution in the fluorine-containing solution [9].

Transition metal halides in a molten state can react with the A layer of the MAX phase as they are electron acceptors (Figure 2.2c). The chloride Lewis acid salts (such as $CuCl_2$ and $CoCl_2$) are generally utilized as etchants to fabricate MXenes with chloride terminations [6]. The molten salt etching method utilizes molten salt as the etchant instead of HF, making it less hazardous than common procedures. The salt composition and temperature are carefully controlled to achieve controlled etching. Ti_3AlC_2, Ti_2AlC, Ti_2AlN, and V_2AlC MAX phases were etched in a mixed $ZnCl_2$/NaCl/KCl molten salt system [10]. $ZnCl_2$ was employed as an etchant for the etching of the MAX phase, while NaCl and KCl with a molar ratio of 1:1 were used for forming the molten salt bath and reducing the melting point of the eutectic system. In the etching process, Zn^{2+} reacts with the A atoms of the MAX phase, where the weakly bonded Al atoms are transformed into Al^{3+}. The reduced Zn atoms subsequently occupy the A layer sites, forming a new Zn-MAX phase, Ti_3ZnC_2. The excessive $ZnCl_2$ then etches the interlayer Zn atoms in the Zn-MAX phase to produce MXenes. The etching process can be described as follows by etching Ti_3AlC_2 as a sample:

$$Ti_3AlC_2(s) + 1.5ZnCl_2(s) = Ti_3ZnC_2(s) + 0.5Zn(s) + AlCl_3(g) \tag{2.5}$$

$$Ti_3ZnC_2(s) + 1.5ZnCl_2(s) = Ti_3C_2(s) + 1.5Zn(s) + AlCl_3(g) \tag{2.6}$$

2.2.2 Bottom-Up Strategies

2.2.2.1 Chemical Vapor Deposition

The chemical vapor deposition (CVD) method was employed to prepare high-quality MXenes, with the advantages of using the MAX phase, low cost, large scalability, and full controllability. A 2D ultrathin Mo_2C crystal was prepared using the CVD approach [10], where methane and copper/molybdenum foil were used as the carbon source, substrate, and molybdenum source, respectively. The Mo_2C crystals had a thickness of several nanometers and a stable structure at ambient conditions. The buckle MXenes, such as Ti_2CCl_2, Ti_2NCl_2, Zr_2CCl_2, and Zr_2CBr_2, are prepared with metal (Ti or Zr foil), metal halides ($TiCl_3$, $TiCl_4$, $ZrCl_4$, or $ZrBr_4$), and carbon or nitrogen sources (graphite, methane, or N_2), respectively [7]. To be specific, Ti_2CCl_2 was synthesized through a hermetical reaction between Ti, graphite, and $TiCl_4$ (Figure 2.2d). Pure Ti_2NCl_2 can even be obtained with Ti foil, $TiCl_4$, and N_2. Atomic-resolution aberration-corrected STEM images reveal that Ti_2CCl_2 has an almost ideal 1:1 stoichiometric ratio, and –Cl groups are fully covered on the surface.

In conclusion, the current methods for preparing 2D MXenes can be mass-produced, but the large number of functional groups generated during the etching process seriously affects the performance of MXenes. For example, MXenes without functional groups exhibit metallic properties, while MXenes with functional groups are semiconductors. Therefore, how to prepare pure MXenes is an important research direction in the future. Moreover, the structure and morphology of MXenes prepared by different methods are also different, so appropriate methods should be selected according to research requirements.

2.3 The Properties of MXenes

2.3.1 Morphologies and Surface Chemistries

The etching method and etchant concentration heavily influence the resultant MXenes morphologies. The types of MXenes include multilayer (accordion), few-layer, and single-layer (SL) MXenes. Concerning functional groups, studies have confirmed that there is a random distribution of MXenes surface terminations, rather than having a specific type of functionality in one region [11]. According to the density functional theory (DFT) calculations, –OH and –F are directly bonded to the surface of MXene flakes, and water molecules are hydrogen-bonded to the –OH groups [12].

2.3.2 Mechanical Properties

Delaminated MXenes possess excellent mechanical flexibility, particularly for the MXenes monolayer. MXenes' mechanical properties strongly depend on their surface terminations. The O-terminated MXenes have very high stiffness, but MXenes terminated by other groups (F and OH) show lower elastic stiffness than their O-terminated counterparts. This may be related to the different lattice constants of MXenes with different terminations: usually, the O-terminated MXenes have smaller lattice parameters than the F- or OH-terminated MXenes. Compared with bare MXenes, the surface-functionalized MXenes have larger flexibility. The functionalization would reduce Young's modulus of Ti_2C, but the functionalized Ti_2C can withstand a larger strain than bare Ti_2C [13]. The surface-terminated groups act as a buffer layer for Ti_2C under tensile deformation, slowing down the collapse of Ti layers and increasing the value of the critical strain where Ti_2C fractures (Figure 2.3).

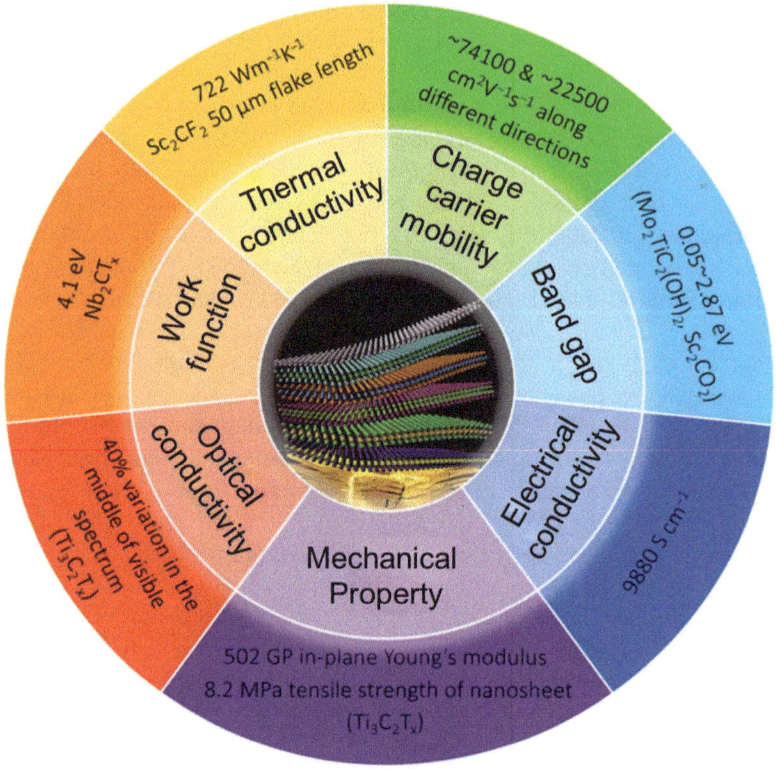

Figure 2.3 Summary of the properties of MXenes: thermal conductivity, work function, band gap, mechanical property, electrical conductivity, optical conductivity, charge carrier mobility. Source: Zhang et al. [2]/with permission of John Wiley & Sons.

2.3.3 Electronic, Transport, and Band Gap Properties

MXenes exhibit metallic behavior with a substantial electron density near the Fermi level. The metallic conductor properties can be regulated not only by the formation of additional Ti—X bonds but also by changing to narrower band-gap semiconductors by adjusting the species and orientations of the surface termination on MXenes. Increasing the heat treatment temperature could further increase the conductivity of Ti_3C_2 due to the decrease in surface functional groups and the formation of shorter conduction paths [14]. Moreover, by controlling the Ti-vacancy concentration in $Ti_3C_2T_x$, the measured conductivity of SL $Ti_3C_2T_x$ was up to $6.76 \times 105\,S\,m^{-1}$, since the defective –OH-terminated SL Ti_3C_2 was still metallic [15]. Additionally, the conductivity and transport of MXenes can be adjusted via plasma treatment or by introducing conductive materials to form a cross-linked and electrically conducting network [16].

Band gap can be engineered by surface termination modification, metal atom substitution, strain modulation, and hybridization. The surface groups of Ti_2CT_x can adjust the band gap of the Ti_2C layer without affecting the basic framework [17]. The dominant charge transport of activated carriers over the narrow energy gaps of the transition metal carbides leads to less sensitivity to the measured transport gap for field-effect mobility at room temperature. Therefore, these excellent electronic, transport, and band-gap properties make MXene-based materials feasible in MA applications.

2.3.4 Thermal Stability Properties

The thermal stability of MXenes is strongly dependent on their chemical composition and the environment. The $Ti_3C_2T_x$ (T_x = F or OH) was stable at temperatures of up to 800 °C in Ar atmosphere. On the contrary, it could be oxidized into anatase TiO_2 partly at 200 °C and fully transformed into rutile TiO_2, CO_2, and H_2O at 1000 °C in an oxygen atmosphere [18]. Owing to a large portion of exposed metal atoms on the surface, the MXenes were usually thermodynamically metastable with high surface energy, thereby suffering from poor oxygen resistance in the air [19]. $Ti_3C_2T_x$ can be converted into carbon-sheet-supported TiO_2 of different crystal structures, particle sizes, and morphologies with the control of temperature, heating rate, and oxidation time [20]. Those properties make MXenes a viable support or template for synthesizing MXene-based hybrids or derivatives.

2.4 Electromagnetic Wave Absorption Performance of Pure MXenes

The dielectric properties of Ti_3AlC_2, $Ti_3C_2T_x$, and annealed $Ti_3C_2T_x$ in a wax matrix with a loading of 50 wt% were compared [21]. Acid etching leads to the Al layers being replaced by surface groups and imparts higher conductivity to MXenes, which results in a significant increase in permittivity and tan δ values. The heat treatment modified the active surface of $Ti_3C_2T_x$ to form amorphous carbon with high permittivity, and the average values of the real and imaginary parts of the permittivity were increased to $\varepsilon' = 14.55$ and $\varepsilon'' = 5.42$, respectively. The electromagnetic wave absorption (EWA) performance of Ti_3C_2 MXenes with different etching states was investigated [22]. The morphology of $Ti_3C_2T_x$ gradually changed from an accordion structure and finally to ultrathin dispersed sheets with increasing etching. The effective absorption bandwidth (EAB) of the accordion microstructure reaches 3.92 GHz at 1.27 mm, while for the $Ti_3C_2T_x$ sheets, the EAB is increased to 4.40 GHz at 1.26 mm.

The etching time can adjust the morphology, structure, surface termination, and dielectric properties of MXenes [23]. The layered structure of $Ti_3C_2T_x$ becomes more obvious as the etching time increases, and the dielectric properties are enhanced. However, the crystal structure and surface morphology of $Ti_3C_2T_x$ were gradually destroyed, and more C was exposed with growing etching time. As time prolongs, the interlayer spacing increases, and the defects, conductivity, surface groups, and surface oxide layer first increase and then decrease. The permittivity and attenuation ability of $Ti_3C_2T_x$ can be significantly improved but decrease the impedance matching.

2.4.1 Content

Wu's team prepared SL $Ti_3C_2T_x$ and studied its EM response under different filler loadings [24]. With the increasing loadings of $Ti_3C_2T_x$, the reflection coefficient S11 and transmission coefficient S21 both show an upward trend, which means that excessive conductivity leads to enhanced surface reflection. Among the SL-$Ti_3C_2T_x$-22% sample, it has the lowest S11 value and the highest S21 value at high frequency due to the well-matched impendence and obtains an RL_{min} of −45.5 dB and EAB of 6.88 GHz at 1.88 mm. In addition, SL-$Ti_3C_2T_x$ delivered superior MA performance compared to multilayer $Ti_3C_2T_x$ because of enhanced dielectric loss resulting from conductive loss, dipolar polarization, and interfacial polarization.

Cao and coworkers investigated the EWA performance of the delaminated $Ti_3C_2T_x$ [25]. All $Ti_3C_2T_x$/wax composites with different concentrations showed excellent EWA. Especially for 40 wt% composites, an RL_{min} value of−47.9 dB and EAB of 3.6 GHz were achieved at 2.5 mm.

2.4.2 Functional Groups and Defects

The surface chemistry affects the electronic structure and electrical properties of MXenes and may enable the conversion from metallic to semiconducting properties [26]. The work function, as an indicator of the magnitude of the conductor's ability to bind electrons, may provide positive insight into understanding the effects of $Ti_3C_2T_x$ termination. The O terminal group exhibits the highest work function, whereas an F-containing group decreases the work function [27]. Thus, surface functional group modulation can affect the Fermi energy level and the electrical properties of MXenes. The surface functional groups and defects of MXenes can serve as the polarization centers that increase dipole polarization. However, the abundant surface-terminating groups can lead to prominent intrinsic defects and decreased electrical conductivity. The surface structure of MXenes can be adjusted by the etching time and annealing treatment, controlling the surface structure and dielectric loss of MXenes [28]. Compared with Ti_3AlC_2 powder, the optimized $Ti_3C_2T_x$ exhibits excellent EWA performance because of appropriate conductivity and abundant defects and functional groups. Interestingly, $Ti_3C_2T_x$-36h achieves an RL_{min} of −65 dB, and the EAB is 5 GHz at 1.4 mm.

Layered MXene, network-like MXene nanoribbons, porous MXene monolayers, and porous MXene layers are rationally constructed with favorable microstructures and surface states for electromagnetic wave (EMW) absorption [29]. The proper electrical conductivity, a large specific surface area, and abundant defects can be achieved in MXene-based nanostructures by tuning the microstructure (layer, ribbon, monolayer, sheet) and surface state (F^-, OH^-, Cl^- terminals). Zhao et al. compare terahertz electromagnetic shielding of micrometer-thick $Ti_3C_2T_x$ on polyimide prepared by two methods: etched using LiF/HF or HCl/HF from the MAX phase [30]. The latter process shows a larger absorption coefficient as well as higher electrical conductivity. The T_x termination of MXene type A is fluorine, while MXene type B is oxygen. Thus, the oxygen termination in type B facilitates Ti–O–Ti bridging between adjacent MXene flakes, resulting in a larger area of continuous flake region.

The removal of –OH and –F surface groups can improve the electrical conductivity and decrease impedance matching, which is confirmed by annealed $Ti_3C_2T_x$ [31]. Besides, the formation of pores and cracks in the flakes destroys the MA mechanism. The concentration of Ti vacancies in MXenes can be controlled after chemical etching in a sulfuric acid solution for improved shielding performance [32]. The enhancement of electromagnetic interference shielding efficiency (EMI SE) originates from the formation of electric dipoles induced by the accumulation of surface charges near the positions of Ti vacancies (Figure 2.4c).

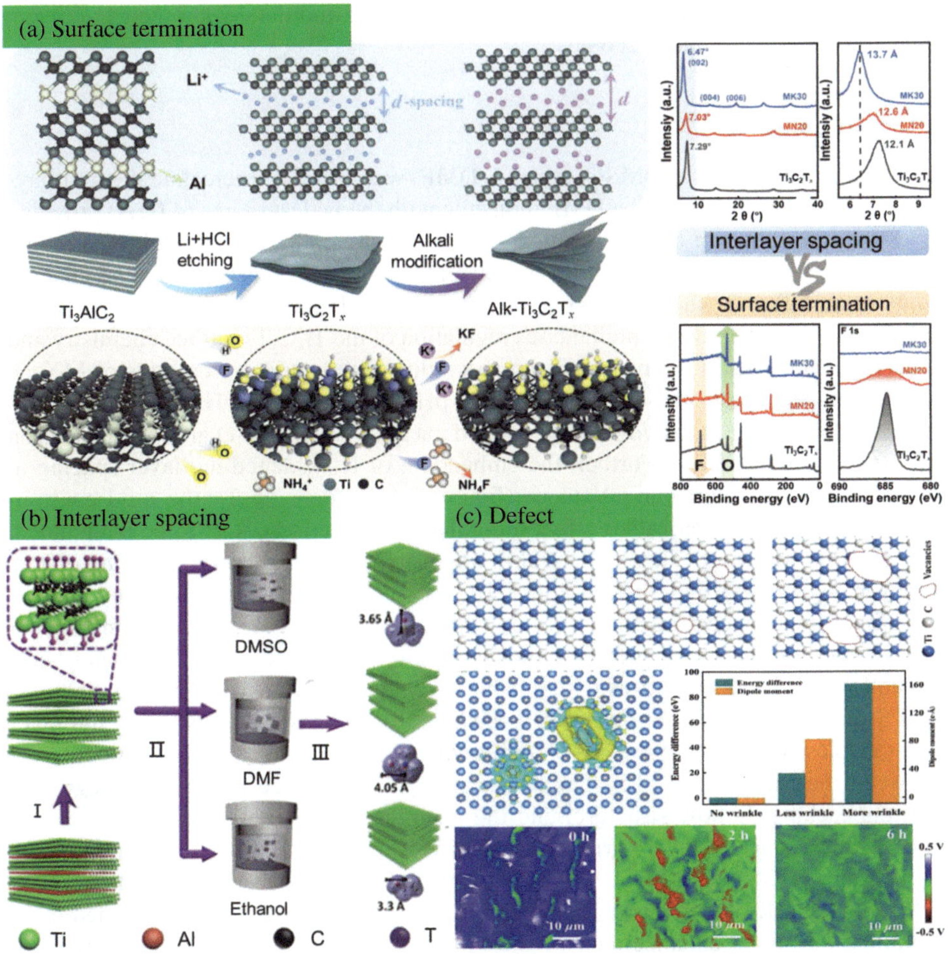

Figure 2.4 (a) Design concepts for the simultaneous adjustment of layer spacing and termination in alk-$Ti_3C_2T_x$. Source: Du et al. [27]/with permission of John Wiley & Sons. (b) Schematic illustration of the preparation of MXene nanosheets with solvent-regulated properties. Source: Xu et al. [33]/ with permisssion of IOP Publishing. (c) Formation of electric dipoles in wrinkled MXene films. The schematic diagram illustrates the ideal crystal structure, single-atom Ti vacancy, and Ti-vacancy cluster. It also depicts the charge density distribution near the positions of Ti vacancies, the energy difference between pristine and wrinkled MXenes, and the dipole moments induced by Ti vacancies. Additionally, it shows the surface potential distribution in the wrinkled MXenes under chemical etching. Source: Ying et al. [32]/with permission of John Wiley & Sons.

2.4.3 Size

The flake size of MXenes plays an important role in the MA property, which is determined by the precursor MAX. The differently sized Ti_3AlC_2 precursors were etched in an HF solution to convert to $Ti_3C_2T_x$ flakes with different lateral dimensions and thicknesses [31]. The large $Ti_3C_2T_x$ flakes derived from the 120-minutes-prepared Ti_3AlC_2 precursor have an attractive RL value of −28.11 dB. The main mechanism of $Ti_3C_2T_x$ is ascribed to the dielectric loss and internal multiple reflection. The richer the functional groups and the thicker their flakes are, the better the MA performance becomes.

2.4.4 Interlayer Spacing

Different intercalating agents (DMSO, ethanol, DMF) were used to intercalate the multilayered $Ti_3C_2T_x$ [33]. Among them, the DMF molecule with the widest radius (4.05 Å) provided the largest layer spacing for the intercalated $Ti_3C_2T_x$ (Figure 2.4b). Expanded layer spacing results in stronger polarization behaviors and increased dielectric loss capability, so that the DMF-intercalated $Ti_3C_2T_x$ shows the RL_{min} of −41.9 dB at 1.1 mm.

Che and coworkers achieved simultaneous regulation of the $Ti_3C_2T_x$ surface chemistry and layer spacing using alkali treatment, which can modify the electrical properties of MXene and optimize impedance matching [27]. The high pH generated by KOH drove the conversion of –F to –OH, whereas dehydration generated many unfavorable –O groups for electron transport (O/F ratio of 33.84). Further, the embedding of K^+ widened the layer spacing to 13.7 Å, increasing the interlayer resistance (Figure 2.4a). Interestingly, the mild environment caused by the weak ionization effect of ammonia allowed the –OH to be maintained, favoring electron transport. Thus, the weakly alkali-treated $Ti_3C_2T_x$ optimized the effects of interlayer engineering and surface chemistry on electrical properties, thereby achieving an optimal impedance matching, excellent MA (−49.1 dB) at 1.7 mm and an EAB of 3.9 GHz. In addition, they used electrostatic adsorption to integrate 2D $Ti_3C_2T_x$ nanosheets into 3D hollow bowl-like structures with tunable layer stacking thickness [34]. On increasing the PS-to-MXenes mass ratio, the wall thickness of the MXene bowls (i.e. the layer stacking thickness of the MXenes) decreased, owing to the increased electrostatic adsorption between the MXenes and PS. The Ti vacancies (V_{Ti}), lattice distortions, degree of aggregation/stacking, hollow structure, and the resultant conductivity/permittivity can be regulated with the stacked-layer variations. Moreover, DFT calculations indicated that the density of states (DOS) from the Ti d orbital near the Fermi level and the dipole moment of the MXene nanosheets increased upon reducing the stacked layers. This provided theoretical validation of the enhancement of dielectric loss ability via structural design and stacked-layer regulation.

The EWA properties of Nb_2CT_x after a solvothermal/hydrothermal treatment with different solvents, including ethanol and DI water, were studied [35]. The results of XRD demonstrated enlarged layer spacing because the typical peak of (0 0 2) of Nb_2CT_x at 10.34° moved to a lower degree at 8.89° and 9.42° for Et-Nb_2CT_x and DI-Nb_2CT_x, respectively. The EWA performance of Et-Nb_2CT_x was superior; the RL_{min} value could reach −52.2 dB at 2.90 mm.

2.4.5 Doping

Element doping can induce lattice distortion and thus affect material properties (e.g. conductivity and polarization behaviors). For S-doped $Ti_3C_2T_x$, the doped-S mainly exists in the Ti—S bonds and surface-adsorbed SO_2 [36]. With the increasing annealing temperature, the S—O bond strength decreases, and the Ti—S bond content increases. Meanwhile, the formation of TiO_2 and amorphous carbon leads to conductivity enhancement. Thus, the conductivity and polarization capability can be regulated, determining the dielectric properties of S-doped $Ti_3C_2T_x$. N-doped $Ti_3C_2T_x$ is prepared using radio-frequency N_2 plasma treatment at room temperature, avoiding unnecessary high-temperature products [37]. Based on the

Stöber criterion, the doped-N atoms with lone-pair electrons bring about the excess DOS and provide $Ti_3C_2T_x$ with a certain magnetic loss capacity, which can be proved from the permeability imaginary curves. In addition, the doped-N atoms also can balance the impedance matching, contributing to the RL_{min} value of −59.2 dB (1.9 mm) of the N-doped $Ti_3C_2T_x$.

Although these works have fully proved that pure MXene nanosheets have substantial potential in the preparation of thin layers and broadband EMW-absorbing materials, the EMW-absorbing material of pure MXenes, due to its non-magnetism, only has the dielectric loss mechanism of EMW absorption without magnetic loss. At the same time, due to the higher dielectric constant, the impedance is not matched so that more EMW are reflected, which limits the further improvement of EMW absorption performance.

2.4.6 Self-Transformation

MXene is easily oxidized under certain conditions, and the oxidation product is TiO_2. The oxidation process generates TiO_2 and C in situ, and the MXene/TiO_2/C can be formed, optimizing the dielectric properties and MA performance. To moderate the excessive conductivity and enhance the dielectric loss of pure MXene, researchers have turned their attention to studying derivatives produced through its self-transformation. Within MXene 2D flakes, three Ti layers alternate with two C layers, rendering it a precursor for achieving an optimal 2D carbon layer structure. The outer transition metal layer and functional groups are susceptible to oxidation, leading to the formation of planar TiO_2 particles. Therefore, the derivatives commonly contain embedded TiO_2 nanoparticles within the carbon layer, which prevents carbon layer agglomeration and introduces more interfaces. Furthermore, the composites feature an abundance of defects and residual oxygen functional groups like oxygen vacancies, grain boundaries, hydroxyl groups, and carbonyl groups, which can serve as polarization centers. Currently, the main self-transformation strategies for MXenes include calcination, hydrothermal, and oxidation methods.

The calcination treatment is widely used for MXene self-transformation. The effect of the atmosphere on structures, compositions, and MA of MXene derivatives was studied [38]. After heat treatment at 800 °C with a CO_2 atmosphere, TiO_2 particles grow in situ on the carbon layer with a large number of defects. Yin and coworkers prepared the annealed Ti_3C_2 and investigated its microstructure and dielectric properties [21]. Owing to the outward migration and oxidation reaction of Ti atoms at a high temperature, new amorphous carbon and defective TiO_2 nanoparticles appear on the surface of $Ti_3C_2T_x$, facilitating interfacial polarization, optimizing the impedance matching, and resulting in increased dielectric parameters. Ultimately, the annealed $Ti_3C_2T_x$ achieves an RL_{min} of −48.4 dB at 1.7 mm. Additionally, Cao and coworkers reported the MA performances of $Ti_3C_2T_x$/TiO_2 composites obtained by calcining $Ti_3C_2T_x$ powder [39]. At 450 °C, the composite exhibited optimal impedance matching and heterogeneous interfaces, resulting in an EAB of 3.04 GHz. In short, $Ti_3C_2T_x$-derived C/TiO_2 hybrids cannot be achieved in an oxygen-free atmosphere. For example, only local oxidation was achieved when $Ti_3C_2T_x$ was annealed in an Ar or N_2 atmosphere [19, 21].

The hydrothermal method is also an approach for the in situ derivatization of MXenes [40]. The anatase-type TiO_2 phase forms through the utilization of Ti atoms within $Ti_3C_2T_x$ sheets as both Ti sources and nucleation centers. By regulating the oxidation extent through reaction duration and temperature, TiO_2 nanosheets can be uniformly dispersed around the layered $Ti_3C_2T_x$. Additionally, under acidic hydrothermal conditions, titanium atoms within

$Ti_3C_2T_x$ might transform into hydrated Ti^{3+} ions, which subsequently oxidize to TiO_2 at the defects of $Ti_3C_2T_x$.

To sum up, MXenes are generally in a thermodynamically metastable state with high surface energy due to the high proportion of exposed metal atoms on the surface, leading to poor antioxidation performance in the air. Thus, air can trigger the oxidation of MXenes at ambient temperatures to obtain their derivatives. However, this oxidation usually occurs at the surface of the MXenes and leads to incomplete oxidation, which is not effective in obtaining composites with suitable dielectric properties. Therefore, some external stimuli like microwave energy and surfactants are necessary. Microwave-assisted preparation of MXene-derived TiO_2 has the advantages of simplicity, good controllability, good uniformity, and high yield. Furthermore, surfactants such as cetyltrimethylammonium bromide (CTAB) and PDDA feature an amphiphilic molecular structure comprising hydrophobic and hydrophilic groups, which can effectively facilitate the oxidation of MXenes.

2.5 Classification of MXenes in EMW Absorbing Materials

MXenes, with the merits of tunable dielectric properties, abundant surface terminations, and controlled structural assembly, provide a positive prospect in the field of MA. Nevertheless, the pristine MXenes absorber suffers from a series of problems, such as self-agglomeration, easy oxidation, single-loss mechanism, and impedance mismatch, which limit their MA properties and practical application. To further improve the MA performance and environmental adaptability of $Ti_3C_2T_x$, $Ti_3C_2T_x$-based MA materials obtained by component optimization and structure manipulation strategies have been extensive.

2.5.1 Component Optimization

The component optimization of MXene-based materials is a common means to adjust EM parameters and obtain good attenuation capacity and impedance matching. The combination of MXenes and other lossy materials (magnetic and/or dielectric mediums) paves an effective way to obtain MXene-based absorbers with tunable EM properties and enhanced interfacial effects.

2.5.1.1 Dielectric Materials

Dielectric materials can be used to adjust the electrical conductivity and dielectric parameters of MXenes to achieve the harmonization between impedance matching and attenuation capacity of the MXene-based MA materials.

Carbon-Based Materials Carbon-based materials include carbon black, graphite, carbon nanofibers, carbon nanotubes, and graphene, which have the merits of high electrical conductivity, lightweight, large specific surface area, environmental friendliness, and chemical stability. Benefiting from MXene's layered structure and evident abundance of functional groups, carbon materials can be uniformly dispersed without aggregation in MXenes. For instance, $Ti_3C_2T_x$ was self-assembled on the surface of carbon fiber and subsequent anchoring of MoS_2 [41]. By the synergistic effects from the MXene sheath increasing the conductive loss and MoS_2 at the outermost layer improving the impedance matching, the MA performance can be effectively regulated and optimized: the optimal RL is −61.51 dB at 3.5 mm, and EAB attains 7.6 GHz at 2.1 mm (Figure 2.5a). $Ti_3C_2T_x$/CNTs hybrids were prepared by CVD. CNTs can be distributed within the interlayer of $Ti_3C_2T_x$. $Ti_3C_2T_x$ is interconnected by curved CNTs, gradually forming a network structure [44]. The CNTs build conductive paths for charge carriers and interface polarization. Charge migration toward heterogeneous interfaces, nanoparticles, and defective areas such as CNTs results in the accumulation of free charges, creating dipoles and boosting polarization loss. $Ti_3C_2T_x$/CNT hybrid exhibits an RL_{min} of −53 dB.

3D seed germination like MXene with in situ growing CNTs/Ni heterojunction was designed for enhanced MA via polarization and magnetization [42]. The MXene-CNTs/Ni composite holds the enhanced MA performance (−56.4 dB) at only 2.4 mm. Compared to traditional magnetic agglomeration, the MXene-CNTs/Ni hybrids have a highly spatial dispersed magnetic architecture. The suspended Ni nanoparticles in a massively orientated CNT framework can form a 3D magnetic coupling network (Figure 2.5b). The 0D-2D structures by in situ polymerization and pyrolysis were prepared to coat the PDA-derived carbon nanoparticles as "nano-clothes" on MXenes (PCM) [45]. The EWA performance of PCM was optimized by adjusting the DA loading and pyrolysis temperature. The PCM shows an RL_{min} of −46.92 dB and an EAB of 7.01 GHz at 2.8 mm. The appropriate PCM creates sufficient defects to induce dipole polarization (functional groups and crystal defects from MXenes and N doping), triggers inhomogeneous space charge and electron density distribution to enhance interfacial polarization, and moderates interfacial contacts to provide good impedance matching.

Conductive Polymer Conjugated conductive polymers (PANI and PPy) offer controlled conductivity through mobile π electrons within double bonds, while also possessing low density, mechanical strength, corrosion resistance, and absorption capabilities.

MXenes/PANI composites with heterogeneous interfaces were designed by electrostatic self-assembly, and PANI acted as an impedance matching modifier [46]. The composites resulted in suitable impedance matching, abundant interfacial polarization, and dielectric loss, thus enabling efficient EWA. The RL_{min} of −60.6 dB can be obtained for the composite with 10 wt% PANI loading, while the EAB reaches 8.64 GHz at 2.3 mm. $Ti_3C_2T_x/TiO_2$/PANI composites with sandwich structures were designed [47]. The growth of TiO_2 particles on the surface of $Ti_3C_2T_x$ through a hydrothermal reaction enhances the interface polarization, and then PANI is doped on the surface of $Ti_3C_2T_x$. Due to the excellent dielectric properties and synergistic effects, the $Ti_3C_2T_x/TiO_2$/PANI composites showed an RL_{min} of −65.61 dB at 2.18 mm and an EAB of 5.92 GHz at 2.10 mm.

$Ti_3C_2T_x$/PPy hybrids with multilayer hierarchical architecture were prepared by in situ polymerization [48]. The composites consisting of 25 wt% $Ti_3C_2T_x$/PPy hybrids exhibit an RL_{min} of −49.2 dB at 3.2 mm and an EAB of 4.9 GHz at 2.0 mm. PPy spheres were also firmly anchored on MXenes while maintaining the layered structure of MXenes [49]. The EAB of the $Ti_3C_2T_x$/PPy composite achieved 6.63 GHz at 2.7 mm, and RL_{min} reached −49.5 dB. PVDF/Ti_3CNT_x@PPyNF composites with multiple heterogeneous interfaces were constructed due to the conductivity difference [50]. Because of the self-doping of N and the coating process, composites generate multiple polarization losses, enabling excellent EWA performance of the material. Specifically, the PVDF/Ti_3CNT_x@PPyNF composite shows an RL_{min} of −65.5 dB and an EAB of 6.95 GHz.

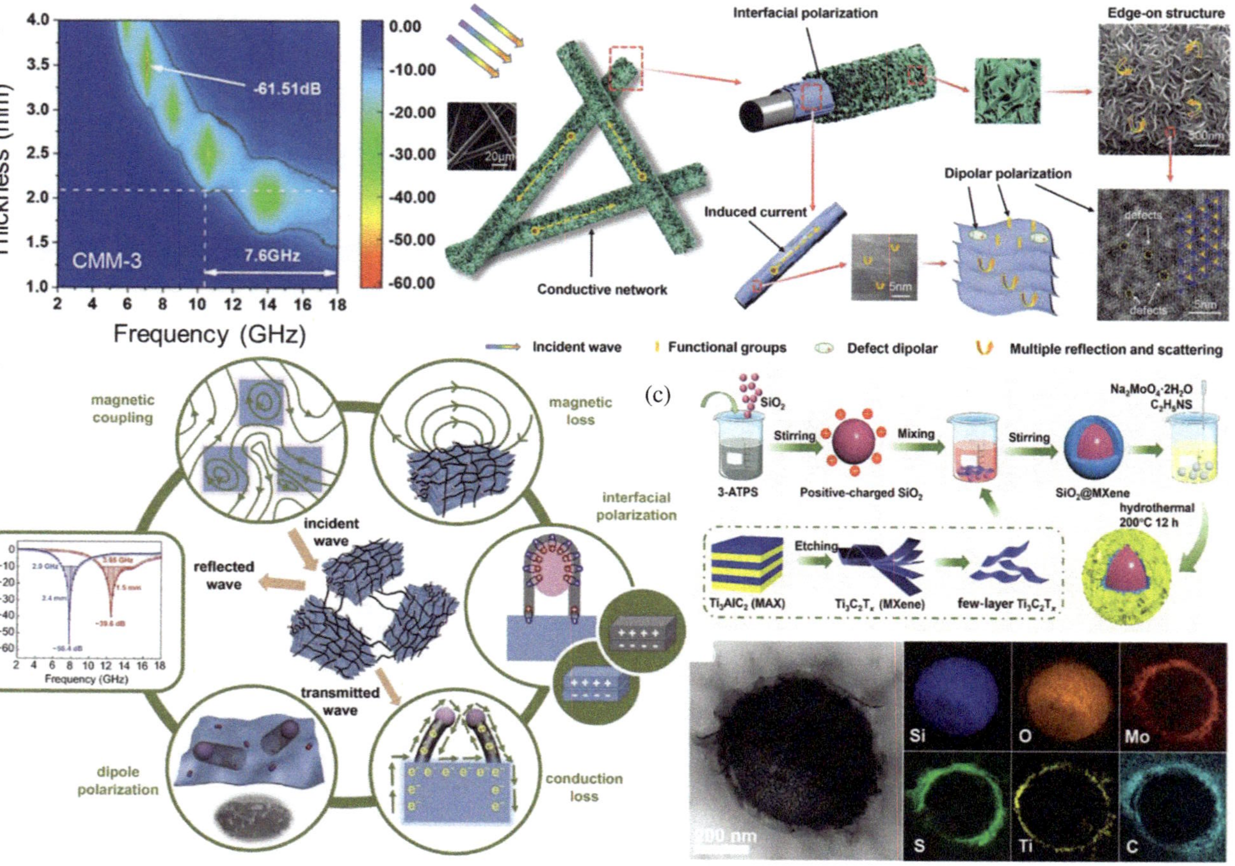

Figure 2.5 (a) Real-life performance and schematic of the MA mechanisms of carbon fiber@MXene@MoS_2 core sheath. Source: Wang et al. [41]/with permission of John Wiley & Sons. (b) Microwave absorption mechanism of MXene-CNTs/Ni. Source: Li et al. [42]/Springer Nature/CC BY 4.0. (c) Schematic diagram of the process of the preparation and TEM elemental mapping images of core–shell structured SiO_2@MXene@MoS_2 nanospheres. Source: Jiang et al. [43]/John Wiley & Sons/CC BY 4.0.

Ceramic Materials The 1D defect-rich SiCnws with a moderate dielectric constant were introduced to balance the impedance matching and conductive loss [51]. The SiCnws were wrapped by 2D MXenes that could not connect and form a 3D network structure. And MXenes could not be agglomerated because of the introduction of SiCnws. SiC/C@MXeneNFs with a core–sheath structure were prepared by electrostatic self-assembly [52]. The ingenious wrapping structure of the composite maximized the advantages of each component and enhanced EWA performance. In addition, the self-restacking of MXenes and the impedance matching of $Ti_3C_2T_x$ can be improved. A RL_{min} of −70.9 dB and a wide EAB of 6.32 GHz were achieved by 8 wt% SiC/C@MXeneNFs at 2.63 mm.

Polymer-derived ceramics (PDCs) provide a route to combine $Ti_3C_2T_x$ with ceramics. PDCs are easy to design due to the realization of fine regulation of composition and structure at the molecular level during polymer synthesis. Besides, the uniform dispersion of MXenes in PDCs can be achieved more easily before the pyrolysis of PDCs. $Ti_3C_2T_x$ was introduced into soluble and meltable hyperbranched polyborosilazane to prepare MXene-derived TiC/SiBCN ceramics by annealing [53]. The EWA performance of TiC/SiBCN was optimized by controlling the mass fraction of $Ti_3C_2T_x$ and annealing temperature. TiC nanocrystals and graphite carbon formed after pyrolysis enhance interface polarization. The inherent defects in TiC crystals enhance dipole polarization. The complex containing 5 wt% $Ti_3C_2T_x$ shows an RL_{min} of −45.44 dB at 2.3 mm. In addition, the excellent high-temperature stability may serve as structural parts in a hostile environment.

Semiconductor Oxides An urchin-like ZnO-$Ti_3C_2T_x$ nanocomposite was synthesized through the coprecipitation method [54]. The ZnO-$Ti_3C_2T_x$ nanocomposite delivers an enhanced EWA performance with an RL_{min} of −26.30 dB at 17.4 GHz. Moreover, the EWA performance can be effectively tuned by controlling the growth time of ZnO nanorods.

CeO_2 features relatively low dielectric properties, and the valence state of Ce oxide is prone to transforming between Ce^{3+} and Ce^{4+}, thus producing abundant oxygen vacancies inside the crystal lattice, which contributes to causing an obvious polarization relaxation. For example, a 2D MXene loaded with CeO_2 was synthesized [55]. The nanocomposite with 30 wt% $CeO_2/Ti_3C_2T_x$ has an RL_{min} of −47.27 dB at 1.9 mm and an EAB of 4.22 GHz. The abundant oxygen vacancies brought by CeO_2 nanoparticles, plentiful micro interfaces brought by the sandwich structure, and a large number of functional groups and defects inherent in MXenes allow the composite to have moderate impedance matching and multiple relaxation polarizations. NiCo/$CeO_2/Ti_3C_2T_x$ composite containing a multilayer structure was synthesized [56]. The NiCo alloy particles can provide magnetic loss, and the amount of CeO_2 affects the EM parameters and impedance of the composite. The formation of heterogeneous interfaces enhances the polarization loss capability. The 3D network framework can extend the path of EMW propagation and further dissipate EMWs. The NiCo/$CeO_2/Ti_3C_2T_x$ performed a wide EAB of 6.32 GHz at 1.9 mm and an RL_{min} value of −42.48 dB at 2.0 mm.

Metal Sulfides/Selenides Metal sulfides/selenides attract attention owing to moderate permittivity, controllable dielectric loss, thin matching thickness, and stable chemical properties. Until now, MoS_2, WS_2, CuS, and CoS_2 combined with MXenes have been reported.

To alleviate the impedance mismatch of MXene, a core–shell structured SiO_2@MXene@MoS_2 nanosphere was synthesized [43]. SiO_2 acts as the core, MXene as the intermediate layer, and MoS_2 as the outer shell (Figure 2.5c). The core–shell structure not only provides a variety of loss mechanisms that effectively dissipate electromagnetic energy but also prevents the self-aggregation of MXene and MoS_2 nanosheets. The core–shell structured nanospheres exhibit a remarkable RL_{min} of −52.11 dB and EAB of 6.72 GHz. 2D/2D MXene-MoS_2 heterostructures with a 3D conductive network were prepared [57]. MXene-MoS_2 composites exhibit an RL_{min} of −46.72 dB at 2 mm than MXenes (−23.93 dB) and MoS_2 (−17.75 dB) due to the synergistic effect in heterostructure interface. Electron holography technology proved that the increased visual charge density distribution greatly promotes the dielectric loss, thereby improving the attenuation capability. In addition, the conductive loss, polarization behavior, and improved impedance matching contributed by the vertical growth of MoS_2 nanosheets on MXenes enhance MA performance. The MXene/CuS composite film with a unique sandwich structure is obtained by an ultrasound-assisted and vacuum filtration method [58], avoiding the negative effects of high temperature on MXene structure and properties. The problem of aggregation of MXenes and CuS nanoparticles can be avoided, extending the microwave propagation path. MXenes and sea urchin-like $NiCo_2S_4$ hybrids were synthesized by electrostatic self-assembly [59]. The hybrids, with a loading of 10 wt%, achieve an RL_{min} of −52 dB and an EAB of 5.6 GHz. The excellent MA performance is attributed to the improved impedance matching and the synergistic effects of dielectric loss and interfacial polarization.

$NiSe_2$-$CoSe_2$@C/$Ti_3C_2T_x$ composites with multilayer structures were synthesized by hydrothermal and selenization methods [60]. The $NiSe_2$-$CoSe_2$@C/$Ti_3C_2T_x$ has several advantages, including a conductive network, rich polarization loss, and multiple scattering evoked by the multilayer $Ti_3C_2T_x$, leading to excellent MA performance. The RL_{min} is −60.46 dB with a load of 40 wt% at 2.6 mm. NiS/MoS_2/$Ti_3C_2T_x$ hybrid with multilayer-scale structure was prepared [61]. Spherical NiS particles are decorated on the $Ti_3C_2T_x$, and 2D MoS_2 nanosheets are stacked on the surface-like scales, achieving good impedance matching. NiS/MoS_2/$Ti_3C_2T_x$ hybrid shows a wide EAB of 5.04 GHz at 2.1 mm and an RL_{min} of −58.48 dB at 2.4 mm.

2.5.1.2 Magnetism

The common magnetic loss materials include magnetic metals (Fe, Co, Ni, and their alloys) and their metal oxides. Based on the MA mechanism, the introduction of magnetic materials provides MXenes with a magnetic loss pathway and thus realizes a dielectric-magnetic synergistic loss.

Magnetic Metal Magnetic metals such as Fe, Co, and Ni are promising candidates for EWA due to their high Snoek's limit, high magnetic saturation strength, and high-temperature resistance. Combining the magnetic particles and MXenes would not only improve impedance matching but also rapidly release the converted thermal energy through MXenes. The size and concentration of Ni in Ni@MXene hybrids can be regulated by adjusting the ratio of Ni ions and $Ti_3C_2T_x$ [62]. The PVDF composite with a 10 wt% load shows an RL_{min} of −52.6 dB and an EAB of 6.1 GHz. The MA mechanism is ascribed to (1) the excellent impedance matching; (2) the magneto-coupling effect, natural resonance, and exchange resonance effect; (3) ohmic loss of free electron conduction, heterointerface polarization, and dipole polarization; (4) multiple reflections and scattering of EM waves; and (5) the effect of interference cancellation.

The Ni/TiO_2/C heterogeneous composites were synthesized through a microwave-mediated reaction in a short time, through which a large amount of TiO_2 with a bipyramid morphology could be generated from hydrazine hydrate and CTAB-modified $Ti_3C_2T_x$ [63]. Owing to the synergistic effect of dielectric loss, magnetic loss, and the unique 2D heterogeneity, the Ni/TiO_2/C composites exhibited an RL_{min} of −39.91 dB at 1.4 mm.

Compared to nanoparticles, metal nanochains or nanofibers with a high aspect ratio can form electrical/magnetic conductive networks more easily. In addition, 1D nanochains have high coercivity and achieve obvious magnetic coupling. For example, the MXenes/Ni hybrids achieve an RL_{min} of −49.9 dB at 1.75 mm when the MXenes content is 10 wt% [64]. The excellent EWA stems from the synergistic effect of MXenes and Ni chains, appropriate impedance matching, and good dissipation ability. Multilayer-sandwiched heterostructure MXenes/Co nanochains (MXenes-CoC) composites were obtained by generating magnetic nanochains in situ in MXenes [65]. MXenes-CoC exhibited an RL_{min} of −46 dB at 1.02 mm. The Co nanochains effectively reduce the dielectric of MXenes while inducing magnetic loss, improving impedance matching. Compared with the MXenes/Co nanoparticles (MXenes-CoP), the MXenes-CoC can achieve excellent MA performance at a lower thickness, which is attributed to the interlacing magnetic flux field effect.

Magnetic Alloy Magnetic alloy with high saturation magnetization can offer strong EWA intensities. The MXenes/FeCo films with various morphologies were prepared, which are composed of alternating layers of magnetic FeCo alloys and MXenes nanosheets [66]. The introduction of FeCo not only reduces the ultra-high conductivity of MXenes but also greatly enhances the magnetic loss. Notably, the magnetic coupling loss in interlayers has expanded from the nanometer scale to the centimeter scale. Compared with nanoplate-like FeCo alloy, the rough surface of nanoflower-like FeCo exhibits stronger interface polarization and magnetic loss, which are proven by the off-six electron holography. The MXenes-FeCo nanoflower-like film demonstrated better MA performance with an RL_{min} value of −43.7 dB, while the RL_{min} reached −36.2 dB of the MXenes-FeCo nanoplate-like film.

The CoFe@$Ti_3C_2T_x$ composites with sandwich-like structures were prepared in DMF medium to enlarge the interlayer space of $Ti_3C_2T_x$, followed by in situ reduction of Co^{2+} and Fe^{2+} ions to CoFe nanosheets [67]. The RL of the composites reaches −36.29 dB at 2.2 mm. The modification of CoFe endows the composites with enhanced magnetic loss, impedance matching, interface polarization, and multiple reflection and scattering of sandwich-like structures.

Magnetic Oxides Ferrite MFe_2O_4 (M = Fe, Co, Ni, Mn, Zn, and so on) possesses a spinel crystalline structure, and different cation distributions have a great impact on EM properties. The loading of Fe_3O_4, $CoFe_3O_4$, and $NiFe_2O_4$ on MXenes can obtain great MA performance.

Fe_3O_4 nanoparticles grew on the $Ti_3C_2T_x$ surface and interlayer (Fe_3O_4@$Ti_3C_2T_x$), thus balancing the impedance matching, introducing more loss mechanisms, and enhancing MA performance [68]. Interestingly, the solvothermal process provides a reducing environment that protects $Ti_3C_2T_x$ from oxidation, endowing $Ti_3C_2T_x$ with high stability compared to the hydrothermal process. The sample containing 25 wt% Fe_3O_4 exhibits an RL_{min} of −57.2 dB at 4.2 mm. The room temperature assembly process prevents oxidation of MXenes. Fe_3O_4/MXene hybrids were synthesized by electrostatic assembly of negatively charged MXene with positively charged hollow Fe_3O_4 nanoparticles [69], which not only enhanced the polarization process by forming a "micro-capacitor"-like structure and magnetic loss but also improved the impedance matching of MXene (Figure 2.6a). The MXene/Fe_3O_4 hybrids exhibited a strong absorption ability of −63.7 dB at 1.56 mm. The sandwich-like m-Ti_3C_2/Fe_3O_4 (MTF) composites were prepared via electrostatic assembly [70]. MTF-1 exhibits an

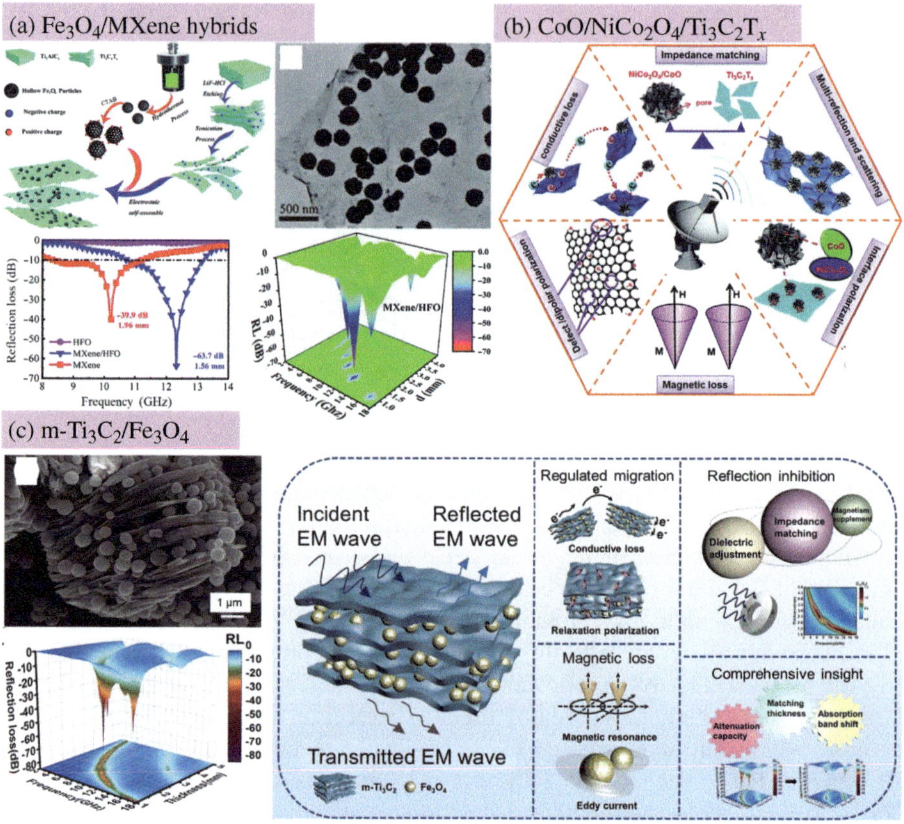

Figure 2.6 (a) Schematic diagram of the preparation process, SEM images, and MA mechanism of Fe_3O_4/MXene hybrids. Source: Deng et al. [69]/Royal Society of Chemistry. (b) Schematic illustration of the microwave absorption mechanism of $CoO/NiCo_2O_4/Ti_3C_2T_x$. Source: Deng et al. [69]/Royal Society of Chemistry. (c) Schematic illustration of the regulation mechanism for m-Ti_3C_2/Fe_3O_4. Source: Yang et al. [70]/Royal Society of Chemistry.

RL_{min} of −77.5 dB at 2.96 mm. The high EWA performance is attributed to the Fe_3O_4 particles inhibiting the electronic hopping behaviors, avoiding EM wave reflection, as well as introducing magnetic loss (Figure 2.6c). $Ti_3C_2T_x/Co_3O_4$ composites are fabricated by hydrothermal treatment in a concentrated strong base followed by an annealing process [71]. OH terminations and intercalation of Na^+ ions can impede the Ti atoms in $Ti_3C_2T_x$ from contacting with water molecules and O_2, ensuring a stable synthesis.

MOF derivatives/MXene ($CoO/NiCo_2O_4/Ti_3C_2T_x$) cross-linking networks were successfully synthesized through connecting flower-like MOF derivatives with 2D flexible MXene flakes [72]. The combination of the enhanced interfacial polarization derived from the flower-like structure and the conductivity loss originating from the network structure endows the composites with remarkable MA performance (Figure 2.6b). The heterostructure shows an RL_{min} value of −72.3 dB at 1.7 mm. MXene nanosheets with homogeneously anchored $NiFe_2O_4$ nanoparticles ($NiFe_2O_4$/MXene) were prepared [73]. Interestingly, the RL_{min} of the composite with only 3% $NiFe_2O_4$/MXene reached −41.83 dB. The MA mechanism ascribes to the existence of $NiFe_2O_4$ that can prevent the self-stacking of MXene nanosheets, causing improved conduction loss. Besides, large amounts of heterogeneous interfaces between $NiFe_2O_4$ and MXenes can also produce additional interface polarization loss. The magnetic loss arising from the magnetic $NiFe_2O_4$ is beneficial for good impedance matching, enabling more EMW to enter into the interior of composites to be dissipated.

2.5.1.3 Multiple Loss Materials

MXenes directly compounding with other constituents inevitably leads to the aggregation of MXenes and weakens their superiority. To achieve the dielectric-magnetic synergistic effect and balanced impedance matching, dielectric-magnetic-MXene systems have been used for MA.

Magnetized MXene-rGO/CoNi film was fabricated via electrostatic self-assembly [74]. Inside the composite film, rGO decorated with highly dispersed CoNi is intercalated into MXenes, suppressing the self-restacking of MXenes, resulting in a reduction of high permittivity (Figure 2.7a). In addition, owing to the strong magnetic coupling between FeCo, the film exhibits a strong magnetic loss. Moreover, the local dielectric polarized fields exist at

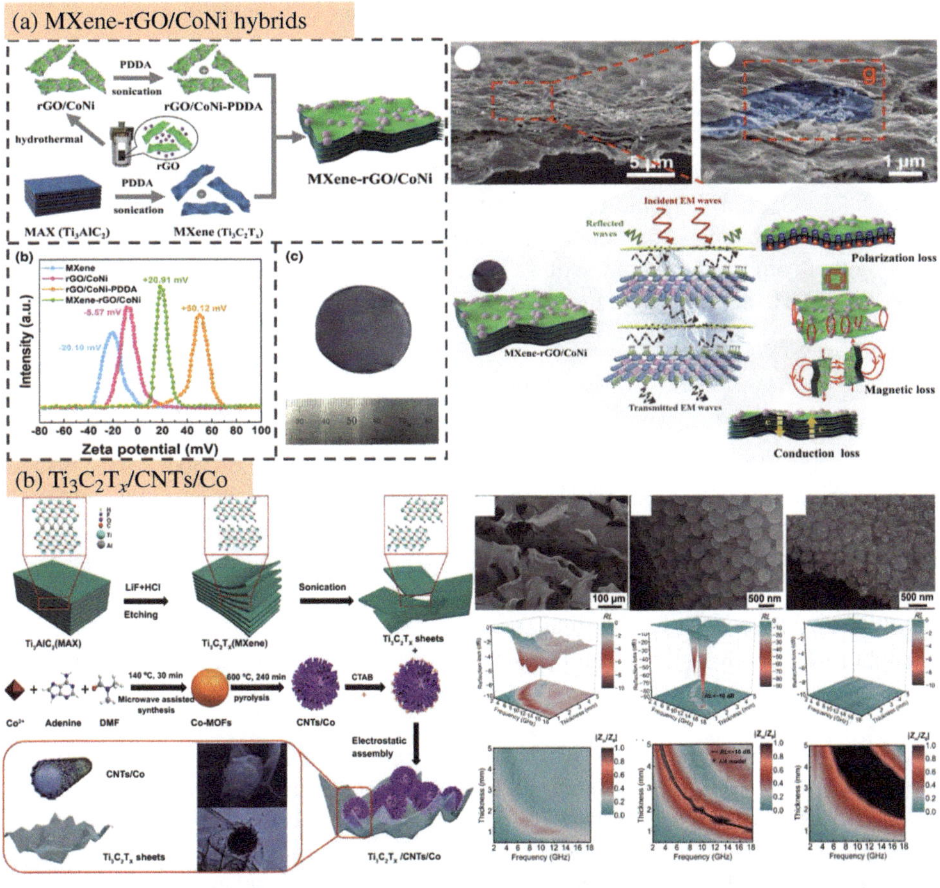

Figure 2.7 (a) Schematic illustration for the synthesis, zeta potential, photograph, SEM images, and MA mechanism of the MXene-rGO/CoNi hybrids. Source: Li et al. [74]/Springer Nature/CC BY 4.0. (b) Schematic diagram of the preparation, SEM images, and RL of $Ti_3C_2T_x$/CNTs/Co composites. Source: Xiang et al. [75]/Springer Nature/CC BY 4.0.

the continuous heterointerfaces between MXenes and rGO. Hence, the composite exhibits an RL_{min} value of −54.1 dB. MXene decorated with SiC_{NWs}@Co/C was prepared by electrostatic self-assembly [76]. The EWA performance of MXene@SiC_{NWs}@Co/C with a bird-nest-like structure could be optimized by changing the proportion of MXenes and SiC_{NWs}@Co/C. The hybrids reveal superior impedance matching, complementary dissipation mechanism, and plentiful heterointerfaces. The hybrids exhibit an RL_{min} of −76.5 dB at 3.9 mm and an EAB of 6.2 GHz at 2.0 mm.

The sea urchin-like CNTs/Co were introduced onto 2D $Ti_3C_2T_x$ sheets by an electrostatic assembly approach to form laminated $Ti_3C_2T_x$/CNTs/Co nanocomposites (Figure 2.7b). A strong RL of −85.8 dB and an ultrathin thickness of 1.4 mm were achieved. The MA performances were enhanced by the synergistic effects between the conduction loss originated from the electronic transport in the conductive network of 1D CNTs and 2D $Ti_3C_2T_x$, the dielectric loss stemmed from the dipole polarization and abundant interface in the laminated structure, and the magnetic loss derived from the ferromagnetic resonance of 0D Co nanoparticles [75].

1D hierarchical core–shell structure with Co-ZIF array-derived CNTs coupled on cotton fiber (CF)-supported MXenes shell (Co/CNTs-MXene@CF) was synthesized [77]. The well-designed structure can efficiently integrate the advantages of Co-ZIF-derived CNTs and MXene to realize multiple electron transfer paths. Besides, the interlayer MXenes and CNTs grow sequentially from the outermost Co-ZIF arrays with appropriate conductivity and abundant heterointerfaces, improving the conduction loss and interfacial polarization. The 1D Co/CNTs-MXene@CF heterostructures exhibit an RL_{min} of −61.41 dB at 2.52 mm and EAB of 5.04 GHz at 1.5 mm. The layered NiCo/TiC/TiO/CNTs structures were prepared [78], avoiding magnetic particle agglomeration, improving the impedance matching, and reducing the loading of the composites. The composites exhibit an RL_{min} of −51.98 dB and EAB of 7.76 GHz at 1.9 and 2.1 mm, respectively.

2.5.2 Structural Regulation

The micro-/macrostructure manipulation is of equal importance. Structures in different dimensions, such as 3D microspheres, layered structures, yolk shells, and fibers, have gained interest from researchers. The enlarged coupling scale of multidimensional units enhances the attenuation capacity and provides a feasible way to prepare the $Ti_3C_2T_x$ MXene-based MA materials.

2.5.2.1 3D Microsphere

3D microstructures, such as core–shell, flower-like, and hollow structures, can optimize impedance matching and lead to enhanced MA capability.

Template Method The high density and uniform dispersion of magnetic components can enhance magnetic loss. However, both the easily agglomerated magnetic components and the restacked MXenes limit the generation of electromagnetic interfaces, resulting in impedance mismatch. An urchin-like 3D PMMA@MXene@Ni was constructed via a spheroidization strategy [79], where the lamellar MXene is reshaped into 3D microspheres that provide the substrate for oriented growth of Ni. The urchin-like double core–shell structure shows several advantages. First, the 2D MXenes anchored on the surface of polymethyl methacrylate

(PMMA) construct a 3D network, preventing MXenes from accumulation. Second, strong bonding between MXenes and Ni via electrostatic adsorption mitigates the agglomerated magnetic components, introducing high-density Ni nanospikes. Third, the strong magnetic anisotropy of the Ni shell enhances magnetic coupling and improves magnetic loss. Hence, the composite microsphere exhibits an RL_{min} of −59.6 dB and an EAB of 4.48 GHz at 1.5 mm.

MXene@Co hollow spheres were prepared via a self-assembling-etching-anchoring growth method [80]. The hollow structure of MXene microspheres optimizes impedance matching and surface modification. In addition, MXene@Co exhibits abundant heterogeneous interfaces, enhancing the interfacial polarization. Meanwhile, the surface-anchored growth of magnetic Co particles forms a magnetic network, which provides a strong magnetic loss capability. The hollow structure shows an RL_{min} of −57.32 dB and EAB of 5.2 GHz at 2.5 mm.

The hollow spheres of $Ti_3C_2T_x$ with a nanometer-thick wall were prepared through a sacrificial template strategy [81]. The hollow spheres result in improving impedance matching, multiple reflections, and scattering, and polarization relaxation. With only 10 wt%, the RL_{min} of $Ti_3C_2T_x$ hollow spheres can reach −49.2 dB at 2 mm, which is much superior to that of $Ti_3C_2T_x$ (−7.4 dB). Moreover, the MA performance of $Ti_3C_2T_x$ hollow spheres can be adjusted by changing the cavity size and thickness. The biomimetic urchin-like $Ti_3C_2T_x$@ZnO hollow microspheres are rationally designed [82], in which ZnO nanoarrays as the urchin spines are evenly grafted onto the surface of $Ti_3C_2T_x$ hollow spheres (diameter: 4.2 μm) as the urchin spherical photosensitive bodies (Figure 2.8a). The construction of gradient impedance and hierarchical heterostructures enhances the attenuation of incident EM waves. The EM wave loss is revealed by limited integral simulation calculations, which fully highlight the advantages of the urchin-like architecture. The $Ti_3C_2T_x$@ZnO hollow spheres deliver an RL_{min} of −57.4 dB and an EAB of 6.56 GH.

The hollow MXene bowls, via sacrificial polystyrene templates, were prepared, the walls of which were composed of MXene nanosheets with tunable layer stacking thickness [34]. The hollow MXene bowls exhibited an RL_{min} of −53.8 dB with an EAB of 4.2 GHz at 1.8 mm when the loading was only 5 wt%. A core–shell structure of SiO_2@$Ti_3C_2T_x$ was prepared by the electrostatic assembly [85]. SiO_2 templates can not only prevent $Ti_3C_2T_x$ from stacking but also modulate the surface impedance and increase interfacial polarization loss. The SiO_2@$Ti_3C_2T_x$ can deliver an RL_{min} of −58.01 dB at 1.3 mm. The interfacial polarization was proven by electron holography.

A hollow structure, coupled with gradient hierarchical heterostructures, was designed to achieve MA [86]. MoS_2 nanosheets are uniformly grown onto the double-layered $Ti_3C_2T_x$@rGO hollow microspheres through self-assembly and sacrificial templates. Notably, the gradient hierarchical heterostructures, comprising a MoS_2 impedance-matching layer, an rGO lossy layer, and a $Ti_3C_2T_x$ reflective layer, have demonstrated improvements in impedance matching and attenuation capabilities. The gradient hollow heterostructures reach an RL of −54.2 dB at 1.8 mm, and the EAB covers up to 6.04 GHz.

Spray-Drying Method Spray-drying technology uses a nozzle to disperse the solution into small droplets, conducts heat and mass transfer with a hot drying medium, and evaporates the solvent to obtain dry particles. Therefore, the size of MXenes obtained by the spray-drying method is usually on a microscale, and multiscale magnetic coupling can be achieved by compounding it with nanoscale magnetic particles.

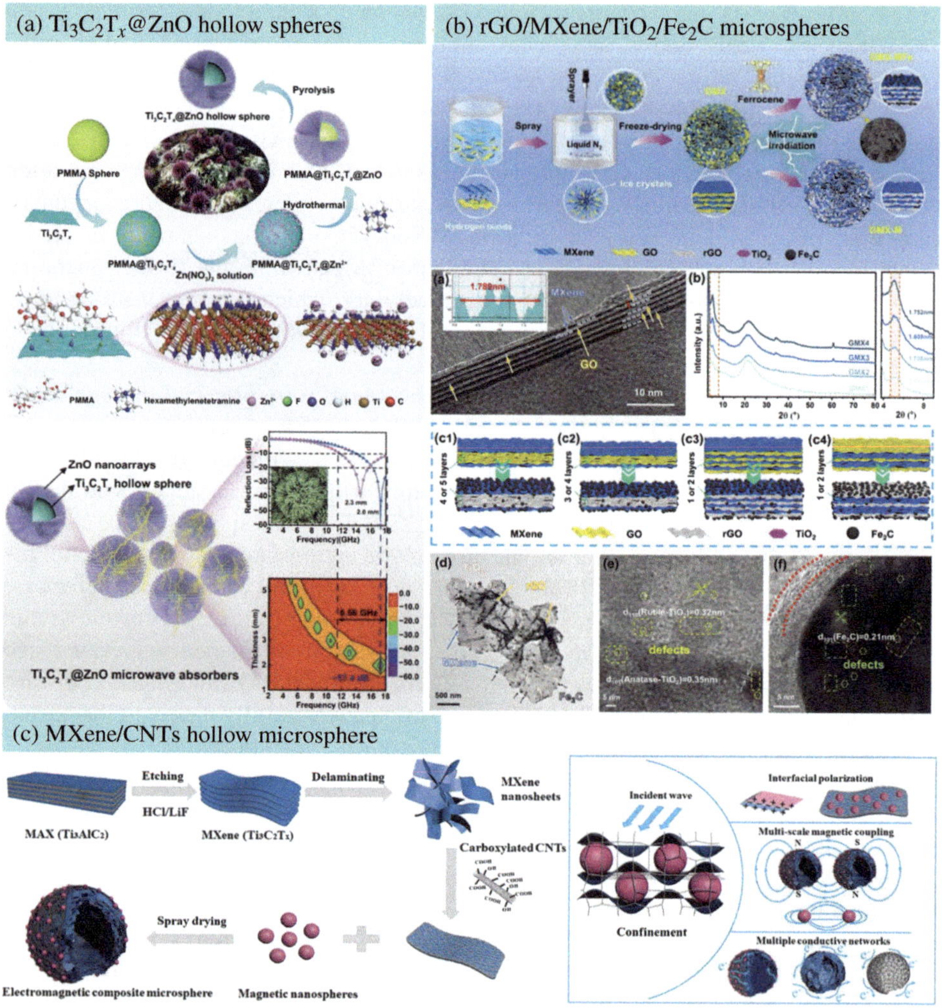

Figure 2.8 (a) Schematic illustration of the synthesis process and MA mechanism of urchin-like $Ti_3C_2T_x$@ZnO hollow spheres. Source: Wang et al. [82]/Springer Nature/CC BY 4.0. (b) Schematic diagram of the preparation process for the 3D rGO/MXene/TiO_2/Fe_2C microspheres and MA mechanisms. Source: Wang et al. [83]/Springer Nature/CC BY 4.0. (c) Schematic illustration describing the preparation process, RL, and MA mechanism of MXene/CNTs hollow microspheres. Source: Zhang et al. [84]/with permission of John Wiley & Sons.

3D rGO/MXene/TiO_2/Fe_2C porous microspheres with periodical intercalated structures and pronounced interfacial effects were constructed [83]. Such an approach reinforced interfacial effects via defects introduction, porous skeleton, multilayer assembly, and a multicomponent system, leading to synergistic loss mechanisms (Figure 2.8b). The abundant 2D/2D/0D/0D intercalated heterojunctions in the microspheres provide a high density of polarization charges while generating abundant polarization sites, resulting in boosted interfacial polarization. By precisely tuning the 2D nanosheet intercalation in the heterostructures, both the polarization loss and impedance matching improve significantly. At a low loading of 5 wt%, an RL_{min} of −67.4 dB can be achieved. The MXenes/Fe_3O_4 microspheres were fabricated [87]. The Fe_3O_4 is uniformly dispersed in the voids of MXenes microspheres, optimizing impedance balance and EM coordination capability. Meanwhile, strong magnetic loss capability with significantly improved magnetic flux line density is achieved from microscale MXenes and nanoscale Fe_3O_4, confirming our 3D multiscale magnetic coupling network. Accordingly, the composites hold an RL_{min} of −50.6 dB and EAB of 4.67 GHz at 2 mm.

The hollow microsphere Fe_3O_4@$Ti_3C_2T_x$/CNTs was assembled using MXenes, CNTs, and Fe_3O_4 [84]. Within the microsphere, MXenes are separated by dispersed CNTs, exposing abundant dielectric polarization interfaces (Figure 2.8c). Besides, Fe_3O_4 nanospheres are uniformly dispersed and confined within nano-cavities between the 1D network and 2D framework. This novel structure promotes interfacial polarization, enhances magnetic loss by microscale and nanoscale coupling network, enlarges conduction loss by MXene/CNTs dual-network, and optimizes impedance matching by hierarchical porous structure. Therefore, Fe_3O_4@$Ti_3C_2T_x$/CNTs composite achieves an RL_{min} of −40.1 dB and an EAB of 5.8 GHz at 2 mm.

$Mo_2TiC_2T_x$, compared with $Ti_3C_2T_x$, contains bimetallic components, which can provide more heterogeneous junctions and thus have stronger dipole polarization. In addition, after high-temperature carbonization, besides $Mo_2TiC_2T_x$ itself, the oxides and carbides of both Mo and Ti metals can be attained at the same time, so that the absorber with abundant heterogeneous components can easily be obtained, which can not only improve the conductive loss but also greatly enhance the interfacial loss capability and improve the impedance matching. Assembled $Mo_2TiC_2T_x$/RGO microspheres were prepared using an ultrasonic spray and annealing treatment [88]. The internal crystalline structure of $Mo_2TiC_2T_x$ was changed at different carbonization temperatures, resulting in the distinction of intrinsic defects and synergistic enhancement effect of $Mo_2TiC_2T_x$/RGO microspheres. Thanks to the good impedance matching and strong MA ability, the optimized $Mo_2TiC_2T_x$/RGO exhibited an RL_{min} of −55.1 dB at only 15 wt% loading, corresponding to the EAB of 5.7 GHz at 2.1 mm. In addition, $Ti_3C_2T_x$@GO hybrid aerogel microspheres (M@GAMS) can be constructed by integrating GO with $Ti_3C_2T_x$ [89]. Driven by the hydrogen bonding interactions, GO and $Ti_3C_2T_x$ layers are assembled into heterogeneous structures with a face-to-face form. Combined with the disparities in conductivity between $Ti_3C_2T_x$ and GO, and the heterointerfaces and abundant surface groups, the M@GAMS with a low loading of 10.0 wt% exhibits an RL_{min} of −49.1 dB at 1.2 mm. More importantly, the M@GAMS reaches an RL of −38.3 dB at 5.0 mm in the S-band.

Nevertheless, porous or hollow structures tend to collapse, thus decreasing the MA capacity. The self-restacking of $Ti_3C_2T_x$ nanosheets unavoidably decreases the number of active sites, specific surface area, and conductivity.

2.5.2.2 Fiber

NiFe LDH/MXene derivatives interconnected with carbon fabric were developed for flexible EWA [90]. The derived $FeNi_3/TiO_2$ inherits the large and flaky features of the MXenes support and acts as the interconnector between the carbon fibers, which facilitates the formation of conductive pathways that allow the free migration of electrons among the fibers. The MA performance shows an RL_{min} of −58.0 dB and an EAB of 7.0 GHz at 2.5 mm.

MXene fibers@CNTs with intertwined 1D heterostructure are prepared [91]. Interestingly, 2D MXenes self-assembled in a strong alkaline to form 1D MXene fibers under the action of shearing force. The growth of CNTs can be controlled by manipulating the loading of MOFs on 1D MXene fibers, thereby realizing the regulation of EM parameters. The 3D cross-linked network constructed by intertwined 1D heterogeneous components features extensive heterogeneous interfaces, hierarchical pore structure, and desirable conductivity, exhibiting superior MA properties. Notably, under a matching thickness of 1.6 mm, the RL_{min} of the composite exceeds −51.6 dB, and the EAB covers 4.5 GHz. The "shearing effect" of alkaline KOH can induce the formation of 1D MXene nanoribbons with a staggered-connected nanoribbon network and microporosity. NiCo layered double hydroxides (LDHs) nanosheets are grown firmly on the MXene nanoribbons and transformed into NiCo embedded in N-doped carbon by heat treatment, forming a porous 0D/1D/2D MXene nanoribbons-NiCo@NC hierarchical network [92]. The synthesized MXene nanoribbons-NiCo@NC enable an RL_{min} value of −57.1 dB at 4.82 mm and an RL_{min} value of −33.19 dB at 1.4 mm.

The $Fe_{0.64}Ni_{0.36}$/MXenes/CNFs composites were prepared by electrospinning method [93]. 2D MXenes with 0D $Fe_{0.64}Ni_{0.36}$ nanoparticles anchored on the surface are embedded in 1D carbon nanofibers, forming a unique heterogeneous network structure. This framework effectively prevents the self-accumulation of MXenes and increases the specific surface area of MXenes, thus contributing to conductive loss and polarization. The $Fe_{0.64}Ni_{0.36}$/MXenes/CNFs composites with a load of 12.5 wt% obtained an RL_{min} of −54.1 dB at 2.7 mm and EAB of 7.76 GHz at 2.1 mm.

In terms of microstructure and surface state, during the functional tailoring of MXene-based nanostructures, MXenes evolve from a layered structure into a 3D skeleton with interconnected 1D nanoribbons and then transform into porous and wrinkled ultrathin nanosheets (Figure 2.9a). The enhanced attenuation ability of P-MXene is mainly attributed to its unique structure, proper electrical conductivity, large specific surface area, and abundant porous defects, which bring about good impedance matching, dipole polarization, conduction loss, and a novel capacitive resistance mechanism [29].

2.5.2.3 Sandwich Structure

LDH were loaded onto MXenes, synthesizing a novel MXenes@LDH composite (Figure 2.9b) [94], which shows a sandwich structure and exhibits excellent EMA performance, thermal conductivity, and flame retardancy. By adjusting the load of LDH, under the synergistic effect of multiple factors such as dielectric and polarization losses, an RL_{min} of −52.064 dB and an EAB of 4.5 GHz were achieved. A NiFe LDH/MXenes composite with a 2D/2D sandwich-like heterostructure was also constructed through electrostatic self-assembly [97]. The composite achieves a broad EAB of 5.1 GHz and an RL_{min} of −58.2 dB at 2.00 mm.

Sandwich-like 2D laminated Fe&TiO_2@C nanocomposites have been synthesized from the MXene/MOF precursors [98]. The formation of Fe and rutile-TiO_2 nanoparticles sandwiched by the two-dimensional carbon nanosheets provided strong electromagnetic energy attenuation and good impedance matching for EMW absorption. The Fe&TiO_2@C nanocomposites exhibited a broad EAB of 6.5 GHz at 1.6 mm and an RL_{min} value of −51.8 dB at 6.6 GHz at 3 mm, respectively.

Graphite/TiC/Ti_3C_2 (G/TiC/Ti_3C_2) hybrids are obtained after Al atoms are etched from G/TiC/Ti_3AlC_2, which are fabricated in a molten salts bath (Figure 2.9c) [95]. Ti_3C_2 sheets

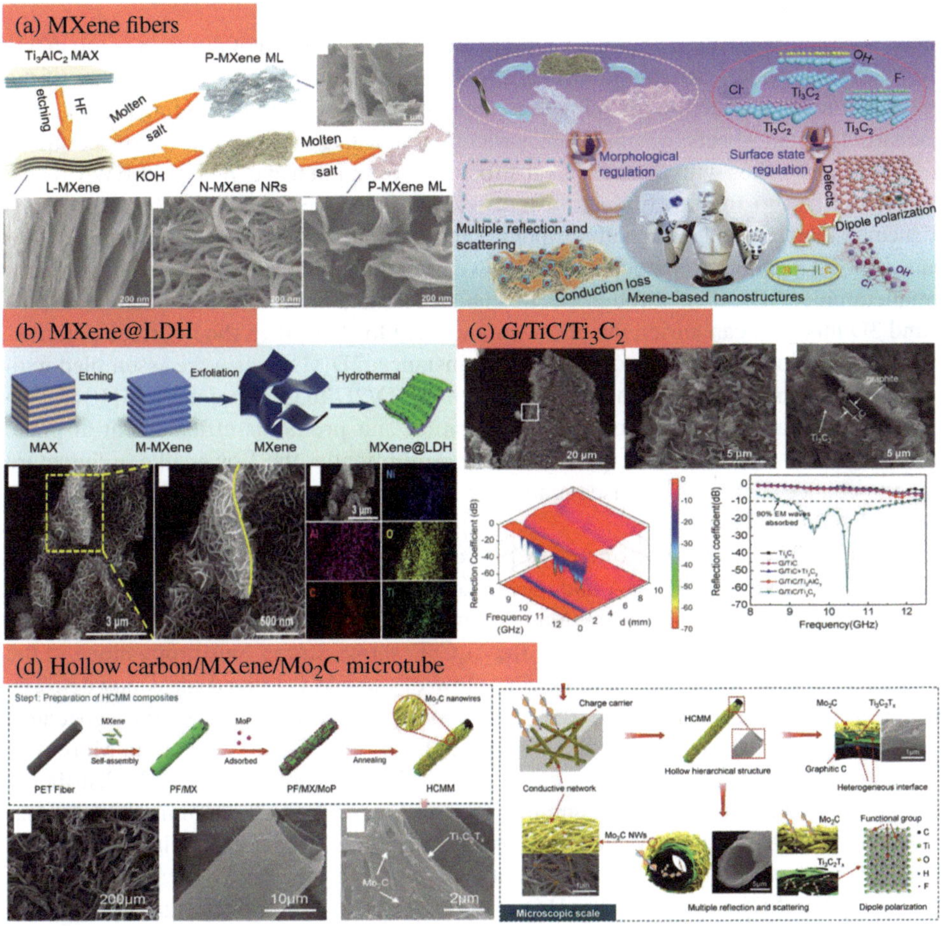

Figure 2.9 (a) Scheme showing the fabrication process and MA mechanism of MXene-based fiber. Source: Zeng et al. [29]/with permission of John Wiley & Sons. (b) Schematic diagram of the preparation process and SEM images for MXene@LDH nanocomposites. Source: Wang et al. [94]/with permission of John Wiley & Sons. (c) The formation, SEM images, EM parameters, and RL of G/TiC/Ti_3AlC_2 and G/TiC/Ti_3C_2 hierarchical structure. Source: Li et al. [95]/with permission of John Wiley & Sons. (d) Schematic illustration of the multiscale design route and MA mechanism of the Hollow Carbon/MXene/Mo_2C Microtube. Source: Wang et al. [96]/with permission of John Wiley & Sons.

are perpendicular to the plane of G/TiC, which is like a fish scale. Due to the special interfacial junction, electron band structure and electron transport properties are changed at the interface, inhibiting the electron transport across it. Space charges accumulated at the interface and aroused interfacial polarization. The RL_{min} of G/TiC/Ti_3C_2 reached −63 dB, and EAB is 3.5 GHz. A novel laminated and magnetic composite NiO&TiO_2@C is synthesized via ion intercalation and a hydrothermal route [99]. The hydrothermal oxidation makes the as-formed atomically thin carbon sheets highly disordered, and the in situ oxidized TiO_2 and NiO nanoparticles are embedded closely into each carbon layer to form a sandwich-like structure. The hybrid exhibits an EAB as high as 11.1 and 9 GHz at 3 and 2 mm, respectively.

2.5.2.4 Hierarchical Structure

Hierarchical structure refers to a complex arrangement or organization of different components at multiple scales, where each part contributes to the overall functionality or properties of the system. A hierarchical structure often involves the integration of various building blocks, such as nanoparticles, nanosheets, or other nanostructures, into larger and more organized architectures. The composites with a hierarchical structure arouse intensive interest in EWA fields, which are ascribed to multipath electron transfer, multiple interfaces, and abundant voids, impedance matching, strong magnetic response, and multiple scattering and reflecting [100]. MXenes, as a 2D negatively charged material, have been integrated with many building blocks to form hierarchical absorbers through electrostatic forces and intermolecular interactions.

2D and 3D modules can enrich the phase contact area to strengthen the interface polarization and cause multiple scattering of EMWs. For instance, 2D MXenes were assembled with 2D carbon shells and 0D MoO_2 nanodots to form a 2D/2D/0D heterojunction [101]. The heterojunction induces strong interfacial polarization and promotes efficient out-of-plane electron flow to dissipate EM energy effectively. Additionally, the flower-like architecture ensures impedance matching. Consequently, the composite exhibits an RL_{min} of −46.7 dB and a wide EAB of 7.7 GHz.

A mesoporous $Ti_3C_2T_x$/Ni/AS composite with a porosity of around 50% was synthesized to simultaneously achieve high MA and mechanical properties [102]. The EMT zeolite was selected as a precursor to realize low-temperature densification for protecting the $Ti_3C_2T_x$ from thermal damage. After densification, the $Ti_3C_2T_x$ shows strong anisotropy alone in the in-plane direction due to the transient viscous flow of the EMT-derived glass powder. The alignment of MXenes combined with the mesoporous glass structure favors the impedance matching because of the low electrical conductivity in the out-of-plane direction. Moreover, taking advantage of the ion-exchanging capacity of EMT, isolated and uniformly distributed Ni nanoparticles can be incorporated into the AS glass matrix as well, attaining high attenuation capability while maintaining good impedance matching. The $Ti_3C_2T_x$/Ni/AS composite exhibited an RL_{min} of −59.5 dB and an EAB of 4.1 GHz.

A hollow carbon/MXene/Mo_2C (HCMM) is fabricated by a multiscale design strategy [96]. The HCMM owns a three-layered hierarchical structure of hollow microtube core, stacked MXene interlayer, and dense Mo_2C nanowire shell. This layer-by-layer structure provides an abundant heterogeneous interface between MXenes, Mo_2C, and graphitic carbon, enhancing polarization loss (Figure 2.9d). Due to the excellent conductivity of MXenes and Mo_2C, the incident waves in MXene interlayer and Mo_2C shell will be scattered in multiple directions and reflected. Meanwhile, the propagation path of the incident waves entering the microtube core will be greatly confined in the cavity, leading to multiple reflections and scattering. Besides, the abundant functional groups on the surface of MXenes can also improve the dielectric loss through dipole polarization. The HCMM can achieve the RL_{min} of −87.28 dB at 1.4 mm and an RL_{min} of −66.30 dB at 1.0 mm.

2.6 The Application Prospects of MXenes in EMW-Absorbing Materials

In conclusion, MXenes offer immense potential in EWA owing to their facile preparation, controllable interface defects and functional groups, large specific surface area, and high conductivity. MXenes are usually coupled with dielectric loss materials or magnetic loss materials to reduce the permittivity or increase the permeability, and thus improve the impedance matching. The synergetic effects between MXenes and the other loss materials will create more loss mechanisms, which benefit the MA performance. The method of coupling MXenes with multiple loss materials is also a popular trend for developing high-performance MAMs. A well-designed structure can optimize contact areas and lengthen EM propagation paths in the absorber without adding thickness to the coating. The rational microstructure design of MXenes can achieve the characteristics of strong absorption, wide band, lightness, and thinness.

Although MXene-based materials have made remarkable achievements in MA, challenges also exist. (1) The preparation methods of MXenes needs further development. The commonly used HF may bring environmental and safety concerns. Despite the fluorine-free synthesis having been developed, large-scale preparation remains a challenge. (2) The MA performance of MXene-based absorbers at low frequencies requires further enhancement. The majority of MXene-based materials exhibit inferior RL at low frequencies and narrower EAB or require greater thicknesses to achieve comparable MA as those observed at high frequencies. (3) The analysis of the microstructure and intrinsic mechanisms of MXene-based absorbers needs further exploration. There is little research on using more advanced characterizations and theories to design the microstructure and explore the internal mechanisms. (4) The multifunctional characteristics of MXene-based absorbers, such as corrosion resistance, flexibility, and flame retardancy, should be considered. Addressing these issues will be crucial in bridging the gap between laboratory research and practical industrial utilization.

References

1 Naguib, M., Kurtoglu, M., Presser, V. et al. (2011). Two-dimensional nanocrystals produced by exfoliation of Ti_3AlC_2. *Adv. Mater.* 23 (37): 4248–4253.

2 Zhang, C., Ma, Y., Zhang, X. et al. (2020). Two-dimensional transition metal carbides and nitrides (MXenes): synthesis, properties, and electrochemical energy storage applications. *Energy Environ. Mater.* 3 (1): 29–55.

3 Verma, C., Zehra, S., Aslam, R. et al. (2022). MXenes as emerging 2D materials for anticorrosive application: challenges and opportunities. *Adv. Mater. Interfaces* 9 (28): 2200579.

4 Salim, O., Mahmoud, K.A., Pant, K.K., and Joshi, R.K. (2019). Introduction to MXenes: synthesis and characteristics. *Mater. Today Chem.* 14: 100191.

5 Li, H., Li, X., Liang, J., and Chen, Y. (2019). Hydrous RuO_2-decorated MXene coordinating with silver nanowire inks enabling fully printed micro-supercapacitors with extraordinary volumetric performance. *Adv. Energy Mater.* 9 (15): 1803987.

6 Bai, Y., Liu, C., Chen, T. et al. (2021). MXene-copper/cobalt hybrids via Lewis acidic molten salts etching for high performance symmetric supercapacitors. *Angew. Chem. Int. Ed.* 60 (48): 25318–25322.

7 Wang, D., Zhou, C., Filatov, A.S. et al. (2023). Direct synthesis and chemical vapor deposition of 2D carbide and nitride MXenes. *Science* 379 (6638): 1242–1247.

8 Chen, J., Chen, M., Zhou, W. et al. (2022). Simplified synthesis of fluoride-free $Ti_3C_2T_x$ via electrochemical etching toward high-performance electrochemical capacitors. *ACS Nano* 16 (2): 2461–2470.

9 Zhang, N., Hong, Y., Yazdanparast, S., and Zaeem, M.A. (2018). Superior structural, elastic and electronic properties of 2D titanium nitride MXenes over carbide MXenes: a comprehensive first principles study. *2D Mater.* 5 (4): 045004.

10 Li, M., Lu, J., Luo, K. et al. (2019). Element replacement approach by reaction with Lewis acidic molten salts to synthesize nanolaminated max phases and MXenes. *J. Am. Chem. Soc.* 141 (11): 4730–4737.

11 Hope, M.A., Forse, A.C., Griffith, K.J. et al. (2016). NMR reveals the surface functionalisation of Ti_3C_2 MXene. *Phys. Chem. Chem. Phys.* 18 (7): 5099–5102.

12 Harris, K.J., Bugnet, M., Naguib, M. et al. (2015). Direct measurement of surface termination groups and their connectivity in the 2D MXene V_2CT_x using NMR spectroscopy. *J. Phys. Chem. C* 119 (24): 13713–13720.

13 Guo, Z., Zhou, J., Si, C., and Sun, Z. (2015). Flexible two-dimensional $Ti_{n+1}C_n$ ($n = 1$, 2 and 3) and their functionalized MXenes predicted by density functional theories. *Phys. Chem. Chem. Phys.* 17 (23): 15348–15354.

14 Wang, H., Wu, Y., Zhang, J. et al. (2015). Enhancement of the electrical properties of MXene Ti_3C_2 nanosheets by post-treatments of alkalization and calcination. *Mater. Lett.* 160: 537–540.

15 Sang, X., Xie, Y., Lin, M.-W. et al. (2016). Atomic defects in monolayer titanium carbide ($Ti_3C_2T_x$) MXene. *ACS Nano* 10 (10): 9193–9200.

16 Yang, Y., Umrao, S., Lai, S., and Lee, S. (2017). Large-area highly conductive transparent two-dimensional $Ti_3C_2T_x$ film. *J. Phys. Chem. Lett.* 8 (4): 859–865.

17 Lai, S., Jeon, J., Jang, S.K. et al. (2015). Surface group modification and carrier transport properties of layered transition metal carbides ($Ti_3C_2T_x$, T: –OH, –F and –O). *Nanoscale* 7 (46): 19390–19396.

18 Li, Z., Wang, L., Sun, D. et al. (2015). Synthesis and thermal stability of two-dimensional carbide MXene Ti_3C_2. *Mater. Sci. Eng. B.* 191: 33–40.

19 Rakhi, R.B., Ahmed, B., Hedhili, M.N. et al. (2015). Effect of postetch annealing gas composition on the structural and electrochemical properties of $Ti_3C_2T_x$ MXene electrodes for supercapacitor applications. *Chem. Mater.* 27 (15): 5314–5323.

20 Ghassemi, H., Harlow, W., Mashtalir, O. et al. (2014). In situ environmental transmission electron microscopy study of oxidation of two-dimensional Ti_3C_2 and formation of carbon-supported TiO_2. *J. Mater. Chem. A* 2 (35): 14339–14343.

21 Han, M., Yin, X., Wu, H. et al. (2016). Ti_3C_2 MXenes with modified surface for high-performance electromagnetic absorption and shielding in the X-band. *ACS Appl. Mater. Interfaces* 8 (32): 21011–21019.

22 Xu, Z., Zhou, R., Ma, Q. et al. (2023). Study on the electromagnetic wave absorption performance of Ti_3C_2 MXene with different etching states. *J. Mater. Sci.* 58 (11): 4824–4839.

23 Tong, Y., He, M., Zhou, Y. et al. (2018). Electromagnetic wave absorption properties in the centimetre-band of $Ti_3C_2T_x$ MXenes with diverse etching time. *J. Mater. Sci. Mater. Electron.* 29 (10): 8078–8088.

24 Zhou, X., Wen, J., Wang, Z. et al. (2022). Broadband high-performance microwave absorption of the single-layer $Ti_3C_2T_x$ MXene. *J. Mater. Sci. Technol.* 115: 148–155.

25 He, P., Cao, M.-S., Shu, J.-C. et al. (2019). Atomic layer tailoring titanium carbide MXene to tune transport and polarization for utilization of electromagnetic energy beyond solar and chemical energy. *ACS Appl. Mater. Interfaces* 11 (13): 12535–12543.

26 Kim, H., Wang, Z., and Alshareef, H.N. (2019). MXetronics: electronic and photonic applications of MXenes. *Nano Energy* 60: 179–197.

27 Du, Y., Yan, Z., You, W. et al. (2023). Balancing MXene surface termination and interlayer spacing enables superior microwave absorption. *Adv. Funct. Mater.* 33 (34): 2301449.

28 Zhang, Z., He, Y., Lv, Y. et al. (2020). Effect of surface structure and composition on the electromagnetic properties of $Ti_3C_2T_x$ MXenes for highly efficient electromagnetic wave absorption. *J. Phys. Chem. C* 124 (36): 19666–19674.

29 Zeng, X., Zhao, C., Jiang, X. et al. (2023). Functional tailoring of multi-dimensional pure MXene nanostructures for significantly accelerated electromagnetic wave absorption. *Small* 19 (41): 2303393.

30 Zhao, Z., Huang, W., Qin, H., and Li, L. (2023). An insight into terahertz electromagnetic interference shielding of two-dimensional titanium carbide thin film: the key role of functional groups. *Appl. Phys. A.* 129 (2): 144.

31 Xu, W., Li, S., Hu, S. et al. (2021). Effect of $Ti_3C_2T_x$ precursor and processing conditions on microwave absorption performance of resultant $Ti_3C_2T_x$ MXene. *J. Mater. Sci.* 56 (15): 9287–9301.

32 Ying, M., Zhao, R., Hu, X. et al. (2022). Wrinkled titanium carbide (MXene) with surface charge polarizations through chemical etching for superior electromagnetic interference shielding. *Angew. Chem. Int. Ed.* 61 (16): e202201323.

33 Xu, G., Wang, X., Gong, S. et al. (2018). Solvent-regulated preparation of well-intercalated $Ti_3C_2T_x$ MXene nanosheets and application for highly effective electromagnetic wave absorption. *Nanotechnology* 29 (35): 355201.

34 Du, Q., Men, Q., Li, R. et al. (2022). Electrostatic adsorption enables layer stacking thickness-dependent hollow $Ti_3C_2T_x$ MXene bowls for superior electromagnetic wave absorption. *Small* 18 (47): 2203609.

35 Jin, Z., Fang, Y., Wang, X. et al. (2019). Ultra-efficient electromagnetic wave absorption with ethanol-thermally treated two-dimensional Nb_2CT_x nanosheets. *J. Colloid Interface Sci.* 537: 306–315.

36 Wu, J., Feng, Y., Xia, Y. et al. (2022). Fabrication of S-doped $Ti_3C_2T_x$ materials with enhanced electromagnetic wave absorbing properties. *J. Alloys Compd.* 891: 161942.

37 Xu, W., Li, S., Zhang, W. et al. (2021). Nitrogen-doped $Ti_3C_2T_x$ MXene induced by plasma treatment with enhanced microwave absorption properties. *ACS Appl. Mater. Interfaces* 13 (41): 49242–49253.

38 Han, M., Yin, X., Li, X. et al. (2017). Laminated and two-dimensional carbon-supported microwave absorbers derived from MXenes. *ACS Appl. Mater. Interfaces* 9 (23): 20038–20045.

39 Liu, T.-T., Cao, W.-Q., Yuan, Q. et al. (2023). Engineering multi-relaxation interfaces in $Ti_3C_2T_x$ for reducing wideband radar cross section. *2D Mater.* 10 (3): 035017.

40 Peng, C., Yang, X., Li, Y. et al. (2016). Hybrids of two-dimensional Ti_3C_2 and TiO_2 exposing {001} facets toward enhanced photocatalytic activity. *ACS Appl. Mater. Interfaces* 8 (9): 6051–6060.

41 Wang, J., Liu, L., Jiao, S. et al. (2020). Hierarchical carbon fiber@MXene@MoS_2 core-sheath synergistic microstructure for tunable and efficient microwave absorption. *Adv. Funct. Mater.* 30 (45): 2002595.

42 Li, X., You, W., Xu, C. et al. (2021). 3D seed-germination-like MXene with in situ growing CNTs/Ni heterojunction for enhanced microwave absorption via polarization and magnetization. *Nano Micro Lett.* 13 (1): 157.

43 Jiang, X., Wang, Q., Song, L. et al. (2024). Enhancing electromagnetic wave absorption with core-shell structured SiO_2@MXene@MoS_2 nanospheres. *Carbon Energy* e502.

44 Li, X., Yin, X., Han, M. et al. (2017). Ti_3C_2 MXenes modified with in situ grown carbon nanotubes for enhanced electromagnetic wave absorption properties. *J. Mater. Chem. C* 5 (16): 4068–4074.

45 Huyan, W., Yao, J., Tao, J. et al. (2022). MXene@C heterogeneous nanocomposites with the 2D-0D structure for ultra-light and broadband electromagnetic wave absorption. *Carbon* 197: 444–454.

46 Yu, M., Huang, Y., Liu, X. et al. (2023). In situ modification of MXene nanosheets with polyaniline nanorods for lightweight and broadband electromagnetic wave absorption. *Carbon* 208: 311–321.

47 Gao, X., Wang, B., Wang, K. et al. (2021). Design of $Ti_3C_2T_x$/TiO_2/PANI multi-layer composites for excellent electromagnetic wave absorption performance. *J. Colloid Interface Sci.* 583: 510–521.

48 Tong, Y., He, M., Zhou, Y. et al. (2018). Hybridizing polypyrrole chains with laminated and two-dimensional $Ti_3C_2T_x$ toward high-performance electromagnetic wave absorption. *Appl. Surf. Sci.* 434: 283–293.

49 Liu, T., Liu, N., An, Q. et al. (2019). Designed construction of $Ti_3C_2T_x$@PPy composites with enhanced microwave absorption performance. *J. Alloys Compd.* 802: 445–457.

50 Liu, Y., Qi, X., Deng, X. et al. (2023). Snake scale-inspired poly(vinylidene fluoride)/Ti_3CNT_x@polypyrrole coatings for ultrawide-bandwidth microwave and visible light absorption. *ACS Appl. Mater. Interfaces* 15 (23): 28491–28502.

51 Li, X., Yin, X., Xu, H. et al. (2018). Ultralight MXene-coated, interconnected SiCnws three-dimensional lamellar foams for efficient microwave absorption in the X-band. *ACS Appl. Mater. Interfaces* 10 (40): 34524–34533.

52 Li, H., Yuan, X., Zhao, P. et al. (2023). A synergistic strategy for SiC/C nanofibers@MXene with core-sheath microstructure toward efficient electromagnetic wave absorption and photothermal conversion. *Appl. Surf. Sci.* 613: 155998.

53 Ding, J., Chen, F., Chen, J. et al. (2021). MXene-derived TiC/SiBCN ceramics with excellent electromagnetic absorption and high-temperature resistance. *J. Am. Ceram. Soc.* 104 (4): 1772–1784.

54 Qian, Y., Wei, H., Dong, J. et al. (2017). Fabrication of urchin-like ZnO-MXene nanocomposites for high-performance electromagnetic absorption. *Ceram. Int.* 43 (14): 10757–10762.

55 Liu, Z., Zhao, Y., Li, S. et al. (2022). CeO_2/$Ti_3C_2T_x$ MXene nanostructures for microwave absorption. *ACS Appl. Nano Mater.* 5 (4): 5764–5775.

56 Sun, C., Jia, Z., Xu, S. et al. (2022). Synergistic regulation of dielectric-magnetic dual-loss and triple heterointerface polarization via magnetic MXene for high-performance electromagnetic wave absorption. *J. Mater. Sci. Technol.* 113: 128–137.

57 Li, X., Wen, C., Yang, L. et al. (2021). Enhanced visualizing charge distribution of 2D/2D MXene/MoS_2 heterostructure for excellent microwave absorption performance. *J. Alloys Compd.* 869: 159365.

58 Zhang, M., Shahnavaz, Z., Yan, X. et al. (2023). Mild synthesis of flower-like CuS modified MXene nanocomposites for high-performance supercapacitors and efficient microwave absorption. *J. Alloys Compd.* 960: 170914.

59 Li, X., Yin, S., Cai, L. et al. (2023). Sea-urchin-like $NiCo_2S_4$ modified MXene hybrids with enhanced microwave absorption performance. *Chem. Eng. J.* 454: 140127.

60 Hou, T., Jia, Z., Wang, B. et al. (2021). Metal-organic framework-derived $NiSe_2$-$CoSe_2$@C/$Ti_3C_2T_x$ composites as electromagnetic wave absorbers. *Chem. Eng. J.* 422: 130079.

61 Chang, M., Jia, Z., He, S. et al. (2021). Two-dimensional interface engineering of NiS/MoS_2/$Ti_3C_2T_x$ heterostructures for promoting electromagnetic wave absorption capability. *Compos. Part B* 225: 109306.

62 Liang, L., Yang, R., Han, G. et al. (2020). Enhanced electromagnetic wave-absorbing performance of magnetic nanoparticles-anchored 2D $Ti_3C_2T_x$ MXene. *ACS Appl. Mater. Interfaces* 12 (2): 2644–2654.

63 Zhou, C., Wang, X., Luo, H. et al. (2020). Rapid and direct growth of bipyramid TiO_2 from $Ti_3C_2T_x$ MXene to prepare Ni/TiO_2/C heterogeneous composites for high-performance microwave absorption. *Chem. Eng. J.* 383: 123095.

64 Liang, L., Han, G., Li, Y. et al. (2019). Promising $Ti_3C_2T_x$ MXene/Ni chain hybrid with excellent electromagnetic wave absorption and shielding capacity. *ACS Appl. Mater. Interfaces* 11 (28): 25399–25409.

65 Pan, F., Yu, L., Xiang, Z. et al. (2021). Improved synergistic effect for achieving ultrathin microwave absorber of 1D Co nanochains/2D carbide MXene nanocomposite. *Carbon* 172: 506–515.

66 Li, X., Wen, C., Yang, L. et al. (2021). MXene/FeCo films with distinct and tunable electromagnetic wave absorption by morphology control and magnetic anisotropy. *Carbon* 175: 509–518.

67 Zhou, C., Wang, X., Luo, H. et al. (2019). Interfacial design of sandwich-like CoFe@$Ti_3C_2T_x$ composites as high efficient microwave absorption materials. *Appl. Surf. Sci.* 494: 540–550.

68 Zhang, X., Wang, H., Hu, R. et al. (2019). Novel solvothermal preparation and enhanced microwave absorption properties of $Ti_3C_2T_x$ MXene modified by in situ coated Fe_3O_4 nanoparticles. *Appl. Surf. Sci.* 484: 383–391.

69 Deng, B., Liu, Z., Pan, F. et al. (2021). Electrostatically self-assembled two-dimensional magnetized MXene/hollow Fe_3O_4 nanoparticle hybrids with high electromagnetic absorption performance and improved impendence matching. *J. Mater. Chem. A* 9 (6): 3500–3510.

70 Yang, Y., Zhao, J., Wang, J. et al. (2023). Regulated electron migration in sandwich-like m-Ti_3C_2/Fe_3O_4 composites derived from electrostatic assembly boosted electromagnetic wave absorption. *J. Mater. Chem. A* 11 (13): 6934–6944.

71 Deng, R., Chen, B., Li, H. et al. (2019). MXene/Co_3O_4 composite material: stable synthesis and its enhanced broadband microwave absorption. *Appl. Surf. Sci.* 488: 921–930.

72 Wang, H.-Y., Sun, X.-b., and Wang, G.-S. (2021). A MXene-modulated 3D crosslinking network of hierarchical flower-like MOF derivatives towards ultra-efficient microwave absorption properties. *J. Mater. Chem. A* 9 (43): 24571–24581.

73 Guo, Y., Wang, D., Bai, T. et al. (2021). Electrostatic self-assembled $NiFe_2O_4$/$Ti_3C_2T_x$ MXene nanocomposites for efficient electromagnetic wave absorption at ultralow loading level. *Adv. Compos. Hybrid Mater.* 4 (3): 602–613.

74 Li, X., Wu, Z., You, W. et al. (2022). Self-assembly MXene-rGO/CoNi film with massive continuous heterointerfaces and enhanced magnetic coupling for superior microwave absorber. *Nano Micro Lett.* 14 (1): 73.

75 Xiang, Z., Shi, Y., Zhu, X. et al. (2021). Flexible and waterproof 2D/1D/0D construction of MXene-based nanocomposites for electromagnetic wave absorption, EMI shielding, and photothermal conversion. *Nano Micro Lett.* 13 (1): 150.

76 Wang, H., Zhao, J., Wang, Z., and Liu, P. (2023). Bird-nest-like multi-interfacial MXene@ SiC_{NWs}@Co/C hybrids with enhanced electromagnetic wave absorption. *ACS Appl. Mater. Interfaces* 15 (3): 4580–4590.

77 Wang, B., Li, S., Huang, F. et al. (2022). Construction of multiple electron transfer paths in 1D core-shell heterostructures with MXene as interlayer enabling efficient microwave absorption. *Carbon* 187: 56–66.

78 Hou, T., Jia, Z., Dong, Y. et al. (2022). Layered 3D structure derived from MXene/magnetic carbon nanotubes for ultra-broadband electromagnetic wave absorption. *Chem. Eng. J.* 431: 133919.

79 Wen, C., Li, X., Zhang, R. et al. (2022). High-density anisotropy magnetism enhanced microwave absorption performance in $Ti_3C_2T_x$ MXene@Ni microspheres. *ACS Nano* 16 (1): 1150–1159.

80 Zhang, Y., Liu, X., Guo, Z. et al. (2024). MXene@Co hollow spheres structure boosts interfacial polarization for broadband electromagnetic wave absorption. *J. Mater. Sci. Technol.* 176: 167–175.

81 Liu, W., Wang, D., Li, K., and Wei, X. (2022). Hollow spheres of $Ti_3C_2T_x$ with a nanometer-thick wall for high-performance microwave absorption. *ACS Appl. Nano Mater.* 5 (12): 18488–18499.

82 Wang, Y.-Q., Zhao, H.-B., Cheng, J.-B. et al. (2022). Hierarchical $Ti_3C_2T_x$@ZnO hollow spheres with excellent microwave absorption inspired by the visual phenomenon of eyeless urchins. *Nano Micro Lett.* 14 (1): 76.

83 Wang, G., Li, C., Estevez, D. et al. (2023). Boosting interfacial polarization through heterointerface engineering in MXene/graphene intercalated-based microspheres for electromagnetic wave absorption. *Nano Micro Lett.* 15 (1): 152.

84 Zhang, C., Wu, Z., Xu, C. et al. (2022). Hierarchical $Ti_3C_2T_x$ MXene/carbon nanotubes hollow microsphere with confined magnetic nanospheres for broadband microwave absorption. *Small* 18 (3): 2104380.

85 Niu, H., Tu, X., Zhang, S. et al. (2022). Engineered core-shell SiO_2@$Ti_3C_2T_x$ composites: towards ultra-thin electromagnetic wave absorption materials. *Chem. Eng. J.* 446: 137260.

86 Wang, Y.-Q., Ding, R., Zhang, Y.-C. et al. (2023). Gradient hierarchical hollow heterostructures of $Ti_3C_2T_x$@rGO@MoS_2 for efficient microwave absorption. *ACS Appl. Mater. Interfaces* 15 (27): 32803–32813.

87 Li, X., Zhang, M., You, W. et al. (2020). Magnetized MXene microspheres with multiscale magnetic coupling and enhanced polarized interfaces for distinct microwave absorption via a spray-drying method. *ACS Appl. Mater. Interfaces* 12 (15): 18138–18147.

88 Ling, M., Ge, F., Wu, F. et al. (2023). Effect of crystal transformation on the intrinsic defects and the microwave absorption performance of $Mo_2TiC_2T_x$/rGO microspheres. *Small* 20: 2306233.

89 Li, Y., Meng, F., Mei, Y. et al. (2020). Electrospun generation of $Ti_3C_2T_x$ MXene@graphene oxide hybrid aerogel microspheres for tunable high-performance microwave absorption. *Chem. Eng. J.* 391: 123512.

90 Wang, Z., Yang, L., Zhou, Y. et al. (2021). NiFe LDH/MXene derivatives interconnected with carbon fabric for flexible electromagnetic wave absorption. *ACS Appl. Mater. Interfaces* 13 (14): 16713–16721.

91 Wu, F., Liu, Z., Wang, J. et al. (2021). Template-free self-assembly of MXene and CoNi-bimetal MOF into intertwined one-dimensional heterostructure and its microwave absorbing properties. *Chem. Eng. J.* 422: 130591.

92 Zeng, X., Zhao, C., Nie, T. et al. (2022). Construction of 0D/1D/2D MXene nanoribbons-NiCo@NC hierarchical network and their coupling effect on electromagnetic wave absorption. *Mater. Today Phys.* 28: 100888.

93 Zhang, S., Jia, Z., Zhang, Y., and Wu, G. (2022). Electrospun $Fe_{0.64}Ni_{0.36}$/MXene/CNFs nanofibrous membranes with multicomponent heterostructures as flexible electromagnetic wave absorbers. *Nano Res.* 16 (2): 3395–3407.

94 Wang, L., Li, X., Qian, Y. et al. (2023). MXene-layered double hydroxide reinforced epoxy nanocomposite with enhanced electromagnetic wave absorption, thermal conductivity, and flame retardancy in electronic packaging. *Small* 20: 2304311.

95 Li, M., Han, M., Zhou, J. et al. (2018). Novel scale-like structures of graphite/TiC/Ti_3C_2 hybrids for electromagnetic absorption. *Adv. Electron. Mater.* 4 (5): 1700617.

96 Wang, J., Wu, Z., Xing, Y. et al. (2023). Multi-scale design of ultra-broadband microwave metamaterial absorber based on hollow carbon/MXene/Mo_2C microtube. *Small* 19 (14): 2207051.

97 Li, Q., Nan, K., Wang, W. et al. (2023). Electrostatic self-assembly sandwich-like 2D/2D NiFe-LDH/MXene heterostructure for strong microwave absorption. *J. Colloid Interface Sci.* 648: 983–993.

98 Deng, B., Xiang, Z., Xiong, J. et al. (2020). Sandwich-like Fe&TiO_2@C nanocomposites derived from MXene/Fe-MOFs hybrids for electromagnetic absorption. *Nano Micro Lett.* 12 (1): 55.

99 Feng, W., Luo, H., Wang, Y. et al. (2019). MXenes derived laminated and magnetic composites with excellent microwave absorbing performance. *Sci. Rep.* 9: 3957.

100 Deng, W., Li, T., Li, H. et al. (2023). In situ construction of hierarchical core-shell SiC_{nws}@ SiO_2-carbon foam hybrid composites with enhanced polarization loss for highly efficient electromagnetic wave absorption. *Carbon* 202: 103–111.

101 Wu, Z., Yang, Z., Jin, C. et al. (2021). Accurately engineering 2D/2D/0D heterojunction in hierarchical $Ti_3C_2T_x$ MXene nanoarchitectures for electromagnetic wave absorption and shielding. *ACS Appl. Mater. Interfaces* 13 (4): 5866–5876.

102 Luo, W., Wang, M., Wang, K. et al. (2022). A robust hierarchical MXene/Ni/aluminosilicate glass composite for high-performance microwave absorption. *Adv. Sci.* 9 (4): 2104163.

3

High-Entropy Electromagnetic Wave Absorption Materials

Shengchong Hui

MOE Key Laboratory of Material Physics and Chemistry Under Extraordinary Conditions, School of Physical Science and Technology, Northwestern Polytechnical University, Xi'an, China

CHAPTER MENU

3.1 The Concept and Features of High-Entropy Materials

3.1.1 The Definition of High-Entropy Materials

The first research results on multi-principal alloys were published in 2004, and the motivation of this work was to research the central region of the phase space of multi-principal alloys [1]. The results indicate that more principal elements might be more favorable for generating single-phase solid solutions (SS) rather than alloys containing intermetallic compounds (IM), but the concept of entropy is not involved at this time. The definitions of high-entropy alloys (HEA) are diverse, and these definitions commonly used in discussions and possible controversies are summarized as follows:

(1) Elemental component-based definition. A pioneering work defined HEA as "a substance composed of five or more major elements in equimolar ratios," at which time the

Electromagnetic Wave Absorbing Materials: Fundamentals and Applications, First Edition.
Edited by Hongjing Wu, Jun Luo, and Meiyin Yang.

concentration of each element in HEA was confined [2]. The other researchers expanded this concept to "between 5% and 35% of each element," and based on this definition, the number of HEA was significantly enhanced [3]. The elemental component-based definition does not limit the entropy magnitude, nor does it require the alloy to be characterized by single-phase SS.

(2) Configurational entropy-based definition. Assuming that the atoms are randomly occupied in the lattice, based on the Boltzmann equation, we can estimate the value of the configurational entropy of a single alloy: $\Delta S = -R\sum_{I=1}^{n} c_i \ln c_i$, where R is the gas constant, c_i is the mole fraction of the element, and n is the total number of elements composing the material. This definition fully considers the disorder of atoms and distinguishes them by configurational entropy (low entropy: $\Delta S<0.69\ R$, medium entropy: $0.69\ R<\Delta S<1.61\ R$, high entropy (HE): $\Delta S>1.61\ R$) [4]. However, this definition requires the material state to be at a high temperature or to provide sufficient thermal energy gain to exclude perturbations to atomic occupation caused by temperature changes. The above issues represent the limitations and challenges of using this definition.

(3) Phase-based compromise definition. The two aforementioned definitions are divergent. For example, consider a five-elemental alloy with the following element proportions: 5%, 5%, 20%, 35%, and 35%. In this case, the configurational entropy $\Delta S=1.36\ R$. This is considered to be a HEA according to the component-based definition but is not recognized as possessing HE characteristics according to the configurational entropy-based definition. Some researchers have defined HEA in terms of the phases that make up the material: "a single-phase stabilized SS consisting of five or more elements in which the contents of the atoms are relatively uniform." With this definition, there is no IM-biased structure in HEA, and there is no requirement for the state of the material. This new "HE" concept opens up a vast and unexplored alloy field. In this field, due to numerous fundamental questions, unexplained experimental findings, and surprising material properties, more and more scholars actively participate in relevant academic investigations. With the further development of research, scholars tend to seek single-phase stable SS by controlling the configurational entropy, which means that phase-based compromise definitions are more commonly used [5–8]. Therefore, in this section, we define "HE" by adopting the phase-based compromise definition. Nevertheless, there is no right or wrong definition of HEA, and the choice of definition depends on the intention of the work being performed. So far, the "HE" concept has not only been used for alloy systems but has also been successfully extended to other material systems. These types of materials have almost infinite elemental combinations and unexplored physicochemical properties and have been named "high-entropy materials" (HEMs) by researchers. This section summarizes and discusses HEM, including their special properties, preparation methods, classification of HE absorbers, and future trends, to accelerate the development of novel electromagnetic wave (EMW) absorption materials and microscopic loss mechanisms.

3.1.2 Broaden Elemental Combination and Microstructure

The primary focus of early publications on multi-principal component alloys was to explore the novel alloy system provided by the internal regions of the multi-principal component phase diagram, offering abundant opportunities for the discovery of new alloy materials with scientific significance, as shown in Figure 3.1. To further develop this concept, additional

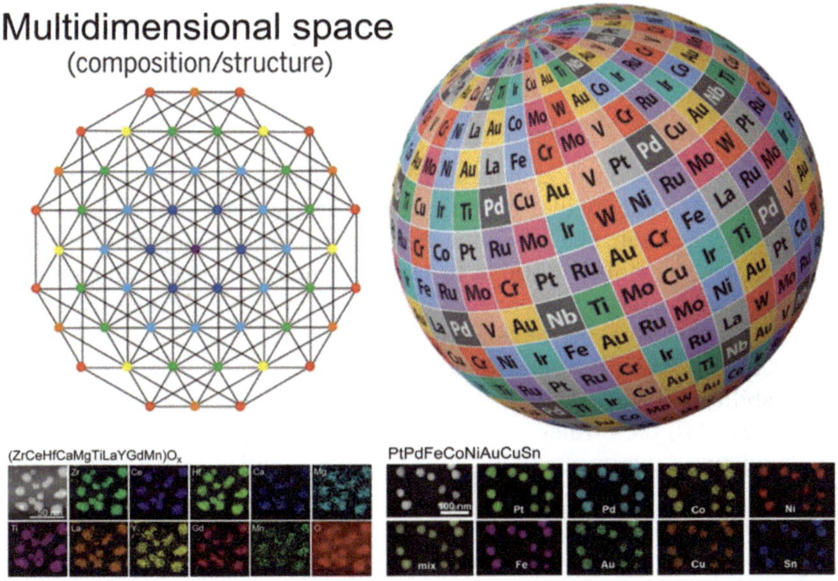

Figure 3.1 Emerging high-entropy nanoparticles feature multi-elemental mixing within a large compositional space and can be used for diverse applications.

structural and functional properties have been documented, such as one multi-principal component alloy exhibiting a fracture toughness that surpasses that of most metal alloys and pure metal materials [9]. Furthermore, a variety of nanostructures have been synthesized, including single-phase SS, multiphase alloys, nanocrystalline microstructures, and amorphous structures. Early research also revealed the appealing multiphase microstructures of borides [10] or nitrides [11], with several alloying elements (such as Ti and Al) yielding unconventional microstructures. Despite the wide array of microstructures provided, the field of HEA has now shifted its focus toward the exploration of single-phase SS.

3.1.3 Dialectical View of Single-Phase Solid Solution Properties

Numerous studies have asserted that the presence of the IM phase leads to substantial embrittlement of alloy materials, rendering the machining process challenging. Conversely, the single-phase SS is regarded as robust due to its favorable ductility and damage tolerance [12]. Consequently, researchers in the field of HEA are focusing on strategies to discover single-phase SS structures. However, this is not universally applicable. In the case of non-alloy materials, such as semiconductors, the phase structure is intricately linked to the electron transport capacity, which directly influences the various physical and chemical properties of the material. While it is undeniable that single-phase or multiphase structures can enhance material properties to a certain extent, their effects are system-specific. Consequently, some researchers are directing efforts toward achieving a balanced distribution of SS and IM phases. The aim of this discussion is not to challenge the overarching conclusion advocating for the pursuit of single-phase SS but rather to offer a more equitable viewpoint on the influence of SS and IM phases on material properties. The typical approach in the development of HEM is to define the requisite performance for a given application and subsequently design components and microstructures capable of delivering the necessary performance [13, 14].

3.1.4 The Theoretical Approach of Phase Selection in HEM

The equilibrium thermodynamic method aims to determine the phase stability of a HEM by calculating the Gibbs free energy of different phases that may form in the HEM. This direct method can predict accurate phases, although it requires data on all possible phases of the alloy. Furthermore, it is often difficult to simulate multicomponent alloys using simulation methods based on physical potentials, such as classical molecular dynamics simulations [15, 16]. Considering metastable phases, such as metallic glass, more challenges are introduced. In addition, empirical and semi-empirical rules were proposed to understand phase stability and guide the design of HEA [17–20]. Compared with the direct calculation method [21–24], these rules are mainly established by comparing the theoretical model with existing experimental data, so they can be applied to equilibrium and metastable phases.

3.1.5 Four Core Effects of HEM

In comparison to conventional materials, HEM exhibits unique features in dynamics, structure, and properties. This is attributed to the HE nature of these materials, typically comprising more than five elements, leading to complex interactions among different elements. The specific behaviors of HEM are often described through the "four core effects." While three of these effects are postulated based on early experimental data, the "cocktail effect" provides a comprehensive description of the diverse properties of HEM. In the following sections, we will expound on the specific content of these hypotheses and their potential implications.

3.1.5.1 "High-Entropy" Effect

The HE effect is a concept that characterizes HEM and proposes that in materials containing five or more elements [25]. HEM typically exhibit reduced Gibbs free energies ($\Delta G_{mix}=\Delta H_{mix}-T\Delta S_{mix}$), wherein higher ΔS_{mix} promotes the reduction of ΔG_{mix} at elevated temperatures, thereby enhancing system stability. Notably, a high ΔS_{mix} increases the disorder of the system and enables the random distribution of constituent atoms on the lattice, thereby impeding the formation of IM compounds and phase separation phenomena. Therefore, enhancing the configurational entropy may be more advantageous for the formation of SS phases.

3.1.5.2 "Lattice Distortion" Effect

The lattice distortion in HEM arises from variations in atomic sizes within the lattice, resulting in the formation of complex and concentrated phases, which are more pronounced compared to traditional materials. The displacement of each lattice position is contingent upon the type of atoms occupying that position and the atoms in the immediate vicinity. The uncertainty in atomic positions caused by these displacements leads to heightened configurational entropy, which can diminish the intensity of X-ray diffraction peaks [26], elevate hardness, reduce conductivity and thermal conductivity, and mitigate temperature dependence [27].

Furthermore, computational simulations allow researchers to precisely characterize the extent of lattice distortion and positional deviation resulting from static and dynamic disorder sequences, which are anticipated to offer theoretical predictions for the properties of HEM. This will be elaborated upon in subsequent content.

3.1.5.3 "Sluggish Diffusion" Effect

Researchers believe that diffusion in HEM is extremely slow because interactions between different atoms and irregular atomic arrangements can affect diffusion behavior. Specifically, due to the influence of high disorder, the lattice potential energy at different positions exhibits significant differences, leading to high diffusion activation energy and thus suppressing atomic diffusion [28].

3.1.5.4 "Cocktail" Effect

Professor S. Ranganathan coined the term "cocktail effect" to describe a synergistic mixture that yields an outcome greater than the sum of its parts, initially inspired by the concept of a pleasant mixture [29–31]. However, it later evolved to signify a synergistic mixture with unpredictable and amplified effects. The cocktail effect is a distinctive phenomenon resulting from the interaction of multicomponent elements, representing more than just a synergistic effect. To some extent, this cocktail effect may confer new characteristics upon HEM that are not present in any constituent materials. This combined effect ultimately manifests in the performance of HEM materials, impacting their mechanical properties, corrosion resistance, and oxidation resistance. In each instance, material properties are intricately and sensitively dependent on composition, microstructure, electronic structure, and other properties. Unlike other "core effects," the "cocktail effect" does not rely on a hypothesis and does not necessitate evidence. The cocktail effect serves as a reminder that unique material properties often arise from unexpected synergistic effects.

3.2 The Synthesis Approach and Advanced Characterization of HEM

From a thermodynamic perspective, the formation of HEM is the result of the competition between enthalpy and entropy. The configurational entropy of HEM increases with the quantity of elements, driving single-phase mixing. The enthalpy (ΔH) of multi-element interactions is largely influenced by the properties of the constituent elements, which directly impact the phase produced under near-equilibrium conditions. For instance, element combinations with highly positive ΔH values (i.e. repulsive forces) can result in immiscibility and phase segregation, while elements with highly negative ΔH values (i.e. attractive forces) can promote structural ordering, such as IM formation. If all ΔH values in a multi-element composition are close to zero, it indicates that there is almost no mutual attraction or repulsion between these elements, and the entropy term dominates, promoting uniform random element mixing and HEM formation. However, due to significant physical and chemical differences among different elements (i.e. the wide range of ΔH values), natural single-phase mixing is often challenging and rare [32, 33]. When synthesizing multi-element compounds using near-equilibrium methods (such as wet chemistry), phase separation structure is more typical [34, 35]. Therefore, the synthesis of HEM is often challenging.

3.2.1 HEM Synthesis

3.2.1.1 Traditional Template Sintering Method

Sintering is a commonly used method to prepare HEM. Traditional solid-state sintering typically involves the use of ball milling technology to achieve uniform mixing of precursor oxides, followed by prolonged high-temperature calcination for sintering. The compression step can be performed either before or during sintering. The resulting materials are typically large or dense particles with a high grain size and low specific surface area. The compression step further reduces porosity and increases crystallinity, which is particularly important for the application of bulk materials [36]. Taking high-entropy oxide (HEO) (MgCoNiCuZn)O as an example, the minimum sintering temperature for the oxide precursor to form a phase through solid-state sintering is approximately 900 °C. Its single phase will be disrupted when annealed at around 700 °C. Salts such as carbonates, nitrates, acetate salts, and acetylacetone salts, as well as metal complexes, metal–organic frameworks (MOFs), and various other types of HEM (such as alloys [37] and hydroxides [38]), can also serve as precursors, as shown in Figure 3.2 [39]. Depending on the precursor, the synthesis temperature of HEM can be as low as 500 °C. Xing et al. prepared metal PVP precursors through electrospinning and subsequent calcination, resulting in all rare earth HEO $(YYbSmEuEr)_2O_3$ with a bixbyite structure [40]. However, products obtained through solid-state sintering are not always suitable for certain research fields that require structured nanoparticles and narrow

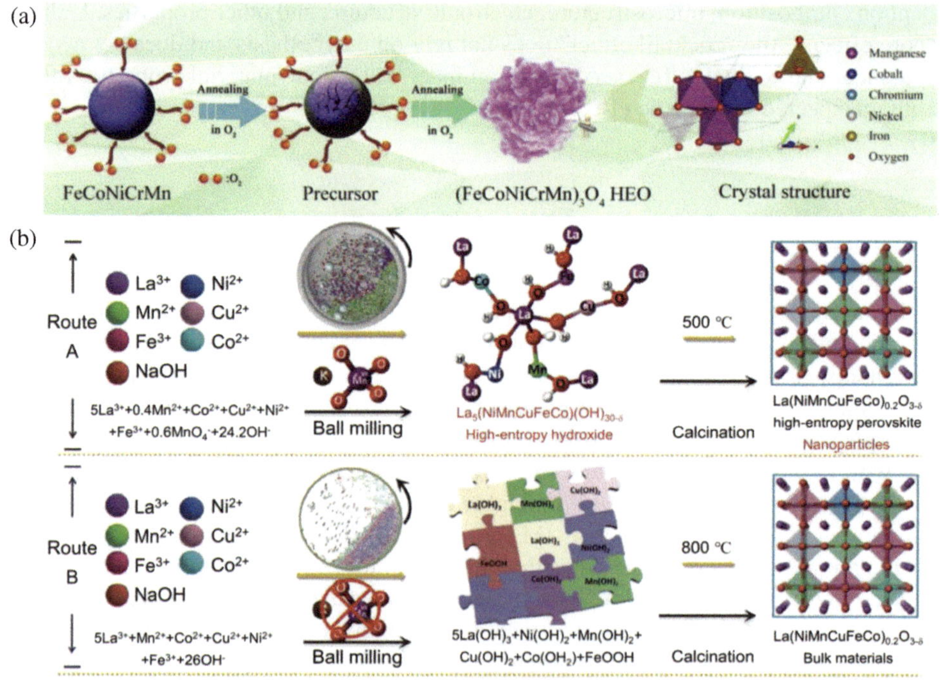

Figure 3.2 (a) A schematic of the preparation process for $(FeCoNiCrMn)_3O_4$ HEO. (b) An illustration of the synthesis of high-entropy LaMnO. Routes A and B corresponded to the process synthesizing Redox. HE-LMONPs and Bulk-HE-IMO, respectively.

distributions [41]. To make HEM suitable for these fields, modifying the synthesis process by incorporating removable templates to separate particles is a viable approach.

Dai's research team selected SiO_2 as a removable hard template, which can be dissolved using a NaOH aqueous solution. Smoked Sil-ICA was subjected to ball milling with metal salts and then pyrolyzed at 900 °C. Following the etching process, a high porosity and atomic dispersion of Pd were achieved, resulting in the formation of fluorite-type HEM with a relatively high specific surface area of 162.1 $m^2\ g^{-1}$ [42]. In the synthesis of perovskite-type $Zr_{0.5}(NiFeCuMnCo)_{0.5-x}O_x$, they still utilized soluble NaCl as a hard template. NaOH and NaCl were added to chloride salts to exchange anions during balling and subsequent calcination at 500 °C. Subsequently, the sodium salts were removed through washing, leading to the retention of pores with a specific surface area of up to 84 $m^2\ g^{-1}$ [43]. The post-processing of template removal adds complexity to the synthesis of porous HEM. Furthermore, the template materials must remain stable during the calcination process (for instance, SiO_2 reacts with many transition metal oxides at high temperatures), while being easily removable or washable, thereby limiting the selection of etchants and templates. This is necessary to maintain the similarity of solid-state synthesis of precursors that form self-templates and enable the fabrication of nanostructures.

Zhou's research team opted for molten glucose as the self-template, drawing inspiration from the "black powder" phenomenon. They introduced additional ammonium nitrate simultaneously with the metal nitrate. Upon heating, the metal nitrate generated a substantial amount of gas, causing the molten glucose to expand and form a foam structure. Through subsequent carbonization and oxidation, the carbon was completely eliminated, resulting in the formation of ultrathin porous oxide sheets [44]. This approach can be utilized to synthesize a broad range of single-component oxides and can be extended to two types of HEO, including cubic bixbyite type $(ErGdYbSm_{1.2}Y_{0.8})_2O_3$ and spinel type $(CoMnNiCrFe)_3O_4$, while retaining their single-phase properties. Dong's team also devised a similar approach, employing a deep eutectic solvent composed of glucose and urea. Due to its low melting point, the liquid is capable of dissolving metal salts directly at manageable temperatures. In addition to the carbonization of glucose, the urea component serves as a foaming agent akin to ammonium nitrate. The precursor was carbonized into black foam using direct microwave irradiation and subsequently calcined to yield HEM [45]. The solvents aid in achieving a more uniform mixing of metals compared to ball milling. Furthermore, by adjusting the metal concentration, the different morphologies of nanowires and nanosheets can be tailored. This method is applicable to various HEMs, encompassing rock salt type, spinel type, and perovskite type.

Soluble salt precursors can also be incorporated into porous templates through straightforward solution-based processes. Qiao's research team synthesized porous sheet-like rock salt-type HEM utilizing graphene oxide sheets as cationic adsorbents [46], resulting in a relatively high specific surface area of 42 m^2g^{-1}. These techniques essentially preserve the original structure, circumvent sintering, and consequently create more interfaces and abundant surface defects. The combustion method represents a variation of the thermal synthesis approach. Combustion agents such as citric acid and glycine exhibit high fuel oxidation rates and excellent coordination performance. By forming a gel with the metal salt precursor, these molecules aid in the uniform distribution of ions and serve as a template. Subsequently, through low-temperature heating, the self-combustion process vigorously completes the reaction. With the assistance of glycine, Gautam and Ahmad proposed the synthesis of rock salt-type HEM at a low ignition temperature of 185 °C, eliminating the need for subsequent annealing [47].

3.2.1.2 High-Temperature "Thermal Shock" Method

Typically, the conventional template sintering method is limited to the synthesis of five or six elemental HEM and cannot be applied to a broader range of elemental HEM. As the number of elements increases, increasingly extreme conditions are required to uphold the exceptional crystallinity of HEM. The initial advancement in HEM synthesis, encompassing a diverse range of compositions (including many immiscible combinations) and a high number of elements (>8), was accomplished through the high-temperature "thermal shock" process pioneered by Hu and coworkers at the University of Maryland, as shown in Figure 3.3 [48]. The cooling rate of this synthesis method is a critical parameter, as it influences the extent of imbalance and structural order achievable by the constituent elements, as elucidated by the temperature-time transition diagrams commonly employed in physical metallurgy and polymer curing [49]. The resulting structures may consist of metal glass nanoparticles (randomly dispersed in a disordered lattice), irregularly shaped HEM particles (randomly dispersed in the lattice), IM particles (ordered between sublattices but randomly dispersed within each sublattice), and heterogeneous particles (phase separation) [50]. Furthermore, the brief duration and rapid quenching of thermal shock synthesis also contribute to the production of small and uniform particles [51], which can be further tailored through defect engineering and suitable substrates [52].

3.2.1.3 HEM Synthesis Strategy Under Mild Conditions

The synthesis methods of HEM typically necessitate high-temperature processes surpassing 1000 °C and highly intricate processing techniques, and such extreme conditions are frequently incompatible with the stability of other materials in the equipment architecture. To address this issue, the pivotal components of the HEM synthesis system should incorporate mild reaction conditions and special lattice configurations tailored to these conditions. Yang and coworkers formulated a synthesis approach suitable for metal halide perovskites in mild environments based on ion bond networks and low cohesive energy crystal structure design, as shown in Figure 3.4a [53]. Due to the easily adaptable lattice and good processability of perovskite materials, this method can effectively produce HEM in room temperature (20 °C) and low-temperature (80 °C) solutions. Due to the soft and ionic lattice properties of metal halide perovskites, HEM is designed based on a vacancy-ordered double perovskite structure

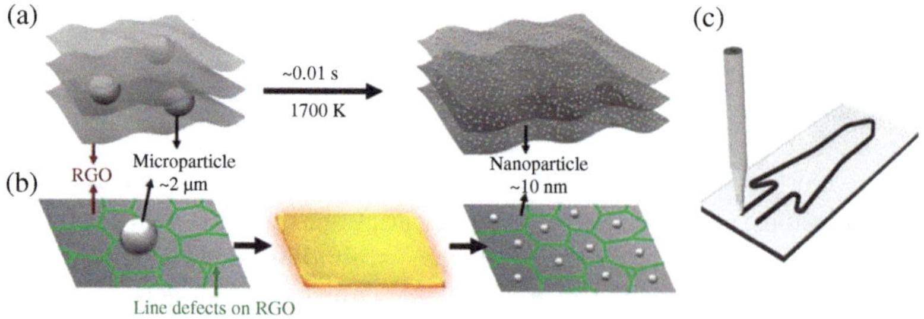

Figure 3.3 (a) Micro-sized particles in a conductive reduced graphene oxide network matrix (left) are self-assembled into nanoparticles (right), driven by direct Joule heating for ~10 ms. (b) Proposed mechanism for nanoparticle formation, where a micro-sized particle melts on heating and self-assembles into nanoparticles due to confinement by the defects of the RGO sheet. (c) The raw materials are 3D printable, wherein nanoparticles can be formed in a conductive RGO matrix.

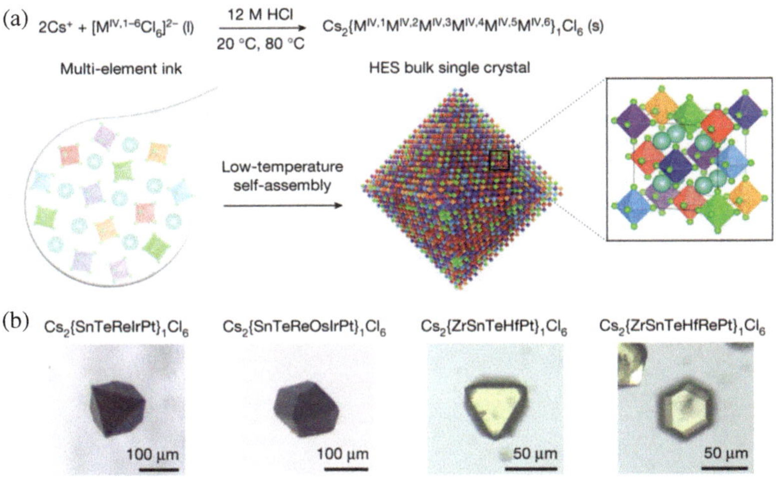

Figure 3.4 (a) Schematic representation of the room- and low-temperature-driven self-assembly processes in solution from multi-element inks to HEM perovskite single crystals. (b) Optical microscope images of HE five-element and six-element single crystals.

of cubic Cs_2MCl_6 (M = Zr^{4+}, Sn^{4+}, Te^{4+}, Hf^{4+}, Re^{4+}, Os^{4+}, $Ir4^{+}$, or Pt^{4+}), which is obtained from the stable self-assembly of complexes during the reaction process. The resulting HEM occupies the M site of six elements in a random alloy form close to an equal molar ratio, preserving the overall crystal structure and stoichiometry of Cs_2MCl_6, as shown in Figure 3.4b. The achievement of synthesizing HEM at room temperature is unprecedented and constitutes a fundamental necessity in HEM applications. This gentle solution synthesis technique has the potential for widespread application to other HEM systems, particularly those characterized by relatively low cohesive energy.

3.2.2 Advanced Characterization of HEM

Based on the definition of HEM, the material is expected to demonstrate a single-phase structure, characterized by uniform and random mixing of constituent elements. Researchers contend that describing this disordered mixing of multiple elements and their synergistic effects is highly challenging. Conventional techniques such as powder X-ray diffraction (XRD, $\lambda = 0.2113$ Å), X-ray photoelectron spectroscopy (XPS), scanning electron microscopy, and transmission electron microscopy (TEM) can aid in identifying the fundamental phase structure, morphology, element distribution, and valence state of materials. Yet they may lack the required high resolution for disentangling multi-element mixing. With the advancement of material characterization technology, synchrotron X-ray technology employing shorter wavelengths (e.g. $\lambda = 0.2113$ Å) can offer a high resolution to better comprehend the atomic arrangement, bonding, and coordination, as well as electronic properties of HEM [35]. Specifically, synchrotron XRD can discern the overall phase structure of HEM with greater precision, along with potential immiscible phases and impurities [54], to verify the attainment of high-entropy mixing. X-ray absorption spectroscopy (XAS) is an element-specific technique that can be employed to investigate the local atomic coordination environment of each element, which is critical for understanding multi-element mixing and potential short-range or local sequences in HEM [55–57]. Furthermore, hard X-ray photoelectron spectroscopy (HAXPES) can unveil the electronic structure of HEM (such as valence band and d-band centers), closely linked to the adsorption and binding energies of key reaction intermediates, thereby streamlining the chemical process of the reaction (Figure 3.5a) [58]. Although X-ray-related techniques can provide comprehensive statistical analysis, electron microscopy-based techniques are essential for directly visualizing particle size and distribution, phase, structure, composition, and chemical environment. For instance, in situ TEM has been utilized to examine the formation process of nanoparticles synthesized by the high-temperature shock method, elucidating their dispersion and stability on defective carbon substrates [59]. Another advanced technique that can fulfill the requirements for higher flux

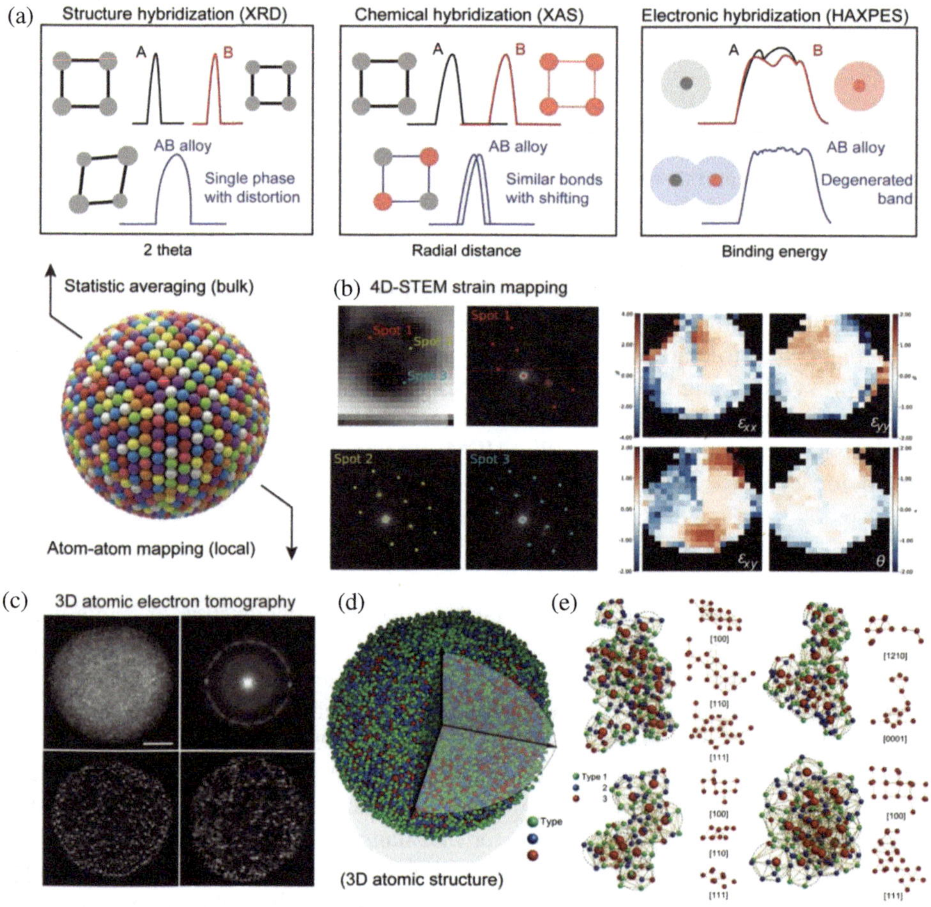

Figure 3.5 Advanced characterization of HEM. (a) X-ray–based techniques, including XRD, XAS, and HAXPES, particularly using synchrotron X-ray sources that provide higher resolution. (b) 4D-STEM and strain mapping of HEM and local lattice strain distribution, including tensile (red) and compressive (blue). (c) Determining the 3D atomic structure of a HE metallic glass nanoparticle by AET. (d) Experimental 3D atomic model of the HE metallic glass nanoparticle. (e) Identification of four types of crystal-like medium-range order that coexist in the metallic glass nanoparticle based on AET results.

and resolution is four-dimensional scanning transmission electron microscopy (4D-STEM) (Figure 3.5b) [60]. 4D-STEM uses a small probe (~1 nm) to scan an area of ~1 × 1 μm^2 in area. A large geometric region allows for rapid and high-resolution characterization of local lattice distortion, structural heterogeneity, and short-range ordering in HEM [61]. Local diffraction patterns can be obtained on HEM, and by comparing the differences between the local phase structure and the average phase structure, corresponding strain maps inside HEM can be obtained, indicating potential lattice distortion and strain. Atomic electron tomography (AET) has been demonstrated to be the preferred method for higher-level characterization of three-dimensional atomic structures [62]. Recently, AET technology has become sufficiently advanced to analyze the three-dimensional atomic structure of HE metal glass nanoparticles containing eight elements: Co, Ni, Ru, Rh, Pd, Ag, Ir, and Pt [63]. Due to the fact that the contrast of AET images depends on the atomic number, the current sensitivity of AET can only classify eight elements into three types: Co and Ni are type 1 (green); Ru, Rh, Pd, and Ag are type 2 (blue); and Ir and Pt are type 3 (red). Figure 3.5c–e show the three-dimensional atomic structure of HE metal glass nanoparticles, where types 1, 2, and 3 atoms are uniformly distributed. The three-dimensional atomic structure reveals four different crystal-like intermediate ordered structures, including a face-centered cubic structure, a hexagonal dense arrangement structure, a body-centered cubic structure, and a simple cubic structure coexisting in high-entropy nanoparticles. These results provide direct experimental evidence that supports the general framework of efficient cluster-filling models for metallic glasses [64] and demonstrate how AET technology enables researchers to study the three-dimensional structure of HE nanoparticles at the single-atom level.

3.3 High-Entropy Electromagnetic Wave Absorption Materials

The distinctive "cocktail effect" of HEM results in unexpected electromagnetic performance and appealing microstructural changes, leading to its widespread application in the field of EMW absorption (including alloys, oxides, and sulfides). Researchers have elucidated a range of microscopic EMW loss mechanisms of HEM through experimental and simulation approaches. However, the impact of high configurational entropy on microscopic factors is not unique across different systems, potentially resulting in varying EMW loss mechanisms. For instance, the influence of configurational entropy on the electron transport characteristics of oxides and sulfides differs (a detailed explanation will be provided in this section). Consequently, a meticulous analysis of specific HEM systems is essential to unveil their EMW absorption performance and the corresponding microscopic loss mechanisms. This section categorizes different HEM systems and discusses their contributions to EMW absorption. Our objective is to enhance readers' comprehension of the design principles governing EMW absorbers and to achieve a comprehensive understanding of the intricate interplay between configurational entropy and EMW absorption in diverse HEM systems.

3.3.1 High-Entropy Alloy

Due to the constraints posed by element compatibility in the fabrication of HEM, incompatible elements are prone to phase separation during the reaction process. Typically, only specific combinations of elements can facilitate the creation of stable single-phase species. Therefore, in this section of the discourse, the emphasis will not be on the rationale behind the selection of specific elements in these studies but rather on examining the interaction mechanism between HEM and EMW. The common feature of all alloy absorbers is their strong EMW absorption ability (the very low reflection loss), enabling them to achieve effective attenuation of over 99% of incident EMW energy at specific frequencies. However, these absorbers often exhibit high intrinsic conductivity, leading to significant reflection of microwaves and hindering their penetration into the material, thereby failing to achieve stealth functionality. The discovery of HEA is anticipated to address the above challenges associated with alloy absorbers. The pronounced lattice distortion induced by the increased configurational entropy exerts a substantial influence on the microscopic properties of the materials. The lattice distortion significantly reduces the conductivity of HEA, thereby mitigating microwave reflection resulting from the skin effect. Additionally, the disordered arrangement of multiple atoms inevitably leads to the enrichment of point defects, triggering defect-induced polarization loss. Consequently, HEA is considered a promising type of EMW absorber. In order to conduct in-depth research on the potential performance regulation patterns within HEA, a large amount of work has been carried out. It is widely recognized that microstructure plays a pivotal role in determining the electromagnetic properties of materials with specific components [65]. Even minute alterations in microstructure can result in substantial variations in electromagnetic properties. Thus, for HEA, the numerous defects, low crystallinity, intricate element distribution, and elevated internal stress induced by HE effects present not only opportunities but also challenges for enhancing microwave absorption performance.

The introduction of magnetic elements can endow absorbers with excellent magnetic loss function, and combined with the inherent dielectric properties of alloy materials, it is expected to achieve magnetic-electric coupling/synergistic effects. Duan and coworkers [66] conducted a study on the impact of alloying methods on the microstructure and microwave absorption performance of $FeCoNiSi_{0.4}Al_{0.4}$ HEA. Through pretreatment involving melting, strip casting, and grinding, they achieved HEA particles with a large aspect ratio and low crystallinity. This resulted in reduced magnetic crystal anisotropy and stress anisotropy, but increased shape anisotropy and conductivity, ultimately leading to higher magnetic resonance losses and GHz conductivity losses. Meanwhile, Hou et al. [67] presented findings on the microwave absorption performance of flake $Sm_{1.5}Y_{0.5}Fe_{17-x}Si_x$ powder. They observed that the primary magnetic loss mechanism at GHz frequencies, natural resonance, is influenced by particle size and is sensitive to the Si content. With increasing Si content, the resonant frequency and dielectric attenuation also increased, leading to an enhanced effective absorption bandwidth of 3.7 GHz. Additionally, Kou and coworkers [68] fabricated porous HEA EMW absorption materials using ball-milled HEA as a template and acid etching. Through precise control of the microstructure, they transitioned from a spherical structure to a porous hollow structure, greatly enhancing the impedance matching and dielectric attenuation of the HEA, resulting in an effective absorption bandwidth of 4.48 GHz. Although research on the absorption performance of HEA is still in its infancy, studies on its magnetic properties are becoming increasingly advanced, providing a crucial foundation for the development of microwave absorption materials.

Although a series of HEA absorbers have been developed, their lower micro-design degrees of freedom limit the controllability of preparation, leading to an upper limit of HEA absorption performance. A research approach based on macroscopic metasurface construction is expected to overcome the above limitations, as macroscopic parameters are more easily manipulated by humans compared to microscopic scales. Che's research team fabricated a series of HEA metasurface bands for EMW absorption [69]. These bands were uniformly distributed in a double V-shape on plate-like HEA powder. Test results revealed that pure HEA powder absorbers were unable to achieve effective absorption in the microwave frequency band. In contrast, the metasurface HEA bands exhibited an effective absorption bandwidth (reflection loss ≤ -10 dB) ranging from 5.86 to 18 GHz, with an absorption peak reaching −50.88 dB (absorption rate exceeding 99.999%) at 14.1 GHz, for a thickness of d=1.9 mm. This observation demonstrates the efficacy of metasurface construction in enhancing EMW losses, and the absorption mechanism is manifested in the following three aspects:

(1) Enhanced magnetic coupling in layered nanostructures. The absorption of ferromagnetic EMW is primarily influenced by magnetic loss, owing to its significant internal magnetic polarization intensity. In the case of existing HEA strip absorption materials, the hierarchical nanostructures contribute to low coercivity and high saturation magnetization, leading to high magnetic permeability [70]. Additionally, the soft magnetic properties of these materials can create greater inductance in the metasurface compared to other conductive materials [71], thereby facilitating robust storage and attenuation of magnetic fields and consequently enhancing the loss effects. Specifically, the accumulation of strong magnetic fields enables the entire band to exhibit remote coupling capabilities, as evidenced by off-axis electron holography of prominent stray magnetic flux lines. This phenomenon enables magnetic localization between the band and the substrate, resulting in strong magnetic resonance with EMWs.

(2) The substantial ferromagnetic resonance induced by exceeding the Snoek limit. Natural resonance represents the primary mode of EMW attenuation in the GHz spectrum, a phenomenon traditionally constrained by the Snoek limit in conventional magnetic materials. In this layered metasurface HEA ribbon, a significant vertical demagnetization factor and a minor horizontal demagnetization factor lead to parallel alignment of magnetic moments on the surface, resulting in planar anisotropy and surpassing the Snoek limit.

(3) Synergistic EM effects. The simulation model demonstrates that the ribbon unit cells facilitate interaction between the substrate and the electric field (Figure 3.6), dependent on the interplay between the internal free charge and the electric field. The appropriate EM parameters enhance the attenuation constant of the HEA bands, surpassing the performance of powder, indicating that HEA bands possess a superior EMW loss capability compared to powder. Duan and coworkers conducted a similar study based on biomimetic structural design, which also confirmed the role of metasurface structure in enhancing the performance of HEA absorbers [72]. Specifically, drawing inspiration from the anti-reconnaissance eye crystals that cover the compound eyes of dragonflies, they proposed an anisotropic dual V-atom biomimetic metasurface made up of $(FeCoNiSi_{8.9}Al_{8.9})C_{0.2}$ HEA bands, which exhibit multi-degree-of-freedom recognition and broadband absorption capabilities. This design allows for the control and identification of the hidden phase, amplitude, and angular momentum of EMWs by manipulating the rotation angle of atomic elements. The HEA bands, formed by a layered nanostructure consisting of FCC/BCC ultrafine particles, B2 nanoparticles, nano twins, and nanoparticles, possess high saturation magnetization and low coercivity. Overall, these bands promote appropriate electromagnetic parameters and superior impedance matching, thus addressing the issue of EMW absorbers being unable to recognize the concealed degrees of freedom of EMWs [73–75].

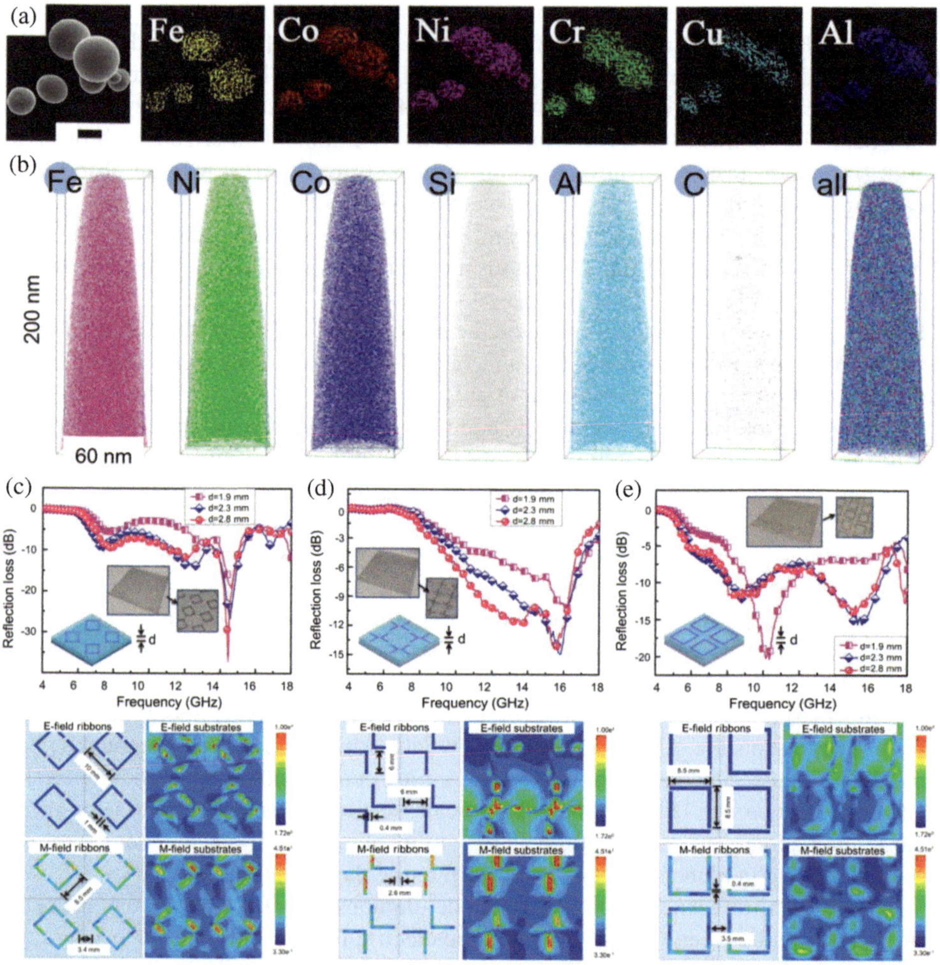

Figure 3.6 (a) Energy dispersive spectrometer (EDS) mapping of FeCoNiCrCuAl HE alloy. (b) Atomic maps for individual and integral elements analyzed by atom probe tomography (APT); the dimensions of the cuboid were $200 \times 60 \times 60$ nm^3. (c–e) Reflection loss of HEA ribbon metasurfaces (200×200 mm) with different unit cells; insets show schematic diagrams and photographs of the metasurfaces; the widths of ribbons were 0.7 mm, and the substrate material was Makrolon. Electric field (E-field) and magnetic field (M-field) distributions of HEA ribbons and Makrolon substrate at the frequency of the absorption peaks corresponding to c–e, respectively.

3.3.2 High-Entropy Oxide

Compared to alloys, common oxides exhibit a relatively low complex dielectric constant, with EMW loss primarily caused by polarization. Due to the absence of high dielectric components, it is commonly believed that the conductivity loss ability is inadequate, which is attributed to the wider bandgap and lower electronic integrated density of states of the oxide. Consequently, researchers are focusing on enhancing the dissipation ability of oxide absorbers for EMW by elevating the level of polarization. Additionally, the high configuration entropy can induce significant lattice distortion, leading to the creation of dense point defects and thereby promoting polarization behavior in HEO. This HE approach aims to improve the overall performance of oxide materials in EMW applications. However, this strategy is still a semi-empirical result, and more theoretical and experimental evidence is still needed.

Considering the limited microwave loss characteristics of single-phase HEO and the constraints posed by semi-empirical laws on the advancement of HEO, Che and coworkers fabricated two types of biphasic HEO through simple sintering, enabling precise control over crystal phase and structural defects [76]. The authors verified that the HEO effect induces various defects in multiphase oxides, which are advantageous for microwave dissipation. In the case of spinel/perovskite HEO, pronounced stress concentration and lattice distortion were directly observed at the interface between spinel (111) and perovskite (110), resulting in a significant accumulation of point defects and stacking faults. The structural defects and multiphase nature of HEO led to the observation of strong polarization phenomena, as confirmed by first-principles calculations. Consequently, the two complex phases of HEO exhibited outstanding microwave absorption performance (Figure 3.7a,b). Similar research has also demonstrated the viability of the dual-phase HEO strategy. Dai et al. enhanced the microwave absorption performance of HEO $(Fe_{0.2x}Co_{0.2}Ni_{0.2}Cr_{0.2}Mn_{0.2})_3O_4$ (x = 1–5) by introducing secondary phases (Figure 3.7c–e) [77]. However, the high-entropy effect of the two-component HEO was not significant, and the improved microwave absorption performance may be attributed to component effects rather than HE effects. Furthermore, a comprehensive understanding of the local atomic characteristics is essential for optimizing the properties of these materials. Therefore, a thorough comprehension of the microstructural evolution of entropy increase is imperative to comprehend the anomalies and unprecedented characteristics of HEO. In order to analyze the lattice distortion effect, the focus was on studying three components: $BaTiO_3$, $(Ca_{1/3}Sr_{1/3}Ba_{1/3})TiO_3$, $(Ca_{0.25}Sr_{0.25}Ba_{0.25}La_{0.25})TiO_3$, and $(Ca_{0.2}Sr_{0.2}Ba_{0.2}La_{0.2}Pd_{0.2})TiO_3$, which transition from low entropy to HE states. The characteristic peaks of XRD gradually broaden from low entropy to HE composite materials, indicating an increase in the lattice strain of the sample. In addition, based on HRTEM images, geometric phase analysis was used to further examine the strain in HE and low-entropy nanoscale perovskites [78]. Compared with low-entropy and medium-entropy perovskites, the lattice edge of HE perovskites ε_{xx} and the strain in the ε_{yy} direction are strong, with green to dark blue and red to bright yellow being related to compressive strain and tensile strain, respectively [79, 80]. In low-entropy compositions, the mismatch in atomic sizes is easily achieved through changes in cell volume, while in HE compositions, the asymmetry of atomic dissociation potential centers and distributions leads to unit cell deformation and large shear strain, which is the main difference between low-entropy and HE compositions. Additionally, electron holography technology was employed to characterize the charge distribution resulting from strain in various perovskites. When compared to $BaTiO_3$ samples, the HE stable $(Ca_{0.2}Sr_{0.2}Ba_{0.2}La_{0.2}Pb_{0.2})TiO_3$ demonstrates significant charge distribution around grain boundaries and areas of strain concentration. Analysis of the quantitative charge density distribution line reveals that the charge density of HE $(Ca_{0.2}Sr_{0.2}Ba_{0.2}La_{0.2}Pb_{0.2})TiO_3$ is more than 10 times higher than that of $BaTiO_3$. The accumulation of positive/negative charges around numerous boundaries and strong strain positions is advantageous for enhancing interfacial polarization loss. The differential charge density plot indicates that as the HE effect gradually intensifies, the polarity region around the oxygen atom significantly expands, signifying a notable electron enrichment effect, as shown in Figure 3.7f–j. This can be attributed to the presence of metal elements with varying valence states in oxides, which give rise to abundant oxygen

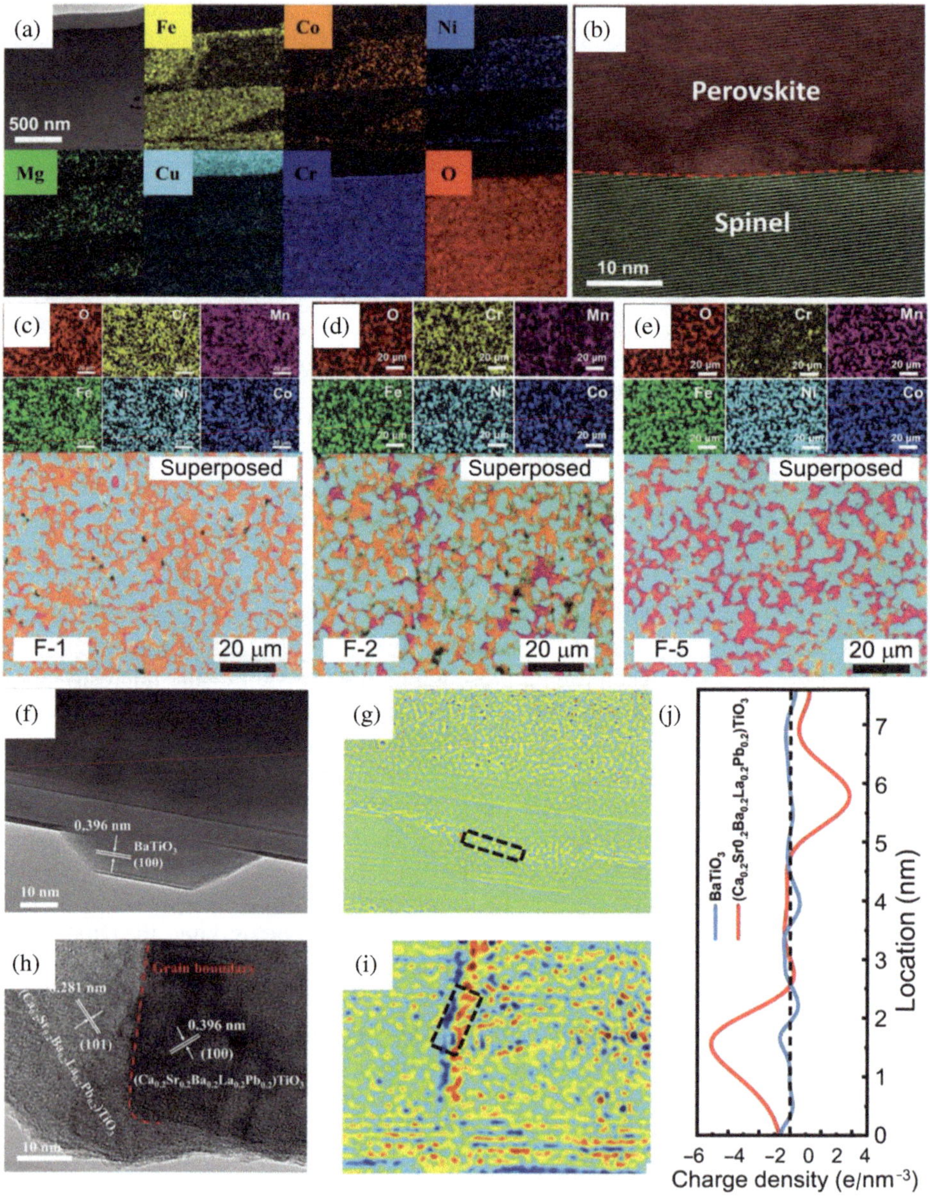

Figure 3.7 (a) Power XRD pattern for HE (Fe, Co, Ni, Cu, Mg)CrO_X prepared at 1300 °C (HE-Cr-1300) after Rietveld refinement. (b) Percentage of perovskite contents and defects in spinel/perovskite hybrid HE oxides. (c–e) Corresponding distributions of O (red), Cr (yellow), Mn (purple), Fe (green), Ni (cyan), Co (blue). (f) HRTEM image, and (g) the reconstructed charge density distribution map of $BaTiO_3$, in which the color reflects the magnitudes of different charge density. (h) HRTEM image and (i) the reconstructed charge density distribution map of HE $(Ca_{0.2}Sr_{0.2}Ba_{0.2}La_{0.2}Pd_{0.2})TiO_3$. (j) The quantitative charge density distribution line profiles in the region of the black dashed frames for the $BaTiO_3$ and $(Ca_{0.2}Sr_{0.2}Ba_{0.2}La_{0.2}Pd_{0.2})TiO_3$.

vacancies, resulting in the structural relaxation of ions around the vacancies. Upon the sudden disappearance of oxygen atoms and the formation of oxygen vacancies, a substantial number of electrons cluster around the A-site atom, leading to electron enrichment. It can be inferred that the electron enrichment effect, caused by the disordered arrangement of local electrons in metal atoms, results in stronger dipole polarization, which is advantageous for EMW attenuation. In short, the abundant presence of boundaries and highly dense, discontinuous lattices leads to pronounced interfacial polarization and dipole polarization. Furthermore, the ultra-dense strain generated by severe lattice distortion in this HE component facilitates robust transport for electron carriers, thereby promoting dielectric loss. Consequently, the EAB of the HE perovskite is nearly twice as high as that of low-entropy samples. This study thus offers valuable insights into entropy engineering for enhanced microwave absorption.

3.3.3 High-Entropy Sulfide

Sulfides are widely utilized in EMW absorption applications due to their outstanding electrical properties. Despite this, there have been few reports on the implementation of the HE strategy in sulfide absorbers, which may be attributed to several factors. On the one hand, in comparison to metal–oxygen (M—O) bonds, the single bond length of metal–sulfur (M—S) chemical bonds is longer, which is not favorable for single-phase stability. In addition, most sulfides undergo thermal decomposition under high-temperature conditions, making the controlled synthesis of HE sulfides a formidable challenge. On the other hand, certain sulfides exhibit metallic properties, and although they are classified as semiconductors, their electron transport capacity far exceeds that of peroxides. The dense defect sites provided by lattice distortion effects may not dominate even if the polarization level is increased.

In order to elucidate the practical role of the HE strategy in sulfide absorbers, our group conducted a pioneering exploration research [81]. We calculated the effect of increased configurational entropy on electron exchange capacity using density functional theory, demonstrating the limitation of HE strategy on sulfide systems, as shown in Figure 3.8a–g. That is to say, in general, the HE strategy is not suitable for enhancing the EMW loss ability of sulfide absorbers. However, to cope with some practical scenarios that require multicomponent sulfides, heterogeneous interface engineering has been proposed to enhance the EMW absorption performance of multicomponent sulfides, which requires careful selection of reaction species based on atomic activity and lattice compatibility. By selecting atoms with different reactivity, multiphase interface components of CuS (1 0 5), $Fe_{0.5}Ni_{0.5}S_2$ (2 0 0), and $CuFe_2S_3$ (2 0 0) were constructed to enhance the interfacial polarization loss of multi-element Cu-based sulfides. Compared with single-phase HE Zn-based sulfides (ZnFeCoNiCr-S), multiphase Cu-based sulfides (CuFeCoNiCr-S) have better EMW absorption performance (EAB of 6.70 GHz at 2.00 mm) because Cu (1 0 5)/$CuFe_2S_3$ (2 0 0) exists at a specific interface under appropriate EM parameters. By heat treatment at 400 °C, single-phase ZnFeCoNiCr-S was decomposed into $FeNi_2S_4$ (3 1 1)/(Zn, Fe) S (1 1 1) heterointerfaces, and strong interface polarization (EAB at 1.45 mm was 4.83 GHz) enhanced the absorption performance of EMW. In addition, photoluminescence (PL) spectra have demonstrated that defect content plays a secondary role in improving the absorption performance of EMW. Therefore, in the system of polysulfides, it can be considered that interface polarization transcends defect-induced polarization and becomes the dominant loss mechanism. Zhao et al.

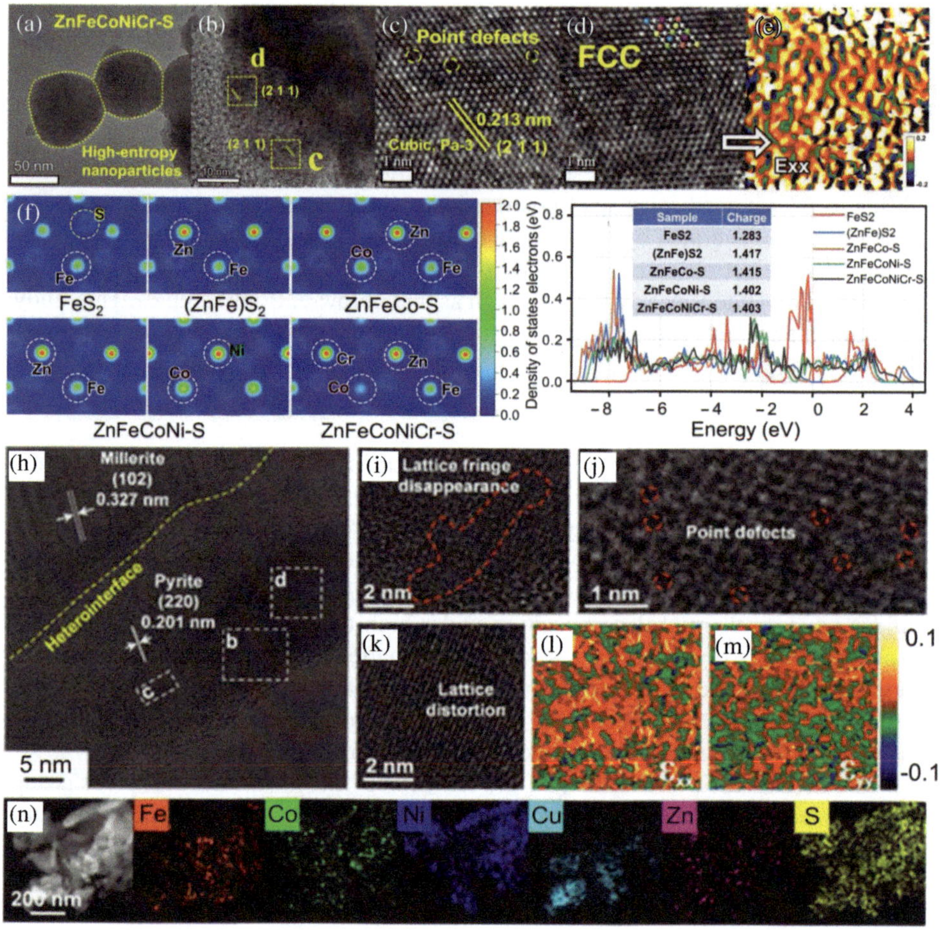

Figure 3.8 The microscopic characterization, simulation calculations, and EMW absorption properties of Zn-based sulfides. (a–d) HRTEM images of ZnFeCoNiCr-S sample. (e) The strain distribution image of ZnFeCoNiCr-S sample. Charge density images of (f1) FeS_2, (f2) $(ZnFe)S_2$, (f3) ZnFeCo-S, (f4, f5) ZnFeCoNi-S, and (f6) ZnFeCoNiCr-S. (g) The density of states of Zn-based sulfides. (h) HRTEM images demonstrating the presence of both millerite and pyrite structural phases. (i–m) Enlarged HRTEM images of marked areas, GPA mappings (strain fields ε_{xx} and ε_{yy}), calculated using a digital micrograph, indicating an overall strain field at the nanoscale. (n) HAADF-STEM image and the corresponding EDS elemental mapping of the (FeCoNiCuZn)S sample.

performed similar research to illustrate the role of phase interface polarization in EMW loss in multi-principal sulfide systems [82]. Specifically, leveraging the HE concept and structural design, they prepared a porous flower-shaped sulfide using a multichannel approach. With an increasing variety of elements, the phase separation of sulfide absorbers becomes more apparent, thereby inducing phase interface polarization and resulting in EMW loss, as shown in Figure 3.8h–n. Hence, it is essential to consider the material characteristics of each system and employ distinct design criteria to enhance EMW absorbers. The semi-empirical design principle can only provide guidance for the initial experimental stage.

3.4 The Challenge and Prospects of HEM

Even though existing research has revealed the EMW attenuation ability and corresponding microscopic loss mechanisms of various HE system materials, HEM absorbers still face enormous challenges. Herein, we have summarized the current challenges and possible solutions, providing forward-looking insights for the future development of HEM absorbers.

(1) Although many effective synthesis methods have been proposed, these methods are often complex and waste energy. Therefore, it is necessary to develop more convenient and green methods to achieve rapid and green synthesis of advanced HE absorbers. In addition, current conventional technologies find it difficult to confine incompatible atoms within a single lattice, resulting in a very limited availability of HEM elements. How to overcome atomic incompatibility and achieve element controllability in HEM is a challenge that researchers will face in the future. The key to solving this problem is to clarify the growth and reaction mechanism of HEM, which may require the development of more advanced in situ characterization techniques.

(2) The understanding of how element composition and synthesis methods affect the microstructure and loss capacity of HEM absorbers by researchers is very limited. The conclusion drawn based on the currently explored small number of HE systems is restricted. The development of future HEM absorbers should expand the concept of HE to other material systems and delve into the role of configurational entropy in various systems, such as HE nitrides, HE-MOFs, HE MXenes, and HE-layered double hydroxides. In addition, the contribution of various elements to the electromagnetic characteristics of absorbers should also be clarified to provide more comprehensive design guidelines.

(3) Although current research mentions increasing the configurational entropy of a system by increasing the types of elements, the influence of the elements themselves cannot be ruled out when studying the regulatory effect of configurational entropy on material properties. That is to say, what is the role of entropy in HEM absorbers? How and when does entropy regulate the micro factors/physicochemical characteristics of HEM absorbers? To fully comprehend the mechanism of HEM absorbers, addressing these issues represents a challenge that researchers must confront. Starting from fundamental theories, such as the thermodynamic/kinetic properties of materials, may be helpful in addressing the aforementioned issues.

(4) At present, the preparation and characterization of HEM absorbers usually adopt a "trial and error" paradigm, which involves complex orthogonal experiments and subsequent performance characterization based only on some empirical or semi-empirical conclusions, resulting in high preparation costs and unpredictable results. Considering the large number of possible combinations of HEM components, it is difficult for researchers to know where to find the next new HEM absorbent. In recent years, high-throughput screening strategies based on machine learning (ML) methods have been widely applied in material discovery and performance prediction. For example, Lu and coworkers constructed an ML-based prediction system for directly predicting the process conditions of carbonyl iron/iron trioxide mixtures with enhanced EMW absorption performance [83]. A series of excellent EMW absorption materials, with the highest correlation coefficient of 0.9844, was recommended based on pattern recognition, high-throughput screening, and back-projection. After manual selection, the maximum absorption efficiency and bandwidth of the optimal mixed spectrum were improved by 207% and 360%, respectively, compared to the original database.

The standardized program greatly shortens the research cycle to several weeks, compared to several months of manual orthogonal experiments. This is a highway that accelerates the discovery and industrialization of high-performance EMW absorption materials. HEM has almost infinite possible elemental composition spaces, and predicting its potential EMW absorption performance through ML is a promising approach. Based on the current research status, a database of primary HEM absorbents, including process conditions and physical properties, can be established to explore the contribution of various elements to EMW attenuation. Train ML algorithms to improve prediction accuracy as much as possible, while returning the output dataset to the database for cyclic learning, which is expected to shift the discovery of HEM absorbers from the traditional "trial and error" paradigm to a "data-driven" paradigm.

References

1 Cantor, B., Chang, I.T.H., Knight, P., and Vincent, A.J.B. (2004). Microstructural development in equiatomic multicomponent alloys. *Mater. Sci. Eng. A* 375–377: 213–218.

2 Yeh, J.W., Chen, S.K., Lin, S.J. et al. (2004). Nanostructured high-entropy alloys with multiple principal elements: novel alloy design concepts and outcomes. *Adv. Eng. Mater.* 6 (5): 299–303.

3 Yeh, J.W. (2006). Recent progress in high entropy alloys. *Ann. Chim. Sci. Mat.* 31: 633–648.

4 Tomlin, I.A. and Kaloshkin, S.D. (2015). 'High entropy alloys' 'semi-impossible' regular solid solutions? *Mat. Sci. Technol.* 31 (10): 1231–1234.

5 Senkov, O.N., Wilks, G.B., Miracle, D.B. et al. (2010). Refractory high-entropy alloys. *Intermetallics* 18 (9): 1758–1765.

6 Senkov, O.N., Wilks, G.B., Scott, J.M., and Miracle, D.B. (2011). Mechanical properties of $Nb_{25}Mo_{25}Ta_{25}W_{25}$ and $V_{20}Nb_{20}Mo_{20}Ta_{20}W_{20}$ refractory high entropy alloys. *Intermetallics* 19 (5): 698–706.

7 Miracle, D.B., Miller, J.D., Senkov, O.N. et al. (2014). Exploration and development of high entropy alloys for structural applications. *Entropy* 16 (1): 494–525.

8 Miracle, D.B. (2015). Critical assessment 14: high entropy alloys and their development as structural materials. *Mat. Sci. Technol.* 31 (10): 1142–1147.

9 Gludovatz, B., Hohenwarter, A., Catoor, D. et al. (2014). A fracture-resistant high-entropy alloy for cryogenic applications. *Science* 345: 1153–1158.

10 Hsu, C.-Y., Yeh, J.-W., Chen, S.-K., and Shun, T.T. (2004). Wear resistance and high-temperature compression strength of FCC $CuCoNiCrAl_{0.5}Fe$ alloy with boron addition. *Metall. Mater. Trans. A* 35: 1465–1469.

11 Chen, T.K., Shun, T.T., Yeh, J.W., and Wong, M.S. (2004). Nanostructured nitride films of multi-element high-entropy alloys by reactive DC sputtering. *Surf. Coat. Technol.* 188: 193–200.

12 Guo, S., Ng, C., Lu, J., and Liu, C.T. (2011). Effect of valence electron concentration on stability of fcc or bcc phase in high entropy alloys. *J. Appl. Phys.* 109: 103505.

13 Li, H., Zhu, H., Zhang, S. et al. (2020). Nano high-entropy materials: synthesis strategies and catalytic applications. *Small Struct.* 1: 2000033.

14 Miracle, D.B., Majumdar, B., Wertz, K., and Gorsse, S. (2016). New strategies and tests to accelerate discovery and development of multi-principal element structural alloys. *Scr. Mater.* 127: 195–200.

15 Zhang, Y., Zuo, T.T., Tang, Z. et al. (2014). Microstructures and properties of high-entropy alloys. *Progr. Mater. Sci.* 61: 1–93.

16 Chen, Y., Zhang, G.F., Li, Y.Y. et al. (2013). Quantitative analysis of dynamic behaviours of rural areas at provincial level using public data of gross domestic product. *Entropy* 15 (1): 10–31.

17 Ye, Y.F., Wang, Q., Lu, J. et al. (2015). The generalized thermodynamic rule for phase selection in multicomponent alloys. *Intermetallics* 59: 75–80.

18 Takeuchi, A., Amiya, K., Wada, T. et al. (2013). Entropies in alloy design for high-entropy and bulk glassy alloys. *Entropy* 15 (9): 3810–3821.

19 Yang, X. and Zhang, Y. (2012). Prediction of high-entropy stabilized solid-solution in multi-component alloys. *Mater. Chem. Phys.* 132: 233–238.

20 Yang, Y., Song, B., Ke, X. et al. (2020). Aerosol synthesis of high entropy alloy nanoparticles. *Langmuir* 36: 1985–1992.

21 Mori, K., Hashimoto, N., Kamiuchi, N. et al. (2021). Hydrogen spillover-driven synthesis of high-entropy alloy nanoparticles as a robust catalyst for CO_2 hydrogenation. *Nat. Commun.* 12: 3884.

22 Löffler, T., Savan, A., Meyer, H. et al. (2020). Design of complex solid-solution electrocatalysts by correlating configuration, adsorption energy distribution patterns, and activity curves. *Angew. Chem. Int. Ed.* 59: 5844–5850.

23 Batchelor, T.A.A., Löffler, T., Xiao, B. et al. (2021). Complex-solid-solution electrocatalyst discovery by computational prediction and high-throughput experimentation. *Angew. Chem. Int. Ed.* 60: 6932–6937.

24 Ludwig, A. (2019). Discovery of new materials using combinatorial synthesis and high-throughput characterization of thin-film materials libraries combined with computational methods. *npj Comput. Mater.* 5: 70.

25 Yeh, J.W., Chang, S.Y., Hong, Y.D. et al. (2007). Anomalous decrease in X-ray diffraction intensities of Cu–Ni–Al–Co–Cr–Fe–Si alloy systems with multi-principal elements. *Mater. Chem. Phys.* 103: 41–46.

26 Yeh, J.W., Chen, S.K., Gan, J.W. et al. (2004). Formation of simple crystal structures in Cu-Co-Ni-Cr-Al-Fe-Ti-V alloys with multiprincipal metallic elements. *Metall. Mater. Trans. A* 35: 2533–2536.

27 Li, T., Yao, Y., Huang, Z. et al. (2021). Denary oxide nanoparticles as highly stable catalysts for methane combustion. *Nat. Catal.* 4: 62–70.

28 Tsai, K.Y., Tsai, M.H., and Yeh, J.W. (2013). Sluggish diffusion in Co-Cr-Fe-Mn-Ni high-entropy alloys. *Acta Mater.* 61: 4887–4897.

29 Wang, T., Chen, H., Yang, Z. et al. (2020). High-entropy perovskite fluorides: a new platform for oxygen evolution catalysis. *J. Am. Chem. Soc.* 142: 4550–4554.

30 Cui, M., Yang, C., Li, B. et al. (2021). High-entropy metal sulfide nanoparticles promise high-performance oxygen evolution reaction. *Adv. Energy Mater.* 11: 2002887.

31 Saito, T., Furuta, T., Hwang, J.-H. et al. (2003). Multifunctional alloys obtained via a dislocation-free plastic deformation mechanism. *Science* 300: 464–467.

32 Troparevsky, M.C., Morris, J.R., Kent, P.R.C. et al. (2015). Criteria for predicting the formation of single-phase high-entropy alloys. *Phys. Rev. X* 5: 011041.

33 Guo, R., Yu, L.L., Liu, Z.Y. et al. (2021). Enthalpy induced phase partition toward hierarchical, nanostructured high-entropy alloys. *Nano Res.* 15: 4893–4901.

34 Chen, P.C., Liu, M., Du, J.S. et al. (2019). Interface and heterostructure design in polyelemental nanoparticles. *Science* 363: 959–964.

35 Kwon, S.G., Krylova, G., Phillips, P.J. et al. (2015). Heterogeneous nucleation and shape transformation of multicomponent metallic nanostructures. *Nat. Mater.* 14: 215–223.

36 Sarkar, A., Djenadic, R., Wang, D. et al. (2018). Rare earth and transition metal-based entropy stabilised perovskite type oxides. *J. Eur. Ceram. Soc.* 38: 2318–2327.

37 Xiao, B., Wu, G., Wang, T. et al. (2022). High-entropy oxides as advanced anode materials for long-life lithium-ion batteries. *Nano Energy* 95: 106962.

38 Nie, S., Wu, L., Zhao, L., and Zhang, P. (2022). Enthalpy-change driven synthesis of high-entropy perovskite nanoparticles. *Nano Res.* 15: 4867–4872.

39 Edalati, P., Wang, Q., Khosroshahi, H.R. et al. (2020). Photocatalytic hydrogen evolution on a high-entropy oxide. *J. Mater. Chem. A* 8: 3814–3821.

40 Xing, Y., Dan, W., Fan, Y., and Li, X. (2022). Low temperature synthesis of high-entropy $(Y_{0.2}Yb_{0.2}Sm_{0.2}Eu_{0.2}Er_{0.2})2O_3$ nanofibers by a novel electrospinning method. *J. Mater. Sci. Technol.* 103: 215.

41 Kusada, K., Mukoyoshi, M., Wu, D., and Kitagawa, H. (2022). Chemical synthesis, characterization, and properties of multi-element nanoparticles. *Angew. Chem. Int. Ed.* 61: e202209616.

42 Xu, H., Zhang, Z., Liu, J. et al. (2020). Entropy-stabilized single-atom Pd catalysts via high-entropy fluorite oxide supports. *Nat. Commun.* 11: 3908.

43 Hou, S., Ma, X., Shu, Y. et al. (2021). Self-regeneration of supported transition metals by a high entropy-driven principle. *Nat. Commun.* 12: 5917.

44 Liu, K., Jin, H., Huang, L. et al. (2022). Puffing ultrathin oxides with nonlayered structures. *Sci. Adv.* 8: eabn2030.

45 Wei, J., Rong, K., Li, X. et al. (2021). Deep eutectic solvent assisted facile synthesis of low-dimensional hierarchical porous high-entropy oxides. *Nano Res.* 15: 2756–2763.

46 Feng, D., Dong, Y., Zhang, L. et al. (2020). Holey lamellar high-entropy oxide as an ultra-high-activity heterogeneous catalyst for solvent-free aerobic oxidation of benzyl alcohol. *Angew. Chem. Int. Ed.* 59: 19503–19509.

47 Liu, F., Yu, M., Chen, X. et al. (2022). Defective high-entropy rocksalt oxide with enhanced metal–oxygen covalency for electrocatalytic oxygen evolution. *Chin. J. Catal.* 43: 122–129.

48 Chen, Y., Egan, G.C., Wan, J. et al. (2016). Ultra-fast self-assembly and stabilization of reactive nanoparticles in reduced graphene oxide films. *Nat. Commun.* 7: 12332.

49 L. Hu, Y. Chen, Y. Yao, (2018). "Nanoparticles and systems and methods for synthesizing nanoparticles through thermal shock," US Patent Application 20180369771.

50 Y. Yao, L. Hu, (2021). "Thermal shock synthesis of multielement nanoparticles," US Patent Application 11193191B2.

51 Kube, S.A. and Schroers, J. (2020). Metastability in high entropy alloys. *Scr. Mater.* 186: 392–400.

52 Aronhime, M.T. and Gillham, J.K. (1986). Time-temperature-transformation (Ttt) cure diagram of thermosetting polymeric systems. *Adv. Polym. Sci.* 78: 83–113.

53 Folgueras, M.C., Jiang, Y.X., Jin, J.B., and Yang, P.D. (2023). High-entropy halide perovskite single crystals stabilized by mild chemistry. *Nature* 621: 282–288.

54 Yao, Y., Huang, Z., Li, T. et al. (2020). High-throughput, combinatorial synthesis of multimetallic nanoclusters. *Proc. Natl. Acad. Sci. U. S. A.* 117: 6316–6322.

55 Yao, Y., Huang, Z., Hughes, L.A. et al. (2021). Extreme mixing in nanoscale transition metal alloys. *Matter* 4: 2340–2353.

56 Yao, Y., Liu, Z., Xie, P. et al. (2020). Computationally aided, entropy-driven synthesis of highly efficient and durable multi-elemental alloy catalysts. *Sci. Adv.* 6: eaaz0510.

57 Morris, D., Yao, Y., Finfrock, Y.Z. et al. (2021). Composition-dependent structure and properties of 5- and 15-element high-entropy alloy nanoparticles. *Cell Rep. Phys. Sci.* 2: 100641.

58 Wu, D., Kusada, K., Yamamoto, T. et al. (2020). On the electronic structure and hydrogen evolution reaction activity of platinum group metal-based high-entropy-alloy nanoparticles. *Chem. Sci.* 11: 12731–12736.

59 Huang, Z.N., Yao, Y.G., Pang, Z.Q. et al. (2020). Direct observation of the formation and stabilization of metallic nanoparticles on carbon supports. *Nat. Commun.* 11: 6373.

60 Ding, Q.D., Zhang, Y., Chen, X. et al. (2019). Tuning element distribution, structure and properties by composition in high-entropy alloys. *Nature* 574: 223–227.

61 Savitzky, B.H., Hughes, L., Bustillo, K.C. et al. (2019). py4DSTEM: open source software for 4D-STEM data analysis. *Microsc. Microanal.* 25 (S2): 124–125.

62 Miao, J.W., Ercius, P., and Billinge, S.J.L. (2016). Atomic electron tomography: 3D structures without crystals. *Science* 353: aaf2157.

63 Yang, Y., Chen, C.C., Scott, M.C. et al. (2017). Deciphering chemical order/disorder and material properties at the single-atom level. *Nature* 542: 75–79.

64 Zhou, J.H., Yang, Y., Yang, Y. et al. (2019). Observing crystal nucleation in four dimensions using atomic electron tomography. *Nature* 570: 500–503.

65 Qin, M., Zhang, L.M., and Wu, H.J. (2022). Dielectric loss mechanism in electromagnetic wave absorbing materials. *Adv. Sci.* 2105553.

66 Zhang, B., Duan, Y., Cui, Y. et al. (2018). Improving electromagnetic properties of $FeCoNiSi_{0.4}Al_{0.4}$ high entropy alloy powders via their tunable aspect ratio and elemental uniformity. *Mater. Des.* 148: 173–183.

67 Yang, W.Y., Zhang, Y.F., Qiao, G.Y. et al. (2018). Tunable magnetic and microwave absorption properties of $Sm_{1.5}Y_{0.5}Fe_{17-x}Si_x$ and their composites. *Acta Mater.* 145: 331–336.

68 Lan, D., Zhao, Z., Gao, Z. et al. (2020). Porous high entropy alloys for electromagnetic wave absorption. *J. Magn. Magn. Mater.* 512: 167065.

69 Huang, L.X., Duan, Y.P., Shi, Y.P. et al. (2022). Novel broadband electromagnetic-wave absorption metasurfaces composed of C-doped FeCoNiSiAl high-entropy-alloy ribbons with hierarchical nanostructures. *Compos. Part B* 244: 110182.

70 Li, G., Cui, Y., Zhang, N. et al. (2016). The precipitation in annealing and its effect on permittivity of Fe-Si-Al powders. *Phys. B Condens. Matter* 481: 1–7.

71 Kong, L.B., Li, Z.W., Liu, L. et al. (2013). Recent progress in some composite materials and structures for specific electromagnetic applications. *Int. Mater. Rev.* 58 (4): 203–259.

72 Huang, L.X., Duan, Y.P., Shi, Y.P., and Pang, H.F. (2023). Bioinspired multiple-degrees-of-freedom responsive metasurface by high-entropy-alloy ribbons with hierarchical nanostructures for electromagnetic wave absorption. *J. Colloid Interface Sci.* 636: 1–10.

73 Chen, W.J., Zhan, J.J., Zhou, Y. et al. (2021). Microwave metamaterial absorbers with controllable luminescence features. *ACS Appl. Mater. Interfaces* 13 (45): 54497–54502.

74 Su, X.G., Han, M.J., Liu, Y.N. et al. (2022). In-situ construction of nitrogen-doped reduced graphene oxide@carbon nanofibers towards the synergetic enhancement of their microwave absorption properties via integrating point defects and structure engineering. *J. Colloid Interface Sci.* 628: 984–994.

75 Zhu, H.Z., Li, Q., Tao, C.N. et al. (2021). Multispectral camouflage for infrared, visible, lasers and microwave with radiative cooling. *Nat. Commun.* 12: 1805.

76 Zhao, B., Du, Y.Q., Yan, Z.K. et al. (2023). Structural defects in phase-regulated high-entropy oxides toward superior microwave absorption properties. *Adv. Funct. Mater.* 33 (1): 2209924.

77 Dai, G.H., Deng, R.X., Zhang, T. et al. (2022). Quantitative evaluation of loss capability for in situ conductive phase enhanced microwave absorption of high-entropy transition metal oxides. *Adv. Funct. Mater.* 32 (35): 2205325.

78 Zhao, B., Yan, Z.K., Du, Y.Q. et al. (2023). High-entropy enhanced microwave attenuation in titanate perovskites. *Adv. Mater.* 35 (11): 2210243.

79 Du, Z.G., Wu, C., Chen, Y.C. et al. (2022). High-entropy carbonitride MAX phases and their derivative Mxenes. *Adv. Energy Mater.* 12 (6): 2103228.

80 Du, Z.G., Wu, C., Chen, Y.C. et al. (2021). High-entropy atomic layers of transition-metal carbides (MXenes). *Adv. Mater.* 33 (39): 2101473.

81 Hui, S.C., Zhou, X., Zhang, L.M., and Wu, H.J. (2023). Constructing multiphase-induced interfacial polarization to surpass defect-induced polarization in multielement sulfide absorbers. *Adv. Sci.* 2307649. https://doi.org/10.1002/advs.202307649.

82 Zhao, B., Yan, Z.K., Li, D.P. et al. (2023). Hierarchical flower-like sulfides with increased entropy for electromagnetic wave absorption. *ACS Appl. Mater. Interfaces* 15 (51): 59618–59629.

83 Che, R.C., Wu, Z.C., Quan, B. et al. (2023). Machine learning-directed fast and high-throughput acquisition of high-efficiency microwave absorbents from infinite design space. *Adv. Funct. Mater.* 2023: 2303108.

4

Novel Microscopic Electromagnetic Loss Mechanisms

Hongsheng Liang

MOE Key Laboratory of Material Physics and Chemistry Under Extraordinary Conditions, School of Physical Science and Technology, Northwestern Polytechnical University, Xi'an, China

CHAPTER MENU

The investigation of electromagnetic wave (EMW) absorbers and the underlying loss mechanisms is crucial for elucidating the electromagnetic (EM) attenuation behavior of materials and guiding the design of novel absorbing materials. Presently, research efforts predominantly focus on material synthesis, with mechanisms studies tending toward empirical and qualitative approaches, lacking quantitative exploration. For instance, studies on dielectric materials often merely enumerate interfacial polarization, dipolar polarization, and conductive losses. Research on magnetic materials typically involves a simple analysis of exchange resonance, natural resonance, and Eddy current losses. Moreover, there is a dearth of in-depth exploration into the dominant mechanisms within specific material systems. Therefore, there is a pressing need for an in-depth investigation into the mechanisms of EMW losses. Here, the recent works on novel microscopic loss mechanisms are summarized.

4.1 Novel Dielectric Loss Mechanisms

4.1.1 Synergistic Effects of Selenium-Sulfur Co-Doping-Induced Dielectric Polarization

Anion vacancy engineering can presumably exert a significant impact on dielectric loss and will be a good option for designing and developing high-performance EMW absorbing

Electromagnetic Wave Absorbing Materials: Fundamentals and Applications, First Edition.
Edited by Hongjing Wu, Jun Luo, and Meiyin Yang.

materials. However, this has been rarely explored so far. Besides, it is believed that not only vacancy concentrations, as most reports focus on, but also various anion vacancies as well as combination thereof may significantly matter. If one can incorporate different anion vacancies with a proper amount of concentration into transition metal chalcogenides (TMCs), a larger regulation room for boosting conduction/polarization losses could be realized. Generally, depending on various anion vacancies, they have different formation energies [1]. To achieve it, various methods have been developed, such as etching strategy [2], plasma treatment [3], and F-regulation engineering [4]. However, these approaches always bring limited vacancy concentration, let alone further enrich various anion vacancies. In this regard, exploring a facile and reliable strategy is desired and of prime significance. Moreover, to the best of our knowledge, a clear correlation between vacancy concentrations, various anion vacancies, and EMW absorption performance has not been established yet, which, if realized, will provide valuable insight for the construction of EM response mechanism. Herein, cobalt sulfoselenide is selected for demonstration purposes owing to its adjustable vacancy level [5].

For the first time, Liu et al. [6] proposed anion-doping-induced vacancy engineering to fabricate a series of Se-doped CoS_2 and S-doped $CoSe_2$ with different vacancy concentrations and various anion vacancies. This work has shown great potential for elevating EMW absorption ability by anion-doping-induced vacancy engineering. However, for this vacancy, what combination of vacancies would be better? To clarify this, different vacancy concentrations and various anion vacancies of Se-doped CoS_2, S-doped $CoSe_2$, and other derivatives, as well as their EMW absorption ability, are summarized in Figure 4.1a. For pure oxide (Co_3O_4), it almost exhibits no response to the incident EMW even if its V_O is as high as up to 29.1%. On the contrary, the effective absorption bandwidth (EAB) of metal sulfide (P-7:0) and selenide (P-0:7) are enhanced to some extent. Even so, regardless of high or low concentration of a single anion vacancy, it is still far from the requirement of excellent EMW absorption ability. This means that the values of single anion vacancy do not represent its ability to the EMW dissipation. However, to our surprise, when the V_S and V_{Se} are introduced simultaneously inside the sample, the situation has shown a huge improvement, even if their individual concentrations are relatively low. Also, such cases can be observed when comparing the C-2:5 and P-2:5, in which victory can be established as long as there is a clear advantage of V_{Se} for C-2:5, although its VS is unfavorable (Figure 4.1b). The result shows that the overall vacancy level, which is dependent on both vacancy concentration and various anion vacancies, really matters for the EMW absorption capacity (Figure 4.1c).

Based on these results, the theoretical analysis and elucidation of the unique roles of various anion vacancies in modulating EM parameters and boosting EMW attenuation ability are then conducted. (1) For the perfect ionic crystals, some polarization effects (such as electron/ion polarization) are not applicable in the commonly investigated microwave range, so the material can hardly respond to the external EM field. On the contrary, the anion doping will produce a high concentration of lattice defects and distortions in the ionic crystal, and the external EMWs will interact with the dipoles induced by these defects, resulting in dielectric polarization-relaxation process for consuming EM energy. (2) Taking the $CoSe_2$ crystal as example, when Se site is replaced by the doped S, the electronegativity of S is larger than that of Se, so the electrons are attracted by S atoms, and Se vacancies spontaneously appear around the doped S atoms to balance the local charge (see C-2:5). The generous generation of Se vacancies in C-2:5 sample will break the symmetrical distribution of the local charge and lead to the formation of a large number of dipoles. (3) When the

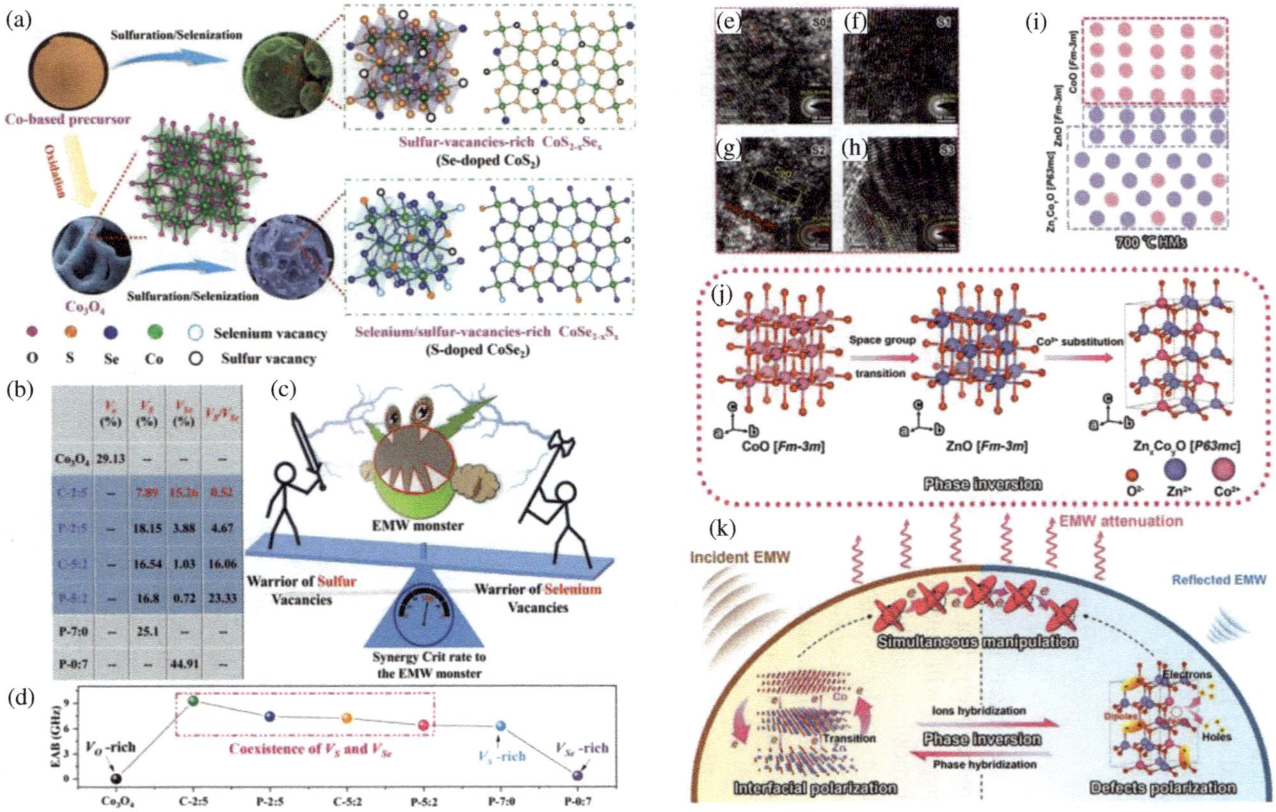

Figure 4.1 (a) Schematic illustration of preparation of selective anion-doped cobalt sulfoselenide with enriched anion vacancy. (b) Vacancy concentrations/various anion vacancies of samples. (c) The corresponding schematic illustration showing the superiority of coexistence of "VS warrior" and "VSe warrior" for improving EMW attenuation. (d) Comparison of EAB of the samples. (e–h) HRTEM images of 700 °C HMs. (i) Schematic multiphase structure of 700 °C HMs. (j) TEA controlling phase inversion mechanism. (k) Schematic representation of simultaneous manipulation of interfacial polarization and defects polarization.

doped S atoms increase in the $CoSe_2$ crystal, the S vacancies also emerge in order to reduce the growing speed of Se vacancies due to the balance of local charge. The appearance of S vacancies together with Se vacancies can further deteriorate the symmetrical distribution of the local charge, and the S-Se vacancies induce more dipoles for dielectric polarization loss in the $CoSe_2$ crystal. (4) The adjacent S-Se atoms would interact with each other. The resulting S-Se dipole-induced polarization loss will be greatly promoted, and this is beneficial to boost EMW absorption ability.

4.1.2 Synergistic Effects of Hybridization State-Induced Dielectric Polarization

In conventional, the effects of ion and phase hybridization on EM properties have been widely studied respectively. For ion hybridization, it has been widely researched that the miscellaneous elements will have a significant impact on dielectric and magnetic properties. For phase hybridization, the EM response is determined by not only phase types but also heterointerface structures. Since the distinctive work functions of dominating interfaces bring about alternate charge density distribution, there will be charge transfer and orientation giving rise to polarization relaxation and local current for EM energy conversion [7]. In addition, variant structural phase boundaries with coherent or incoherent symmetries, indicating the degree of phase separation, have an inevitable influence on the charge transfer kinetics, since more identical crystal symmetry (coherent interface) is usually accompanied by lower interfacial energy [8]. As higher interfacial energy usually generates higher carrier mobility, the heterointerface structures can also be a flexible controller for the storage and loss of EM energy by the regulation of complex permittivity [9]. In view of the two aspects aforementioned, it is extremely indispensable to investigate the synergistic manipulation mechanism of ion and phase hybridization toward bimetal hybrid materials (HMs).

A metal–organic framework (MOF)-controlled deposition and high-temperature pyrolysis crystallization strategy of bimetal HMs with tunable ion/phase hybridization and EM properties were put forward by Gao et al. [10] for the first time. In detail, Zn and Co zeolitic imidazolate frameworks (ZIF-L) were first assembled with the assistance of complexing agent triethanolamine (TEA) with claw-like structures, for further fabrication of Zn/Co HMs. Given the complexation constant difference between Zn^{2+} and Co^{2+} hydrometallatranes ($Zn^{2+} > Co^{2+}$), the polynuclear clusters were thermodynamically and kinetically controlled by adjusting the ratio between metal cations and TEA, resulting in successful formation of precursors with tunable Zn/Co separation. Afterward, the change of pyrolysis temperature made Zn^{2+} and Co^{2+} present different hybrid states, where higher temperature caused single-component phase with atomic modification, while phase separation was generated under lower temperature as the higher temperature enhanced the movement ability of metal ions. As shown in Figure 4.1i, the schematic interfacial structure of polycrystals with tunable phase hybridization revealed that gradually disappearing penetration would induce richer incoherent interfaces. Therefore, high resolution transmission electron microscopy (HRTEM) images were employed to verify the boundary structures by analyzing the polycrystalline particles dispersion and heterointerfaces coherent condition. What's more, it can be found in Figures 4.1e–h that the polycrystals with stronger grain boundary penetration always occupied more distinctive boundaries. In detail, as shown in Figure 4.1e,f, there were almost no clear grain boundaries of S0 and S1, which verified their coherent interface structure. Figure 4.1f showed that hexagonal Zn_xCo_yO [P63mc] transitioned to cubic CoO [Fm-3m] without obvious boundaries, indicating that the space group transition in S0 was accompanied with coherent interfaces. As is marked out by blue background, a local area of phase inversion can be identified. On the contrary, clear grain boundaries can be observed in S2 (Figure 4.1g) and S3 (Figure 4.1h), inducing more incoherent interfaces and leading to the disappearance of grain boundary penetration.

To sum up, the contribution of defects polarization and interfacial polarization on EMW absorption was analyzed. Above all, there were almost only interfacial or defects polarization in 600 and 800 °C HMs, respectively. Fortunately, moderate thermal motion during phase conversion resulted in the coexistence of ion and phase hybridization, giving rise to the simultaneous tailoring manipulation of interfacial polarization and defects polarization. In detail, the TEA leading phase inversion was depicted in Figure 4.1j. As mentioned earlier regarding the controllable stepwise deposition mechanism, the major ZnO and CoO phases will be separated from each other, generating more incoherent interfaces. During the pyrolysis, ZnO [Fm-3m] grew between ZnO [P63mc] and CoO [Fm-3m], while the phase separation performing as the ZnO [Fm-3m] moved into ZnO [P63mc]. Then, owing to the absence of transition phase, Co^{2+} migrated into ZnO [P63mc] more easily, producing more Zn_xCo_yO [P63mc]. However, excessive thermal motion led to high movement and high-entropy mixing of Zn/ Co, so as to make a single-phase Zn_xCo_yO [P63mc]. Finally, as shown in Figure 4.1k, the ε' and ε'' values of 700 °C HMs are far greater than those of 600 and 800 °C HMs. For 600 and 800 °C HMs, the single polarization form limited the increase in dielectric response. For 700 °C HMs, the synergistic effect of defects and interfacial polarization led to their high complex permittivity. According to the structure and electrochemical analysis, the interfacial polarization has strengthened, and the defects polarization was weakened gradually along with the TEA increasing. On the one hand, S0 and S1 performed better defects polarization, while the nonideal EMW absorbability of S2 and S3 was caused by a lack of defects. On the other hand, the excellent EMW absorption performance of S0 and S1 revealed that the interfacial polarization performed at coherent interfaces was more advantageous to EMW dissipation, though the polarization at incoherent interfaces was stronger. It seems that the defects polarization plays an important role on dielectric loss of Zn/Co HMs rather than interfacial polarization and conductive loss. Finally, it can be concluded that the synergistic effects of ion and phase hybridization can be employed on the design of HMs with tunable EM parameters. Meanwhile, the defects polarization and interfacial polarization complement each other, bringing about electromagnetic wave absorbing (EMAs) with controllable EM attenuation mechanism.

4.1.3 Synergistic Effects of Twin Structure-Induced Dielectric Polarization

Metal materials, benefiting from strong electronic transmission characteristics, high ductility, durability, and high mechanical strength, have been widely used in energy, catalysis, environment, and other fields [11]. However, as a potential research object of EMA materials, the disadvantages of large density and serious agglomeration limit the construction of its effective conductive network under the condition of the same mass percentage of filler, resulting in a much lower dielectric loss than carbon at the same level [12]. To make full use of the advantages of metal materials, the balance of impedance matching and EM loss capability is the first consideration. Synergistic optimization of defect level, electronic conductivity, composition, and interfacial structure is an effective strategy to optimize the EMW dissipation capability [13]. In the past research, a series of strategies have been widely used in the modification of metal materials [14]. For those metal-embellished EMW absorbers, various synthesis strategies have been used for developing multicomponent composites, which optimize the intrinsic EM responsiveness of composites. However, few studies have focused on the metal-predominated EMA materials, and little has been done to unequivocally reveal the underlying EM loss mechanisms involved. For the first time, Liang et al. [15] successfully formed metal Cu^0-dominated EM wave absorbers with a twin-crystal structure by adjusting the ratio of copper salt to the reducing agent chitosan to control metal nucleation.

Significant evidence on twin boundaries (TBs) is shown in Figures 4.2a–c. The HRTEM image reveals lattice spacing of 0.208 nm, corresponding to the (111) crystal plane of metal Cu^0 phase. As previously reported, the (111) crystal plane of face-centered cubic metal is the "warm nest" of twins. As shown in Figure 4.2c, the symmetrical crystal planes can be clearly observed, and the deflection angle between the symmetrical (111) crystal planes is 70.5°. Introducing TBs through GBE can also create internal boundaries within the crystal. The atomic arrangement on one side is the specular reflection of atoms on the other side separated by twinned planes. According to the atomic ordering at the twin boundary, the TBs of Cu-NC-10 belong to the coherent twin boundaries, which are lower than the grain boundary barrier of incoherent twin boundaries. Benefiting from its advantages, charge transfer is easier to achieve at TBs. The corresponding selected area electron diffraction (SAED) profile provides further strong evidence for the successful synthesis of the twinned structure of Cu-NC-10 (Figure 4.2d). Moreover, the detailed heterogeneous interfaces and defects of the prepared samples are further investigated. A small amount of (111) plane $Cu_{2+1}O$ and amorphous carbon regions are observed. No lattice fringes of CuO are detected, indicating that $Cu_{2+1}O$ is relatively stable without further oxidation. The high-angle annular dark field (HAADF) and EDX mapping image further explain the uniform distribution of various elements in the Cu-NC-10 (Figure 4.2e).

Subsequently, the EM wave absorption mechanism associated with the twin-crystal structure was investigated. Remarkably, Cu-NC-10 with TBs structure shows an abnormal trend with a sharp rise in its value, which also maps the same law in $\tan\delta_E$ and attenuation constant (Figure 4.2g). The Cu-NC-10 exhibits multiple Debye relaxation processes that may be ascribed to its special dielectric loss mechanism. The polarization states and carrier behaviors are further confirmed by electrochemical impedance spectroscopy (EIS) (Figure 4.2f). The decrease in charge transfer resistance (Rct) of Cu-NC-10 is related to its TBs structure, which implies higher charge mobility, indicating that the special structure of TBs is indeed beneficial for fast interfacial charge transfer. It can be seen that the electric conductivities of Cu-NC-10 > Cu-NC-9 > commercial copper powder > Cu-NC-11 are approximately consistent with their EMA performances, indicating the conduction loss plays a major role in boosting EM wave attenuation. However, the lower EMA performance of commercial copper powder may be due to its relatively low dispersity at the same filler ratio. X-ray photoelectron spectroscopy (XPS) is performed to investigate the electronic structure of Cu-NC-x (x = 9, 10, and 11) (Figure 4.2h). The Cu 2p spectrum of Cu-NC-10 is less red-shifted (−0.25 eV) compared to that of other samples due to the existence of TBs structure in which the Cu atoms might be not easily oxidized. The oxygen vacancies (OVs) in the samples are insufficient for defect-induced dielectric polarization in these Cu-predominated samples. The excellent conduction loss of Cu-NC-10 can be ascribed to the following microscopic mechanisms: (1) The high-dispersive metal Cu^0-predominated nanoparticles (NPs)

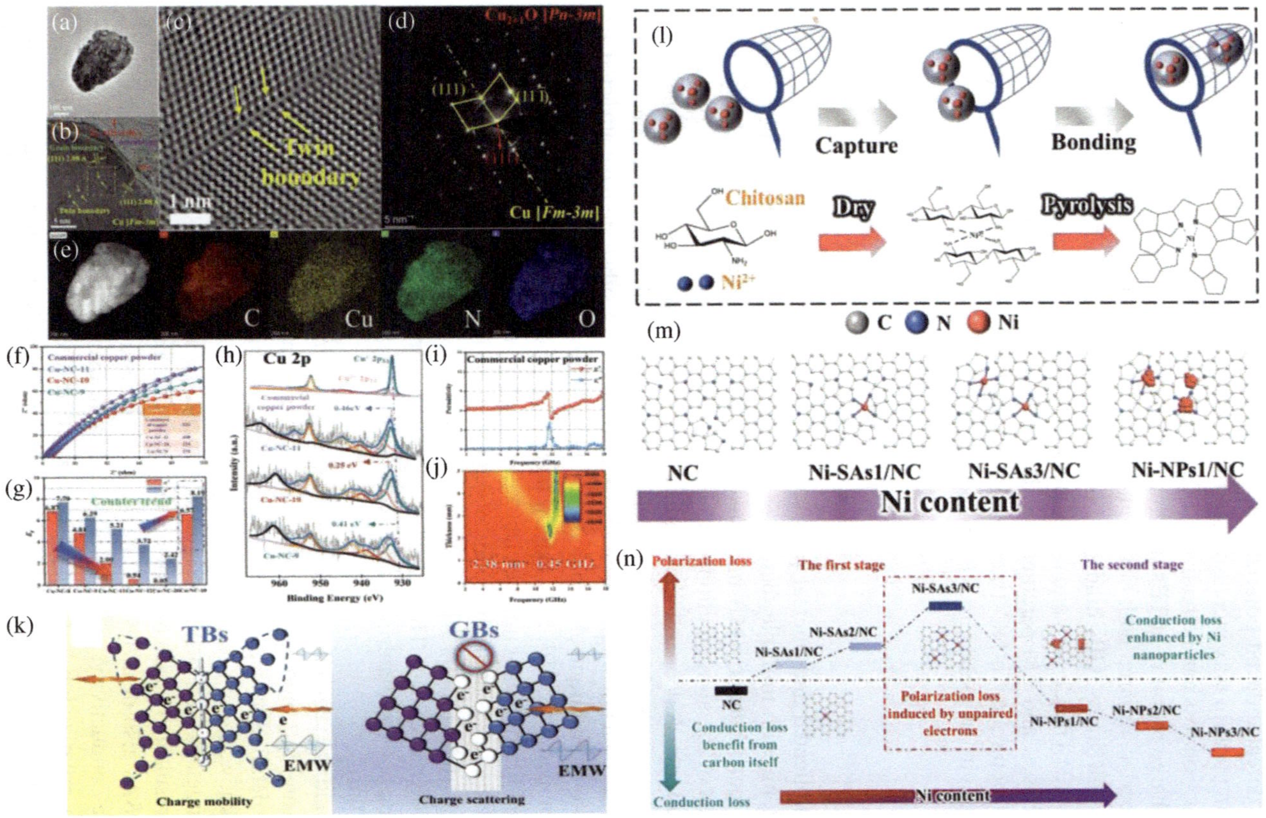

Figure 4.2 Transmission electron microscopy (TEM), high resolution transmission electron microscopy (HRTEM), images, inverse fast Fourier transform (IFFT), selective area electron diffraction (SAED) patterns for (a–d) Cu-NC-10. (e) High-angle annular dark field (HAADF) and corresponding elemental mapping images for Cu-NC-10. (f, h) Nyquist plots and X-ray photoelectron spectroscopy (XPS) Cu 2p spectra of Cu-NC-9, Cu-NC-10, Cu-NC-11, and commercial copper powder. (g) Complex permittivity of the samples. (i, j) 2D plot of reflection loss (RL) values versus frequency at different thicknesses for commercial copper powder. (k) Schematic diagrams of EM wave absorption mechanisms for twin boundaries (TBs) and grain boundaries (GBs). (l, m) Schematic illustration of the fabrication process of the samples. (n) Continuous evolution law of microscopic EM loss based on single-atom model absorbers.

uniformly blend with paraffin wax, and abundant nonmagnetic Cu NPs with relatively weak interaction can contribute to the formation of an isotropic dielectric loss model. This model is the foundation for understanding varied dielectric losses such as dipole polarization, interfacial polarization, defect-induced polarization, conduction loss, and so on. (2) When the density of Cu NPs in paraffin wax increases and the distance between Cu NPs reduces considerably, the conduction loss can be realized in the composites with discontinuous conductive network. The charge mobility at TBs structure between homogeneous interfaces can be significantly enhanced, compared with GBs region. Due to the existence of large amount of crystal defects between grains, GBs hinder the transport of charges, and charge scatterings occur at defect sites. (3) The Cu NPs and commercial copper powder could be oxidized into $Cu_{2+1}O$ and CuO phases, and EM parameters will reduce in these samples (Figure 4.2i,j); however, the TBs structure in Cu-NC-10 can resist the oxidation of metal Cu^0 and keep the high electric conductivity of metal Cu^0 NPs (Figure 4.2k).

4.1.4 Synergistic Effects of 3D Orbitals Unpaired Electron-Induced Dielectric Polarization

At present, the complicated and unclear physical loss mechanism of mixed magnetic and dielectric losses, along with multiple factors as a theoretical guide has limited the further exploration of high-performance EMA materials [16, 17]. Hence, it is urgent to clarify the intrinsic mechanisms of the aforementioned problems but still faces great challenges, especially on the atomic scale. A deeper understanding of the relationship between mechanisms and properties requires rational physical models at the atomic scale. Since the concept of isolated atoms was introduced, single atoms (SAs) dispersed on a substrate have attracted a lot of attention due to their unique electronic structure, quantum size effect, and unsaturated coordination environment [18, 19]. SAs exhibit physical properties different from the corresponding NPs in addition to their unique electronic behavior, which makes them promising for a wide range of applications in the fields of catalysis, environment, and so on. Herein, Liang et al. [20] constructed stable Ni-SAsx/N-doped ($x = 1, 2, 3$) carbon absorbers with the strategy of ligand polymerization. The morphology, composition, conductivity, defects, and electronic interactions of the material can be well tailored by Ni species modulation engineering (Figure 4.2l,m).

To better understand the segmented regular variation of the complex permittivity based on the Ni single-atom and NP models, induced by the intrinsic microscopic EM loss mechanism, the following detailed discussion should be carried out (Figure 4.2n). First, for the first stage (1) as we all know, carbon content is hardly conducive to impedance matching due to ultra-high EM parameters. Part of the carbon as reducing agent was consumed during the introduction of Ni, resulting in a slightly lower carbon content in Ni-SAsx/NC. (2) The initial introduction of Ni SAs also destroys the integrity of the conductive lattice of the support, causing a fall in the conductivity, which is reflected in an overall decrease in the complex permittivity. (3) For polarization loss, the Ni species in Ni-SAsx/NC exist as electron configuration coordination and do not form heterogeneous interfaces, so the interfacial polarization is not taken into account. Moreover, the increased level of unique polars (Ni SAs-N_x) is positively correlated with the high concentration of unpaired excess of Ni species, while the carbon structural defects play a negligible role in Ni-SAsx/NC. Therefore, the unpaired excess of Ni species, synergistically induced by Ni-N_x bonds and abundant Ni SAs, leads to the formation of excess Ni SAs bonded with limited N atoms (i.e. Ni SA_y-N_x species, $y > 1$). This enhances the polarization loss, and the highest polarization loss has been achieved in Ni-SAs3/NC. Second, for the second stage (1) the enrichment of surface carbon enhances the conductivity in Ni-NPs*x*/NC. (2) Emphatically, despite the significantly different band structure between NC and Ni-NPs*x*/NC compared to the Ni-SAs*x*/NC, the lower quality of unpaired electrons on the Ni 3*d* orbital due to the weaker interaction between Ni NPs and carbon atoms, and formation of abundant near-zero-valent Ni NPs, lead to the lower binding energy values of Ni species. This indicates a weaker polarization center (i.e. Ni NPs-C_x species), suggesting a limitation of its polarization loss. Based on the preceding discussion, both the physicochemical properties and electronic interactions of the material synergistically determine the inherent characteristics of the absorbers. Based on the Ni single-atom and NP models, the segmented regular variation of complex permittivity is mainly attributed to the electron density of excess unpaired electrons on Ni 3*d* orbital, which also indicates the related band structures. The abundant excess unpaired electrons on Ni 3*d* orbital of the Ni SAs model could steadily exist due to the excess single-atom confinement effect and stronger electronegativity of N atoms, which would be together with N_x atoms and act as unique polar/defect centers, contributing dominantly to its dielectric losses.

4.1.5 Interpretation of Energy Band Theory in Dielectric Loss

Recent research has highlighted the significance of polarization/conduction loss in semiconductors, particularly arising from OVs in metal–cation-doped MnO_2, leading to the optimization of semiconductor absorber designs [21, 22]. According to theoretical analysis, since the absence of oxygen atoms creates OVs, the periodic underlying field generated by the rigorously periodic arrangement of atoms will be destroyed, and it is possible to introduce the energy levels that electrons can reach in the forbidden band. For the p-type semiconductors, the introduction of OVs can increase the charge density near the valence band (i.e. from Ti^{4+} to Ti^{3+} in Ti_xO_{2x-1} ($4 \leq x \leq 9$)) and change the intrinsic band gap of the semiconductors [23, 24]. The ability of OVs to close the band gap between the Fermi level and valence band dictates the conductivity of p-type semiconductors. The relationship between electrical conductivity (σ) and permittivity (ε'') could be expressed by the free electron theory: $\varepsilon'' = \sigma/2\pi\varepsilon_0 f$, where f and ε_0 are the frequency and dielectric constant in vacuum, respectively. Therefore, the related EM properties, including ε' and ε'', can be indirectly reflected through energy band structures of defective semiconductors. Despite numerous experimental studies addressing defect engineering in semiconductor absorbers, there remains a significant gap in understanding the intricate relationship among charge carriers, defects, heterointerfaces, and EM properties.

Li et al. [25] utilized energy band theory to link charge carrier/defects/interfaces to EM properties; a series of oxygen-deficient multiphase Ti_xO_{2x-1} ($1 \leq x \leq 6$) composites were fabricated by using facile hydrothermal reaction and subsequent hydrogenation of a temperature gradient approach. Benefiting from the charge carrier/heterointerfaces induced strong conduction or polarization loss, oxygen-deficient Ti_xO_{2x-1}/carbon nanotubes (CNT) (TiOC) composites exhibit superior EM performances compared with pristine CNT/TiO_2 material. XRD Rietveld plots and schematic diagrams of the phase transition process between different Ti_xO_{2x-1} phases during hydrogen reduction are shown in Figure 4.3a,b. To depict the charge transfer process, the energy band diagrams of the TiOC composites are explored by

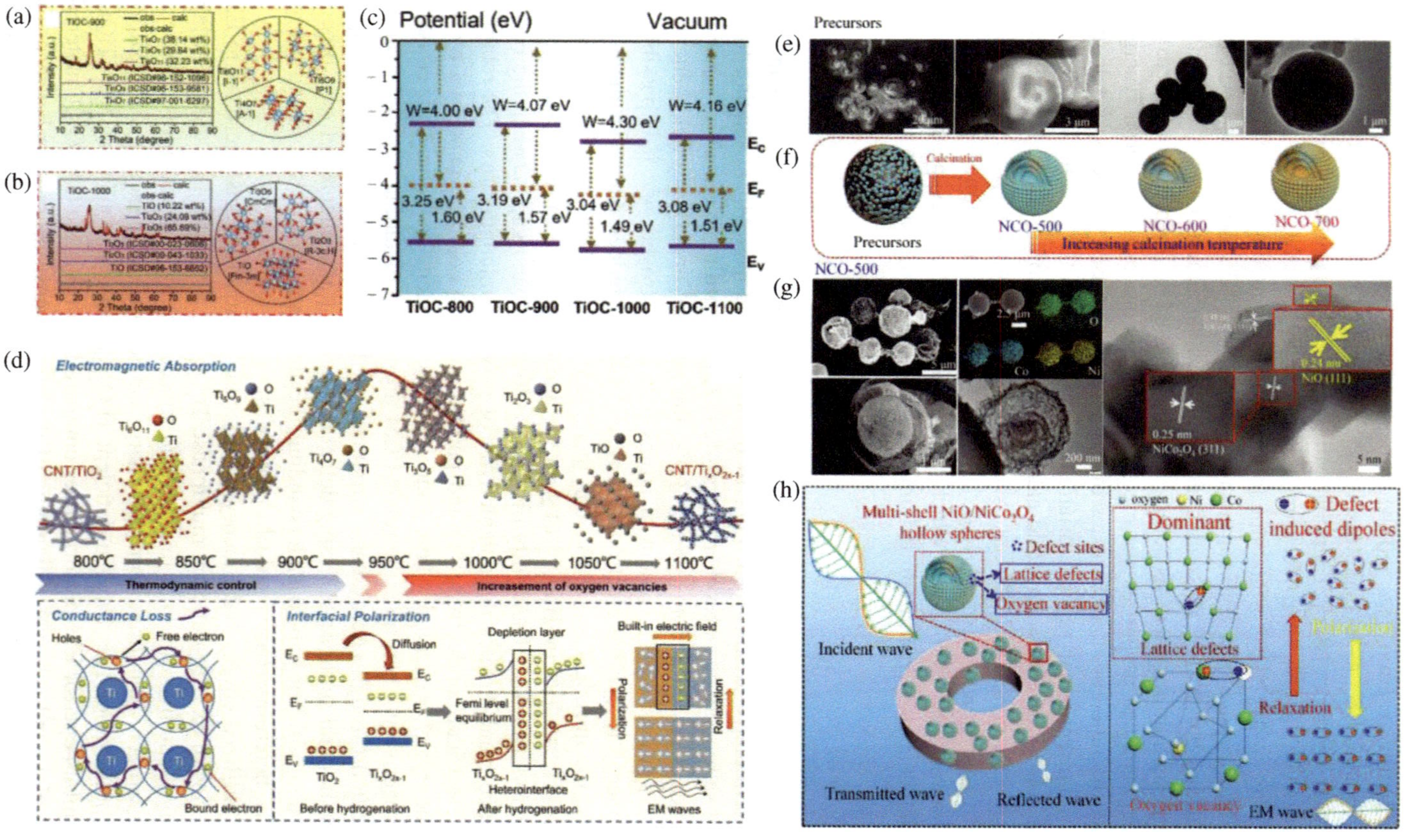

Figure 4.3 (a, b) The XRD Rietveld plots and the schematic diagrams of the phase transition process between different titanium oxides during the hydrogen reduction. (c) Energy band diagram of TiOC composites hydrogenated at different temperatures. (d) The holes induced conductance loss and interfacial polarization caused by interfaces at the atomic level. (e, g) TEM and HRTEM images of the sample and (f) its synthetic mechanism. (h) EM wave absorption mechanism.

ultraviolet photoelectron spectroscopy (UPS) spectra and ultraviolet–visible spectroscopy (UV-vis). Based on the energy band theory, the work function is utilized to describe the electron transfer behavior of the samples, which is determined by UPS spectra. The band gaps of TiOC-800, TiOC-900, TiOC-1000, and TiOC-1100 are 3.25, 3.19, 3.04, and 3.08 eV, respectively. In this heterojunction, CNTs act as electron catchers due to their excellent conductivity, which inhibits carrier recombination on the surface of Ti_xO_{2x-1} and leads to charge separation and transfer. As a result, the Fermi levels of the four samples are closer to the valence band than that of the conduction band, indicating that they are p-type semiconductors (Figure 4.3c).

The conductance loss resulting from an abundance of positive charge holes was further explored based on semiconductor characteristics. All Ti_xO_{2x-1} underwent conversion into typical p-type semiconductors after the hydrogenation reaction, implying that positive charge hole migration was the primary cause of charge movement. In contrast to n-type semiconductors like ZnO, where OVs enhance conductivity and optimize EM wave absorption by shifting the oxide Fermi level upward, shrinking the band gap, and reducing the band gap between the Fermi level and the conduction band, p-type semiconductors in this study experience a downward shift of the Fermi level. This reduction in the band gap between the Fermi level and VB optimizes positive charge hole migration. Positive charge holes migrate more slowly than free electrons in a crystalline structure, acting as large moving electric dipoles in the Ti_xO_{2x-1}. This behavior differs from free electrons inducing conductance loss and is akin to unique dipole polarization with movement ability. A small number of electrons trapped in the OVs cannot move freely in the crystalline, and electron-filled OVs cannot induce local charge nonuniform distribution; thus the electron-induced conductance loss and defect-induced dielectric polarization do not occur in these Ti_xO_{2x-1} composites. Furthermore, many heterointerfaces are formed between Ti_xO_{2x-1} and CNTs, and the difference in Fermi level between the two materials will generate electron transfer across the interface, resulting in a built-in electric field and space charge region, thus affecting the transport of charges and interface polarization. Meanwhile, the difference in the conductivity or polarity among the Ti_xO_{2x-1} can also lead to nonuniform distribution of positive and negative charges, resulting in the formation of a spatial electric dipole moment that enhances the polarization-relaxation process. The microscopic factors such as charge carriers, defects, and interfaces can be reflected by the energy band structures of these multiphase Ti_xO_{2x-1}. In this case, the relationship between charge carriers, defects, interfaces, and energy band has already been investigated by exploiting a series of spectral characterizations such as UPS, UV-vis, EPR, and PL. The energy band indicate the proportion level of charge carriers, defects, or interfaces in these Ti_xO_{2x-1} composites, which can reflect the macroscopic electromagnetic properties (Figure 4.3d).

4.1.6 Defect-Induced Polarization Loss in Multi-Shelled Spinel Hollow Spheres

Co-based spinel-structured metal oxides are regarded as prospective candidates as effective EM wave-absorbing materials. To date, a series of MCo_2O_4 (M = Fe, Ni, Mn, etc.)-based EM wave absorbers have been prepared and exhibit promising absorption behaviors [26–29]. Though these advances are inspiring, the high-density issue of MCo_2O_4-based EM wave absorbers still hinders their practical applications. More importantly, researches on these

MCo_2O_4 absorbers are individual, and a systematic research on these serial EM wave-absorbing materials to reveal their connections and differences on EM wave attenuation has not been reported yet. Based on the previous discussion, it is highly desired to design a specific morphology to reduce the high-density issue of MCo_2O_4-based EM wave absorbers. At the same time, a systematic study on a series of MCo_2O_4 EM wave absorbers to comprehensively disclose their attenuation behaviors and mechanisms is of great significance.

Qin et al. [30] successfully verified that the defects (more precisely, line defects and planar defects induced polarization) were the main source for EM wave absorption in serial multi-shelled Co-based spinel hollow spheres. First, they selected $NiCo_2O_4$ as a representative and changed their defect sites by different calcination temperatures. The morphologies of these samples were viewed by TEM. For the precursors, it displays solid spherical structure with smooth surface (Figure 4.3e). The spheres possess uniform size with a diameter of around 4.7 μm. The uniform geometry and size distribution make it a valid template for the synthesis of multi-shelled Co-based spinel hollow spheres. After heating treatment, the solid carbon spheres transform into multi-shelled hollow sphere structure (Figure 4.3e). Though the products retain the spherical shape, abundant pores and folds are in the presence of the shell. The pores may arise from the release of CO_2 gas, which results from carbon spheres combusting under heating treatment. The magnified TEM images clearly reveal that the outmost shell of these spheres is constructed by considerable loosely packed NPs. The broken spheres in Figure 4.3e confirm that the NCO-500 possess a thin shell and a core. The core has a diameter of ~1.55 μm. As a whole, the diameter of complete spheres of NCO is 2.5 μm, much smaller than the pristine carbon spheres (as schematic in Figure 4.3f). Subsequently, the EM absorption properties were compared above the samples. EM wave loss capacity was found to decrease along with increasing calcination temperature. Thanks to their identical multi-shelled hollow sphere structure, morphology influence is ruled out in this system. With respect to other properties, an in-depth study is carried out. Polar water molecules and functional groups are eliminated during high-temperature calcination; thus dipolar polarization was excluded. Meanwhile, conductivity is found to decline with elevated temperature, and conductive loss is also not the main contribution to EM wave loss. Similarly, more NiO is generated at higher calcining temperature, leading to more interfacial junctions between NiO phase and $NiCo_2O_4$ phase, thus promoting interfacial polarization. This result confirms that interfacial polarization is not the main contribution to dissipate EM energy. Only defects and its induced dipolar polarization are responsible for the variance trend for serial materials. Also, abundant defect sites can be observed in TEM images in $NiCo_2O_4$ samples in Figure 4.3g. With respect to defect sites, OVs and crystal defects such as planar defects exist in spinel-structured materials.

Interestingly, oxygen vacancy is found to increase with higher temperatures. Therefore, it can also be inferred that oxygen vacancy is not dominated loss mechanism. After careful analysis, crystal defects display the same tendency with EM wave performance variation, while oxygen vacancy shows a reverse trend, revealing crystal defect-induced polarization is the main loss mechanism. Notably, an electron spin resonance (ESR) test was conducted to uncover the OVs in serial Co-based ferrite samples. In general, resonate peaks in ESR curve can be discerned due to the electrons trapped on OVs. However, the peaks were absent in curves, but an interesting phenomenon was discovered instead. It should be noted that ESR is known to be a very sensitive method to probe not only paramagnetic centers but also microwave absorption. For a high dielectric loss material, it will distort the microwave field inside an EPR resonator, which results in distortion of the EPR signal from spins inside the material. The samples to be tested in this work have the potential to dissipate the energy of EM wave by dielectric loss, which may influence the test signals. In the test results, the data in test curves are basically negative due to their consumption of EM wave energy, leading to the lack of peak values. Moreover, the negative values are proportional to their EM wave attenuation capacities. Among serial Co-based samples, NCO500 clearly displayed the lowest values, and the curve even changed into arch. This result suggests the strongest EM wave absorption capacity of NCO500, which is in line with the conclusion of this work. In short, ESR technique may be not suitable to investigate the oxygen vacancy in EM wave absorbers but can be a useful tool to validate their absorption abilities. This finding may also provide us a more visualized and facile approach to pick out materials with strong EM wave absorption capacities rapidly. A detailed schematic diagram for defect-induced polarization loss mechanism is present. This result is also suitable for serial multi-shelled MCo_2O_4 hollow sphere spinel materials. In detail, $NiCo_2O_4$ exhibits the highest defect sites in serial spinel ferrites in view of Raman spectra, thus delivering the best EM wave attenuation capacity (Figure 4.3h).

4.2 Novel Microscopic Magnetic Loss Mechanisms

At present, microwave absorption properties of absorbers refer to the polymer composites introduced by absorbing fillers. Because of the random distribution of absorbing fillers in a polymer matrix, the optimization of EM parameters (ε_r and μ_r) to strengthen microwave absorption abilities is a problem that has yet to be solved. Researchers have invested considerable attention in the permittivity-tunable EM wave absorption. Lv et al. [31] fabricated a sandwich electronic device with a conductive layer and a microwave absorption film, regulating the permittivity values using an external electrical field to control the EM wave absorption performance. Song et al. [32] prepared polymer ionic conductive gels capable of optically transparent microwave absorption. By controlling hydrogen bonding networks, complex permittivity can be tuned to obtain manipulatable microwave stealth properties. Ye et al. [33] designed edge-rich graphene grown on porous Si3N4 ceramics via the chemical vapor deposition method, with tunable dielectric properties by the tailored structures. Wen et al. [34] fabricated reduced graphene oxides (rGO) that display strong EMI shielding at an increased temperature as a result of the synergy of dipole polarization and hopping conductivity. The rGO composites reveal various temperature-dependent imaginary permittivity and EMI-shielding performances. Moreover, the complex permittivity of ZnO needles is also temperature-dependent [35]. Li et al. [36] prepared carbon nanocomposites by adding numerous vertical interphases, and the dielectric response is successfully adjusted. The relaxation process in vertical-interphase nanocomposites, indicating the discrepancy in polarization ability among the interphases, enhances microwave absorption properties. By analyzing the previously mentioned dielectric-dependent materials, it can be deduced that the manipulation of EM wave absorption properties for these permittivity-dependent absorbers is limited because of the absence of magnetic loss. The study of magnetic loss, which is much less common in the field of wave absorption than dielectric loss, also plays a key role in the modulation of absorption properties and therefore needs to be emphasized.

4.2.1 Magnetic Losses Induced by the Sequence Structure of Metallic Magnetic Chains

Concerning magnetic absorbers, various 1D magnetic absorbing materials exhibit competitive EM wave absorption capabilities, such as FeNi/C nanofibers, Fe/C nanofibers, Fe nanowires, Fe3O4@carbon nanochain, CoNi chains, Ni chains, and Co chains. These materials are characterized by their lightweight and thin nature, demonstrating robust absorption due to unique 1D structures, orientation polarization, and inherent magnetic loss. Current evidence supports that the magnetic field-induced synthesis primarily yields magnetic metallic 1D nanostructures. Nevertheless, measurements of EM parameters in polymer composites featuring randomly distributed 1D magnetic samples may compromise complex permeability, adversely affecting EM wave absorption properties. The orientation of magnetic 1D absorbing materials within polymers becomes a critical concern. Limited research addresses magnetic composites with oriented structures, focusing on controlling complex permeability to enhance EM wave absorption performances.

Zhao et al. [17] introduced a novel approach, employing a modified galvanic replacement reaction to produce 1D core–shell Cu@Ni and Cu@Co bimetallic magnetic chains under the influence of an external magnetic field. This pioneering method opens avenues for constructing 1D core–shell magnetic composites with noble metals, including Ag@Ni or Co, Au@ Ni or Co, Pt@Ni or Co, and Pd@Ni or Co. Subsequently, they achieved the orientation of these magnetic chains within a polyvinylidene fluoride (PVDF) matrix by employing both an external magnetic field and a hot-pressing process. The fabrication process for 1D core–shell magnetic chains is succinctly illustrated in Figure 4.4a. Typically, core–shell bimetallic magnetic composites are formed through reduction reactions involving various redox potentials using the heterogeneous nucleation method. Figure 4.4a depicts the preparation of oriented core–shell magnetic chains/PVDF composites. Initially, the as-prepared magnetic chains and PVDF are blended in a N, N-Dimethylformamide (DMF) solution. Subsequently, orientated magnetic chains/PVDF composite films are obtained by solvent evaporation with the assistance of an external magnetic field, followed by a hot-pressing treatment. Upon exposure to alternating EM fields, whether in an oriented or random configuration, the magnetic chains/PVDF composites invariably absorb microwave energy or convert it into Joule heating. In the oriented configuration, the parallel alignment of magnetic chains results in the separation of contacts between adjacent chains, leading to decreased electrical conductivity. This, in turn, results in a lower complex permittivity (ε_r) compared to the random configuration.

Figure 4.4b1,b2 present the morphological characterization of the magnetic Cu@Co chains as prepared in this study. It is apparent that all the magnetic samples exhibit 1D structures, self-assembled by spherical NPs under the influence of a magnetic field. Notably, the surfaces of the bimetallic magnetic chains appear coarser and rougher, contributing to the effective attenuation of microwave energy. Particularly in the case of Cu@Co chains, crumpled walls are observed on the surfaces, possibly resulting from the presence of dual phases comprising Cu and Co components. The coarser and rougher surfaces of these bimetallic magnetic chains are advantageous for harvesting more microwave energy. First, owing to their high specific surface area, the rough special lamellar surfaces induce multiple scattering, effectively attenuating microwave energy. Additionally, the magnetic coupling interaction readily occurs between adjacent or connected flakes within the chains, leading to a significant dissipation of microwave energy. The combination of these factors underscores the potential

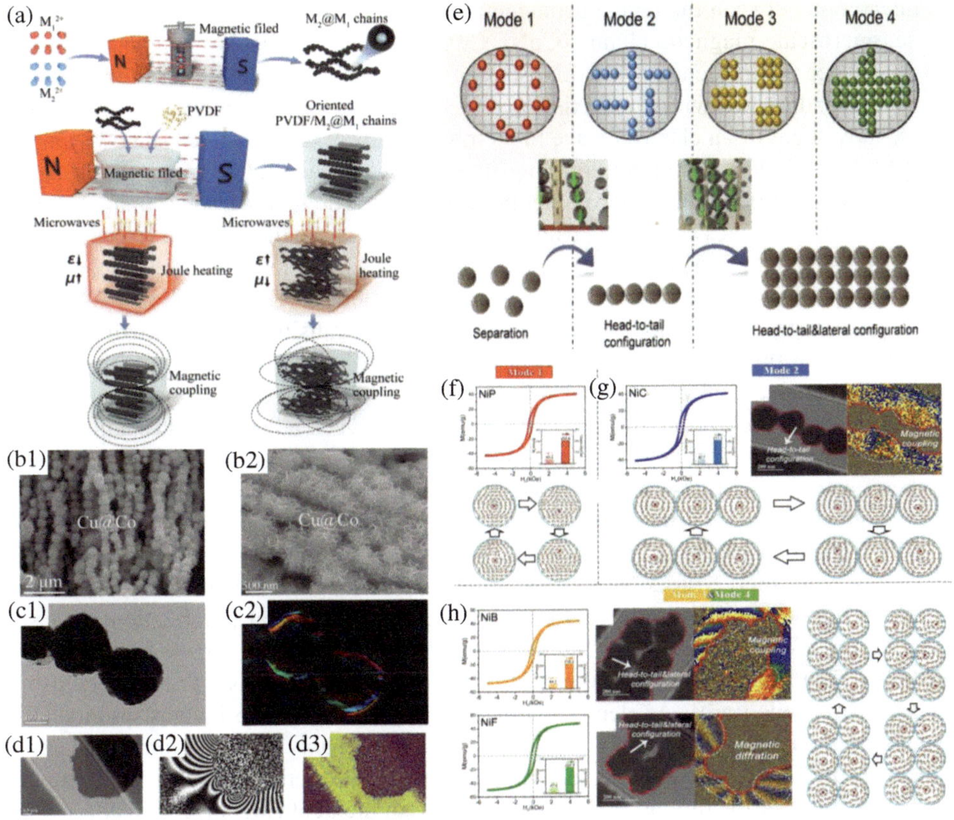

Figure 4.4 (a) Fabrication of orientated magnetic chains/PVDF composite films and microwave absorption. (b1, b2) Morphology characterizations of the magnetic chains. (c1, c2) TEM image and magnetic domains of the magnetic chain. (d1–d3) Electron holograms, magnetic coupling, and charge density distribution of magnetic chains. (e) Four mesoscopic modes of sequential architecture designed with ferromagnetic functional units, assisted by an applied magnetic field. (f–h) Strange magnetic behaviors of different modes.

of the observed morphological features to enhance the absorption of microwave energy by the magnetic chains. Comparing the oriented composites with the random distribution of magnetic chains in PVDF, they observed a simultaneous reduction in complex permittivity and an increase in complex permeability, resulting in enhanced EMW absorption properties. In the case of the oriented core–shell Cu@Co chains embedded in PVDF, they achieved an optimal reflection loss (RL) value of −43.5 dB, along with an effective absorption bandwidth (RL< −10 dB) spanning 7.3 GHz.

To directly elucidate the EMW absorption mechanism involving magnetic coupling interaction and polarization loss, it conducted an off-axis electron holography analysis to observe the magnetic stray field and multipolar charge density emitted from the magnetic chains. This analysis provides a visual representation of the movement of magnetic moments within the magnetic chains, offering insights into the microwave absorption process (Figure 4.4c1,c2). In the fringe area of magnetic particles, the magnetic moments exhibit faint fluctuations due to the interaction between the native demagnetized field and the external magnetic field. In this case, microscale magnetic chains (CoNi, Cu@Ni, Cu@Co) display magnetic vortices movement under the influence of the high-frequency microwave field. To directly observe magnetic responding behaviors, off-axis electron holograms were employed to distinguish the phase movement of an electron wave in its travel pathway (Figure 4.4d1–d3). The alternating EM fields induce a mutual coupling interaction between adjacent magnetic particles, leading to significant microwave energy dissipation. The reconstructed images reveal intense magnetic flux lines emanating from the chains, indicating the inherent ferromagnetic behavior of each chain. Notably, the magnetic flux lines spread into the vacant space constrained by the 1D structure, consistent with findings related to Co nanochains. When the morphological surfaces of monometallic magnetic particles transition from smooth to rough, the orientations of the magnetic lines adjust instantaneously. In essence, the 1D magnetic self-assembled coarser flaky particles exhibit enhanced microwave absorption ability. The coarser flaky surfaces facilitate the magnetic coupling between adjacent flakes, attenuating energy. Additionally, the self-assembled magnetic chains, resembling candied haws on a stick, strengthen magnetic coupling due to the enhanced interaction between adjacent magnetic particles in the direction of orientation. Building upon the aforementioned characterizations, it can note a slight oxidation occurring on the surfaces of the magnetic chains. Particularly in the case of core–shell Ni@CoNi, Cu@Ni, and Cu@Co magnetic chains, the presence of multiple interfaces results in the generation of intensive interfacial polarizations. Consequently, microwave energy can be attenuated through the interfacial polarization induced by the EM field. The distribution of charge density in the magnetic chains was evaluated using off-axis electron holograms. As depicted in the reconstructive phase images, pronounced charge distribution occurs adjacent to the interfaces. Remarkably, clear regions of negative and positive carriers are evident near the interfaces of the magnetic particles, signifying the presence of a polarization dipole. When stimulated by an incident EM wave with alternating vector directions over a short period, both the natural dipole moment and electric field within numerous hetero-joints realign to a confined orientation, generating macroscopic polarization intensity. During this process, it becomes evident that both interfacial polarization and spatial charge polarization contribute to dissipating the energy of the penetrated EM wave. This dual polarization mechanism enhances the overall microwave absorption capabilities of the magnetic chains.

4.2.2 Multi-Model Sequence Structure for Improving Magnetic Loss

Sequential structural engineering, constructed by artificially configuring functional units, has been demonstrated to be a new strategy compared to conventional methods, focusing on the modulation of microstructure through specific spatial configurations and arrangements. The coupled enhancement effect of neighboring functional units leads to strange physical behaviors and shows great potential in improving the macroscopic properties of materials. However, the filler loading (FL) of traditional ferromagnetic microwave absorbers (MAs) is high, in the range of 50–80 wt%, which results in two intractable problems: (1) the obtained MAs with large surface density fail to meet the lightweight requirements in current applications; (2) the mechanical properties are weak due to the poor interface bonding between the ferromagnetic absorbent and the binder matrix. Microstructure design is considered an efficient strategy to optimize the microwave absorption performance of ferromagnetic MAs due to the advantages in the modulation of the micro-electric field and magnetic anisotropy [37–40].

However, the high FL issue of these MAs still exists because the remanence of ferromagnetic filler leads to inevitable aggregation, even at low FL [41]. The disordered aggregation enhances the isolation effect of the binder but inhibits the expression of the improvements. Appropriate models and methods based on sequential architecture are required to solve the problems mentioned previously, considering the effects of dimensions, sizes, morphologies, and components on the strange dielectric-magnetic behaviors of ferromagnetic MAs. Herein, Pan et al. [42] developed a series of ferromagnetic MAs with different sequential architectures via a bottom-up self-assembly method based on four mesoscopic models (particles, chains, bundles, and fibers) of ferromagnetic functional units. In this work, a novel magnetic field-assisted growth method was applied to prepare Ni-based MAs with bottom-up sequential architectures, including Ni particles (NiP), Ni chains (NiC), Ni bundles (NiB), and Ni fibers (NiF), which correspond to different sequential modes (Figure 4.4e). The period of the applied magnetic field is essential to tailor the different sequential architectures.

The different modes show abnormal magnetic behaviors. The head-to-tail and lateral configuration will not only reduce the defect concentration of the Ni but also enhance the magnetic coupling of adjacent functional units: (1) magnetic behaviors of mode 1: the static magnetic properties of mode 1 are detected through the hysteresis loops at room temperature. Besides, micromagnetic simulation is adopted by software at the microwave frequency band to clarify the unusual magnetic behavior. A uniform, coherent rotation mode dominates the magnetization reversal mechanism of a sphere, while the magnetization reversal mechanism is vortex rotation when the radius is larger than the critical value. For mode 1 with dispersed ferromagnetic functional units, the magnetic moment presents a noticeable clockwise vortex motion in the unit period, which relates to the magnetic loss mechanism of domain wall migration (Figure 4.4f). (2) magnetic behaviors of mode 2: when the particles are assembled via head-to-tail, Ms increases due to the contact between particles that reduces the specific surface area of the materials. The Lorentz electron holography was performed further to analyze the magnetic coupling behavior of different modes and observe the stray magnetic field emitted (Figure 4.4g). The magnetic coupling is observed from mode 2. When alternating EM fields irradiate the NiC sample, denser magnetic coupling interactions around the ferromagnetic functional units are triggered, which is beneficial to dissipate the EM energy through magnetic polarization. The triggering coupling interaction between the adjacent ferromagnetic functional units tends to form a closed magnetic-line semicircle from mode 2. Furthermore, the micromagnetic simulation of mode 2 also shows the strange magnetic coupling, different from mode 1. The motion direction of the magnetic domains in adjacent functional units changes counterclockwise, displaying a periodic state where the homodromous chiral vortex occurs in the interval functional units and the heterodromous chiral vortex appears in the adjacent functional units. The flip of the domain represents a reinforcement of both the magnetic coupling and the magnetic loss. (3) magnetic behaviors of modes 3 and 4: the magnetic behaviors further change from mode 2 to modes 3 and 4 due to the formation of the lateral configuration. This phenomenon indicates that magnetic loss gradually enhances due to lateral and fibrous configurations. The configuration transition further reduces the defects and improves the magnetic coupling of the ferromagnetic functional unit. According to the Lorentz electron holography (Figure 4.4h), head-to-tail aggregation and the lateral aggregation induce strong magnetic coupling to the samples. Mode 3 acts as high-density magnetic activation antennas to irradiate out magnetic flux lines and interact with each other, resulting in an interconnected magnetic coupling network. The corresponding micromagnetic simulation also demonstrates the enhanced magnetic coupling of mode 3. A similar inverse change of domain motion occurs between the adjacent chains as the lateral configuration occurs, resulting in a heterodromous chiral vortex between the ferromagnetic functional units in the bundles (Figure 4.4h).

4.2.3 Enhanced Magnetic Coupling in Hollow Porous Carbon Three-Dimensional Magnetic Networks

Recently, the focus of research on MA materials has shifted from individual electrical or magnetic components to innovative EM composites, such as FeCo/C [43], FeCoNi/carbon fiber [44], and Ni/MXene/GO [45]. These composite materials, leveraging the dual properties of electrical and magnetic components for efficient EMW dissipation, coupled with optimized impedance matching through EM synergies, demonstrate satisfactory MA performance. However, many strategies primarily concentrate on densely decorating magnetic NPs onto the electrical matrix [46]. During the compounding process, magnetic NPs tend to agglomerate into bulk structures due to magnetic dipole interactions. This agglomeration results in fewer interfaces and weaker magnetic coupling, ultimately compromising the EMW absorption capacity. Zhang et al. [47] employed a high-throughput, environmentally friendly, and scalable spray-drying method for the fabrication of hierarchically hollow porous spheres (Figure 4.5a). Leveraging the extensive specific surface area, Co^{2+} ions were adeptly anchored onto the surface of these porous spheres, providing a conducive environment for the in situ growth of magnetic carbon nanotubes (CoNC). This simultaneous growth not only confined the magnetic particles within the structure but also enhanced the dispersion of magnetic components. The resultant high-density CoNC, in conjunction with carbon nanosheets, established a compact interface and magnetic network, thereby amplifying interfacial polarization and magnetic coupling. This strategic arrangement contributes to the realization of efficient MA performance.

High-density magnetic dispersion and confinement play pivotal roles in influencing magnetic loss. To gain a comprehensive understanding of the magnetic properties, it conducted an analysis of the measured complex permeability parameters of the samples. The permeability constant of CoNC/HCP microspheres exhibits a strong correlation with the magnetic properties of Co particles. As illustrated in Figure 4.5b, the μ' value increases from 1.08 to 1.13 at 2 GHz with the Co content, indicating an enhancement in magnetic storage ability. Furthermore, the μ'' value varies with the increase in Co content, and the CoNC/HCP-2 microsphere exhibits the highest μ'' value overall (Figure 4.5c), highlighting its superior magnetic loss performance. In general, μ'' is influenced not only by the magnetic particle content but also by the degree of dispersion and particle size. The confinement effect of porous spheres and their excellent magnetic dispersion contribute to the enhanced μ'' value of the CoNC/HPC-2 microsphere. Consequently, appropriate magnetic particle content and size, along with good dispersion, endow CoNC/HPC-2 with outstanding magnetic loss capability. Similarly, the magnetic loss tangent ($\tan\delta\mu$) serves as an indicator for evaluating magnetic loss capability. In this context, CoNC/HPC-2 exhibits the highest $\tan\delta\mu$ value among the EM composite microspheres (Figure 4.5d), aligning with the magnetic parameter results and further confirming its robust magnetic loss capability. Based on the results of Figure 4.5e, the CoNC/HPC-2 microsphere exhibits high saturation magnetization. Calculating a constant value for $C = \mu''(\mu')^{-2}f^{-1}$ indicates the presence of Eddy current loss in a specific frequency range. Notably, Figure 4.5f highlights the occurrence of Eddy current loss at high frequencies (12–18 GHz).

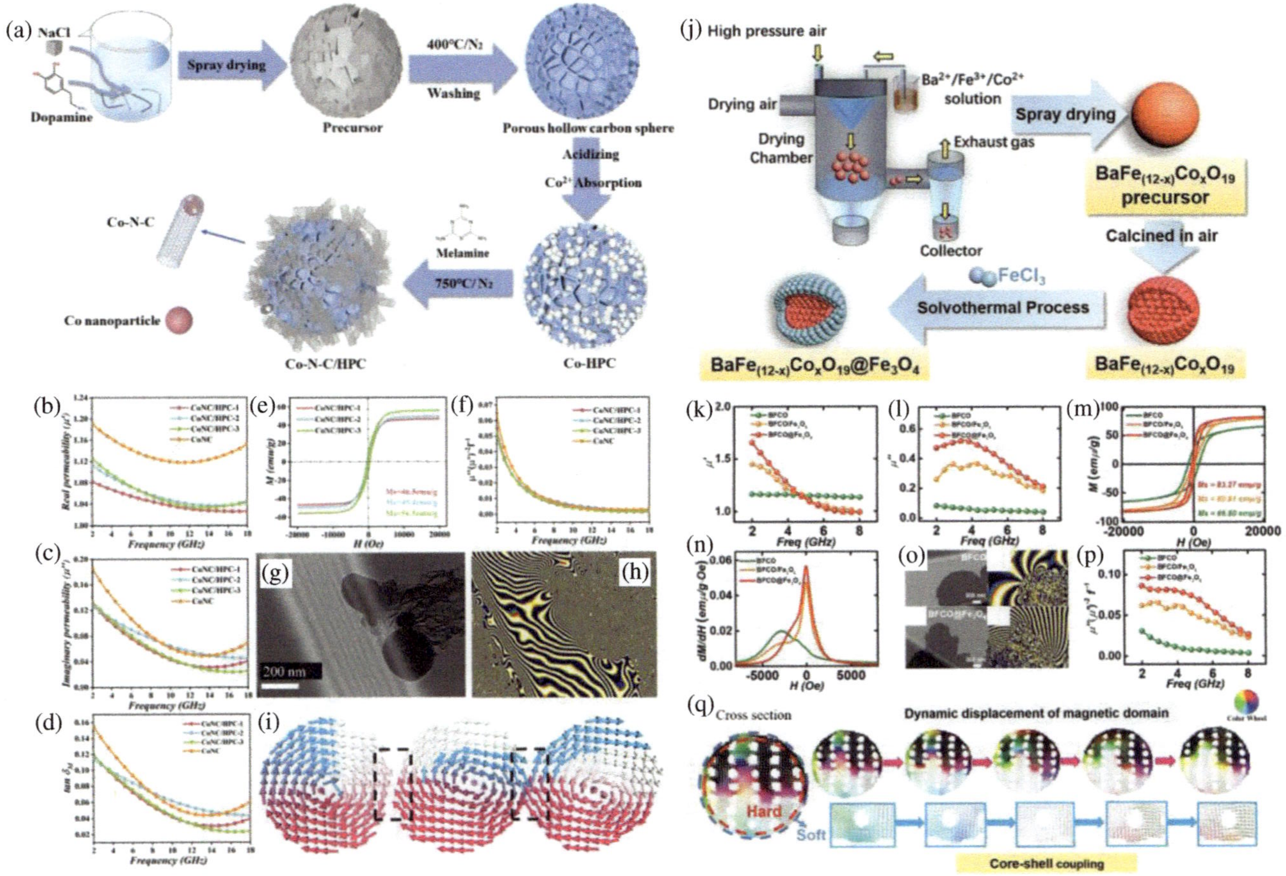

Figure 4.5 (a) Schematic illustration describing the preparation process of the CoNC/HPC microsphere. (b) Real and (c) imaginary permittivity of CoNC/HPC samples. (d) Tangent of magnetic loss of samples. (e) Room-temperature hysteresis loops of CoNC/HPC microspheres. (f) Frequency dependence of $\mu''(\mu')^{-2}f^{-1}$ for CoNC/HPC microspheres. (g, h) Holograms images and magnetic flux lines of CoNC/HPC-2. (i) The simulated magnetic moments of multiple Co NPs under 16 GHz EM field through Mumax3 simulation. (j) Fabrication process for $BaFe_{(12-x)}Co_xO_{19}@Fe_3O_4$ microsphere. (k) Real part of permeability, (l) imaginary part of permeability, (m) M–H curves, (n) SFD plot, (o) electron hologram and corresponding magnetic lines, electron hologram, and corresponding magnetic lines of flux for $BaFe_{11.6}Co_{0.4}O_{19}$. (p) Value of $\mu''(\mu')^{-2}f^{-1}$. (q) The distribution of magnetic moments of the cross-section of an individual hard/soft core–shell magnetic microsphere under 6.4 GHz EM field obtained from the Mumax3 simulation. The color wheel shows the direction of magnetic moments.

The magnetic loss capability of the constructed CoNC/HPC microspheres is significantly influenced by the dispersion degree of Co particles. Within porous carbon spheres, the high specific surface area provides abundant active sites for anchoring magnetic particles, enhancing dispersion and loading density. This results in the formation of a 3D magnetic coupling network, further facilitating the dissipation of EMWs. When EMWs penetrate the microsphere, they induce an alternating magnetic field that causes the magnetic moment of Co particles within the sphere to oscillate periodically. This oscillation extends to the surrounding magnetic particles, leading to friction and energy dissipation. Consequently, a mutual inductive coupling interaction occurs between adjacent nanomagnets. To elucidate the magnetic behavior, off-axis electron holography was employed to analyze the distribution of magnetic force lines in CoNC/HPC-2. The reconstructed electron holographic phase images demonstrated that, despite being enveloped by graphitized carbon (Figure 4.5g,h), Co particles emitted high-density magnetic flux lines, confirming the effectiveness of the limited dispersion strategy. The Co particles dispersed by the CNT matrix exhibited a noticeable magnetic coupling effect, which is advantageous for further dissipating EMWs. For a deeper understanding of the magnetic behavior of CoNC/HPC-2 microspheres, micromagnetic simulations were utilized to unravel the underlying magnetic response and coupling mechanisms. Specifically, Co particle models with sizes of 80 nm were constructed to analyze variations in magnetic moments when subjected to alternating magnetic fields at 16.0 GHz. The occurrence of magnetic vortices inside the sample led to the formation, migration, and disappearance of these vortices with changes in the external magnetic field. Additionally, due to the interplay between internal and external magnetic fields, magnetic moments underwent slight vibrations that are conducive to dissipating EMWs. As illustrated in Figure 4.5i, the simulated magnetic response of multiple Co particles indicates that the magnetic moments at the edges of adjacent particles align either in the same or opposite direction, suggesting the presence of magnetic coupling.

4.2.4 Enhanced Magnetic Coupling Through Core–Shell Structural Design

Ferrite, such as $BaFe_{(12-x)}Co_xO_{19}$, represents a promising magnetic candidate for constructing absorbers effective at lower frequencies due to its robust magnetic loss properties and superior impedance matching compared to magnetic metals [48]. Among ferrites, M-type hexaferrite, in particular, stands out as a high-performing hard ferrite material, possessing significant uniaxial anisotropy and a high ferromagnetic resonance frequency. The introduction of heteroatoms like Ca^{2+}, Co^{2+}, Zr^{4+}, and Al^{3+} into M-type hexaferrite offers a means to finely tune its working frequency and enhance its magnetic/dielectric characteristics [49, 50]. On the other hand, magnetite (Fe_3O_4) nanomaterial is known for its potent magnetic loss capabilities and is widely acknowledged as an excellent absorber, exhibiting natural resonance in the 1–4 GHz range [51]. Given these distinctive properties, M-type hexaferrite and magnetite were selected as the advanced hard and soft magnetic components, respectively, forming a unique and efficient magnetic composite in the core–shell structure. Yuan et al. [39] fabricated a core–shell structured composite material, namely $BaFe_{(12-x)}Co_xO_{19}@Fe_3O_4$, featuring a hard/soft magnetic combination with abundant heterointerfaces (Figure 4.5j). This composite was achieved through a straightforward process involving spray-drying followed by a solvothermal approach. The hard magnetic core consists of $BaFe_{(12-x)}Co_xO_{19}$, while the soft magnetic shell is composed of Fe_3O_4.

The complex permeability (μ', μ'') of the samples is illustrated in Figure 4.5k,l. The real part of permeability (μ') is associated with the energy storage capability, while the imaginary part (μ'') corresponds to magnetic loss. The core–shell BFCO@Fe_3O_4 composite demonstrates both high μ' and μ'', indicating a high magnetic permeability (μ_r). This outcome suggests that BFCO@Fe_3O_4 possesses the strongest energy storage capability and magnetic loss among the examined samples. Due to Snoek's limitation, the initial permeability in most magnetic materials experiences a sharp decline as the frequency increases above GHz, primarily due to the restricted saturation magnetization (Ms). Breaking the Snoek's limitation, and consequently achieving a higher value of complex permeability, is crucial. Analyzing the magnetization versus applied magnetic field curves (M–H curves) of the three samples (Figure 4.5m), notably, the observed Ms in BFCO@Fe_3O_4 (83.27 emu g^{-1}) is higher than that in BFCO/Fe_3O_4 (80.81 emu g^{-1}). This higher Ms indicates that modifying the saturation magnetization value of the material, specifically through the core–shell combination of BFCO and Fe_3O_4, successfully overcomes the Snoek's limitation of the BFCO and Fe_3O_4 composite. Consequently, this modification promotes both energy storage and magnetic loss ability in the resulting composite.

Particularly, for the two hard/soft composites, considering the identical content of magnetic components, the elevated Ms value of BFCO@Fe_3O_4 can be attributed to its distinctive core–shell structure. As indicated in the literature, an increase in Ms can be indicative of an improved degree of exchange coupling interaction within the magnetic system. The exchange coupling degrees of the samples can be assessed through the switching field distribution (SFD) curve, which represents the first derivative of the demagnetization curve (see Figure 4.5n). A single peak around zero external fields indicates a synchronous magnetic switching process. The stronger intensity of the peak signifies a higher degree of exchange coupling induced between the magnetic phases. It is noticeable that, in addition to the main peak, a distinct plateau appears in the curve of BFCO/Fe_3O_4, suggesting a decoupling behavior in the material. Moreover, no plateau is observed in BFCO/Fe_3O_4, and the intensity of the peak in the core–shell composite is higher than that of BFCO/Fe_3O_4, indicating a stronger exchange coupling interaction. Combining the previous analyses, the improved Ms in the core–shell structure arises from its robust exchange coupling. To characterize the magnetic field distribution around the microsphere, off-axis electron holography was employed. The electron holographic reconstructions (Figure 4.5o) vividly illustrate the strong magnetism of the core–shell BFCO@Fe_3O_4. These holographic reconstructions visualize the stray magnetic field outside the microsphere through Fourier and inverse Fourier transforms. From the results, several bold and clear magnetic stray-flux lines emanate from the polar point of the hexaferrite microsphere, leading to a sparse distribution around the sphere, indicative of weak magnetism. In contrast, for the BFCO@Fe_3O_4 composite, there is a noticeable increase in the density of magnetic flux lines radiating from the microsphere. Thin and closely spaced magnetic stray-flux lines extend straight out of the observation region with a scattered spatial distribution, indicating strong magnetism. This outcome suggests that Fe_3O_4 NPs adhered to the surface of the BFCO microsphere enhance the magnetism of the original hexaferrite microsphere. In Figure 4.5p, the $\mu''(\mu')^{-2}f^{-1}$ values of all composite samples exhibit a decrease with the increase in frequency, and no constants are observed in the 2–8 GHz range. Consequently, magnetic loss from Eddy currents can be excluded. Moreover, micromagnetic simulations were conducted to elucidate the magnetization process of the product. As illustrated in Figure 4.5q, magnetic moments in both hard and soft phases flip synchronously, indicating a notable core–shell magnetic coupling. Within the porous microsphere, magnetic vortices form and shift, vanish, and reappear frequently in response to changes in the external field. This behavior corresponds to the mechanism of domain wall migration, contributing to the observed magnetic loss. The micromagnetic simulations provide valuable insights into the dynamic magnetization behavior of the core–shell microsphere, supporting the experimental findings and contributing to a comprehensive understanding of the composite material's EM properties.

4.3 Conclusion and Outlook

In conclusion, we have detailed the loss mechanisms of some representative new EMA materials, including dielectric and magnetic materials. Though tremendous advances have been achieved in the EM wave absorption field, there are still many research aspects that should be taken into consideration. In future research, we will emphasize the mechanism of multi-functionalization, low-frequency EMA materials, and multi-physical field regulation. These outlooks will provide strong guidance and support for the in-depth development of the field of EMA materials.

(1) Multifunctional absorbing materials have become an urgent requirement for the next generation of EMW absorbers. In addition to being lightweight, thin, possessing a wide EAB, and strong absorption capabilities, practical applications demand features such as flame resistance, self-cleaning waterproofing, anti-oxidation corrosion resistance, flexible wear resistance, near-infrared absorption, self-healing, and optical transparency. For instance, traditional microwave-absorbing materials typically have hydrophilic surfaces, making them susceptible to impairment in environments characterized by heavy rain, ice, humidity, and dust. Such contaminants significantly impact normal usage and can lead to the deterioration or loss of absorption performance during operation, thereby greatly reducing their lifespan. Consequently, superhydrophobic absorbing materials, combining outstanding super-hydrophobicity and microwave absorption properties, serve the purpose of self-cleaning, waterproofing, and stain resistance. They effectively address these issues, prolonging the material's longevity and demonstrating wide-ranging prospects for application. Currently, there are two main approaches to prepare superhydrophobic absorbing materials. The first involves constructing micro-nano rough structures using absorbing materials and subsequently modifying them with low-surface energy substances such as perfluorooctanoic acid (PFOS) and polydimethylsiloxane (PDMS). The second approach entails the composite construction of micro-nano rough structures by combining absorbing materials with other nanomaterials, followed by modification with low surface energy substances. These approaches offer diverse pathways for the preparation of superhydrophobic absorbing materials, laying the foundation for optimizing their performance and diversifying their functionalities. Future research can further delve into these two approaches, utilizing precise design and synthesis to enhance the performance of superhydrophobic absorbing materials in the field of EMW absorption. This will contribute to expanding their application scope.

(2) Currently, research on stealth materials in the high-frequency wave range has made significant progress, but there is still a need for further exploration and development of stealth materials in the low-frequency wave range. Therefore, the investigation of absorption performance in low-frequency materials has become one of the key directions in the development

of new absorbing materials, with a particular emphasis on the microscale absorption mechanisms in the low-frequency range. This emphasis is primarily due to the challenge that, as the frequency decreases, achieving a wide effective absorption bandwidth at relatively thin thicknesses requires absorbing materials to possess higher magnetic permeability. Common materials include ferrites, carbonyl iron powders, and metal alloys. Although improvements in magnetic permeability have been achieved through material modifications such as core–shell structure design, lamellar modifications, and the introduction of organic groups, these materials still fall short of meeting the goal of broadband absorption. Consequently, enhancing the intrinsic magnetic permeability of materials to improve low-frequency absorption performance becomes a crucial research direction for the future.

(3) EMW absorption is a passive process, and the absorption performance of EMWs is influenced by the intrinsic properties of materials (defects, interfaces, charge carrier transport capabilities, dipoles, magnetic crystal anisotropy, etc.) and the control of the material's macroscopic structure. Current research on EMW absorption mainly focuses on manipulating the intrinsic properties of materials through various means, thereby regulating microscale absorption mechanisms. This, in turn, allows for the control of the material's EM parameters within suitable ranges, achieving broadband absorption. However, the current degree of control over the intrinsic properties of materials is relatively low, leading to a somewhat monotonous change in control mechanisms. This limitation restricts the expansion of the effective absorption bandwidth of EMW absorption materials. The intrinsic properties of materials are influenced by the crystallization process, and applying physical fields during material synthesis can alter mass and heat transfer processes in the crystallization, causing a drastic change in the material's intrinsic properties and inducing the emergence of novel physical phenomena. Therefore, investigating the control mechanisms of various physical fields, such as ultrasonic fields, electric fields, magnetic fields, and so on, on properties such as content, types, distribution of defects, bandgap width, charge carrier transport capabilities, particle aggregation status, and so on, is of significant importance. This exploration can elucidate new microscopic EM loss mechanisms under the influence of multiple physical fields.

It is worth noting that the development of new EMA materials is significant but still a great challenge. Nevertheless, we believe that further studies of microscopic loss mechanisms will pay great dividends and help to accomplish the design of novel EMW absorbers.

References

1 Li, G., Blake, G.R., and Palstra, T.T. (2017). Vacancies in functional materials for clean energy storage and harvesting: the perfect imperfection. *Chem. Soc. Rev.* 46: 1693.

2 Xu, L., Jiang, Q., Xiao, Z. et al. (2016). Plasma-engraved Co_3O_4 nanosheets with oxygen vacancies and high surface area for the oxygen evolution reaction. *Angew. Chem. Int. Ed. Engl.* 55: 5277.

3 Liu, J., Zhang, L., and Wu, H. (2022). Enhancing the low/middle-frequency electromagnetic wave absorption of metal sulfides through F-regulation engineering. *Adv. Funct. Mater.* 32: 2110496.

4 Fang, L., Li, W., Guan, Y. et al. (2017). Tuning unique peapod-like $Co(S_xSe_{1-x})_2$ nanoparticles for efficient overall water splitting. *Adv. Funct. Mater.* 27: 1701008.

5 Zhu, T., Xiao, Y., Ren, Y. et al. (2021). Unusual formation of $CoS_{0.61}Se_{0.25}$ anion solid solution with sulfur defects to promote electrocatalytic water reduction. *ACS Appl. Energy Mater.* 4: 2976.

6 Liu, J., Zhang, L., and Wu, H. (2022). Anion-doping-induced vacancy engineering of cobalt sulfoselenide for boosting electromagnetic wave absorption. *Adv. Funct. Mater.* 32: 2200544.

7 Lv, H., Wu, C., Tang, J. et al. (2021). Two-dimensional SnO/SnO_2 heterojunctions for electromagnetic wave absorption. *Chem. Eng. J.* 411: 128445.

8 Du, Z., Gong, Y., Zhao, H. et al. (2021). Unveiling the interface structure of the exsolved Co-Fe alloy nanoparticles from double perovskite and its application in solid oxide fuel cells. *ACS Appl. Mater. Inter.* 13: 3287.

9 Zhang, Q., Wang, C., Zhang, H. et al. (2020). Designing ultrahard nanostructured diamond through internal defects and interface engineering at different length scales. *Carbon* 170: 394.

10 Gao, Z., Lan, D., Zhang, L., and Wu, H. (2021). Simultaneous manipulation of interfacial and defects polarization toward Zn/Co phase and ion hybrids for electromagnetic wave absorption. *Adv. Funct. Mater.* 31: 2106677.

11 Xin, Y., Yu, K., Zhang, L. et al. (2021). Copper-based plasmonic catalysis: recent advances and future perspectives. *Adv. Mater.* 33: 2008145.

12 Qin, M., Zhang, L., Zhao, X., and Wu, H. (2021). Lightweight Ni foam-based ultra-broadband electromagnetic wave absorber. *Adv. Funct. Mater.* 31: 2103436.

13 Ji, C., Liu, Y., Xu, J. et al. (2022). Enhanced microwave absorption properties of biomass-derived carbon decorated with transition metal alloy at improved graphitization degree. *J. Alloy. Compd.* 890: 161834.

14 Zhao, H., Xu, X., Fan, D. et al. (2021). Anchoring porous carbon nanoparticles on carbon nanotubes as a high-performance composite with a unique core-sheath structure for electromagnetic pollution precaution. *J. Mater. Chem. A* 9: 22489.

15 Liang, H., Zhang, L., and Wu, H. (2022). Exploration of twin-modified grain boundary engineering in metallic copper predominated electromagnetic wave absorber. *Small* 18: 2203620.

16 Qin, M., Zhang, L., and Wu, H. (2022). Dielectric loss mechanism in electromagnetic wave absorbing materials. *Adv. Sci.* 9: 2105553.

17 Zhao, B., Li, Y., Zeng, Q. et al. (2020). Galvanic replacement reaction involving core-shell magnetic chains and orientation-tunable microwave absorption properties. *Small* 16: 2003502.

18 Wang, A., Li, J., and Zhang, T. (2018). Heterogeneous single-atom catalysis. *Nat. Rev. Chem.* 2: 65.

19 Nie, L., Mei, D., Xiong, H. et al. (2017). Activation of surface lattice oxygen in single-atom Pt/CeO_2 for low-temperature CO oxidation. *Science* 358: 1419.

20 Liang, H., Chen, G., Liu, D. et al. (2022). Exploring the Ni 3d orbital unpaired electrons induced polarization loss based on Ni single-atoms model absorber. *Adv. Funct. Mater.* 9: 2105553.

21 Song, L., Duan, Y.P., Liu, J., and Pang, H. (2020). Assembled Ag-doped α-MnO_2@δ-MnO_2 nanocomposites with minimum lattice mismatch for broadband microwave absorption. *Compos. Part B: Eng.* 199: 108318.

22 Song, L., Duan, Y.P., Zhang, Y., and Wang, T. (2017). Promoting defect formation and microwave loss properties in δ-MnO_2 via Co doping: a first-principles study. *Comp. Mater. Sci.* 138: 288.

23 Gai, Y.Q., Li, J.B., Li, S.S. et al. (2009). Design of narrow-gap TiO_2: a passivated codoping approach for enhanced photoelectrochemical activity. *Phys. Rev. Lett.* 102: 036402.

24 Wang, D., Zou, Y.H., Wen, S.C., and Fan, D.Y. (2009). A passivated codoping approach to tailor the band edges of TiO_2 for efficient photocatalytic degradation of organic pollutants. *Appl. Phys. Lett.* 95: 012106.

25 Li, Y., Qing, Y., Cao, Y. et al. (2023). Positive charge holes revealed by energy band theory in multiphase Ti_xO_{2x-1} and exploration of its microscopic electromagnetic loss mechanism. *Small* 19: 2302769.

26 Li, X., Wang, L., You, W. et al. (2019). Morphology-controlled synthesis and excellent microwave absorption performance of $ZnCo_2O_4$ nanostructures via a self-assembly process of flake units. *Nanoscale* 11: 2694.

27 Qin, M., Liang, H., Zhao, X., and Wu, H. (2020). Filter paper templated one-dimensional NiO/$NiCo_2O_4$ microrod with wideband electromagnetic wave absorption capacity. *J. Colloid Interface Sci.* 566: 347.

28 Qin, M., Lan, D., Wu, G. et al. (2020). Sodium citrate assisted hydrothermal synthesis of nickel cobaltate absorbers with tunable morphology and complex dielectric parameters toward efficient electromagnetic wave absorption. *Appl. Surf. Sci.* 504: 144480.

29 Qin, M., Zhang, L., and Wu, H. (2020). Dual-template hydrothermal synthesis of multi-channel porous $NiCo_2O_4$ hollow spheres as high-performance electromagnetic wave absorber. *Appl. Surf. Sci.* 515: 146132.

30 Qin, M., Zhang, L., Zhao, X., and Wu, H. (2021). Defect induced polarization loss in multi-shelled spinel hollow spheres for electromagnetic wave absorption application. *Adv. Sci.* 8: 2004640.

31 Lv, H., Yang, Z., Wang, P.L. et al. (2018). A voltage-boosting strategy enabling a low-frequency, flexible electromagnetic wave absorption device. *Adv. Mater.* 30: 1706343.

32 Song, W.-L., Zhang, Y.-J., Zhang, K.-L. et al. (2020). Ionic conductive gels for optically manipulatable microwave stealth structures. *Adv. Sci.* 7: 1902162.

33 Ye, F., Song, Q., Zhang, Z. et al. (2018). Self-assembly core-shell graphene-bridged hollow MXenes spheres 3D foam with ultrahigh specific EM absorption performance. *Adv. Funct. Mater.* 28: 1707205.

34 Wen, B., Cao, M., Lu, M. et al. (2014). Reduced graphene oxides: light-weight and high-efficiency electromagnetic interference shielding at elevated temperatures. *Adv. Mater.* 26: 3484.

35 Liu, J., Cao, W.-Q., Jin, H.-B. et al. (2015). Enhanced permittivity and multi-region microwave absorption of nanoneedle-like ZnO in the X-band at elevated temperature. *J. Mater. Chem. C* 3: 4670.

36 Li, Y., Qin, F., Quan, L. et al. (2019). Vertical interphase enabled tunable microwave dielectric response in carbon nanocomposites. *Carbon* 153: 447.

37 Li, X., Qu, X., Xu, Z. et al. (2019). Fabrication of three-dimensional flower-like heterogeneous Fe_3O_4/Fe particles with tunable chemical composition and microwave absorption performance. *ACS Appl. Mater. Inter.* 11: 19267–19276.

38 Xie, X.B., Ni, C., Lin, Z.H. et al. (2020). Phase and morphology evolution of high dielectric CoO/Co_3O_4 particles with Co_3O_4 nanoneedles on surface for excellent microwave absorption application. *Chem. Eng. J.* 396: 125205.

39 Yuan, M., Zhao, B., Yang, C. et al. (2022). Remarkable magnetic exchange coupling via constructing Bi-magnetic interface for broadband lower-frequency microwave absorption. *Adv. Funct. Mater.* 32: 2203161.

40 Yang, Z., You, W., Xiong, X. et al. (2022). Morphology-evolved succulent-like FeCo microarchitectures with magnetic configuration regulation for enhanced microwave absorption. *ACS Appl. Mater. Inter.* 14: 32369–32378.

41 Xiao, W., Lei, W., Gong, M. et al. (2018). Recent advances of structurally ordered intermetallic nanoparticles for electrocatalysis. *ACS Catal.* 8: 3237–3256.

42 Pan, F., Ning, M., Li, Z. et al. (2023). Sequential architecture induced strange dielectric-magnetic behaviors in ferromagnetic microwave absorber. *Adv. Funct. Mater.* 23: 2300374.

43 Liu, R., Du, G., Liao, B. et al. (2022). Simultaneous dual pyrolysis synthesis of heterostructured FeCo/C porous hollow microspheres for highly efficient microwave absorption. *J. Mater. Chem. A* 10: 1547–1559.

44 Yang, B.T., Fang, J.F., Xu, C.Y. et al. (2022). One-dimensional magnetic FeCoNi alloy toward low-frequency electromagnetic wave absorption. *Nano-Micro Lett.* 14: 170.

45 Liang, L.Y., Li, Q.M., Yan, X. et al. (2021). Multifunctional magnetic $Ti_3C_2T_x$ MXene/graphene aerogel with superior electromagnetic wave absorption performance. *ACS Nano* 15: 6622–6632.

46 Wu, Z.C., Pei, K., Xing, L.S. et al. (2019). Enhanced microwave absorption performance from magnetic coupling of magnetic nanoparticles suspended within hierarchically tubular composite. *Adv. Funct. Mater.* 29: 1901448.

47 Zhang, C., Luo, K., Liu, J. et al. (2024). Realizing optimized interfacial polarization and impedance matching with CNT-confined Co nanoparticles in hollow carbon microspheres for enhanced microwave absorption. *J. Mater. Sci. Technol.* 175: 1–9.

48 Akhtar, M.N., Yousaf, M., Lu, Y. et al. (2021). Highly efficient absorber with enhanced magnetoelectric properties based on Y, Gd, and Pr doped NMZ nanoferrites. *J. Magn. Magn. Mater.* 537: 168232.

49 Yang, Y., Wang, F., and Shao, J. (2016). Influence of stoichiometry on the phase constitution and magnetic properties of M-type Sr-Ca hexaferrites. *Optik* 127: 6096.

50 Liu, C., Xu, Q., Tang, Y. et al. (2016). Zr^{4+} doping-controlled permittivity and permeability of $BaFe_{(12-x)}Co_xO_{19}$ and the extraordinary EM absorption power in the millimeter wavelength frequency range. *J. Mater. Chem. C* 4: 9532.

51 Chen, Y.H., Huang, Z.H., Lu, M.M. et al. (2015). 3D Fe_3O_4 nanocrystals decorating carbon nanotubes to tune electromagnetic properties and enhance microwave absorption capacity. *J. Mater. Chem. A* 3: 12621.

5

Bridging Mechanisms Between Micro and Macro

Geng Chen, Tao Zhang, and Yuntong Wang

MOE Key Laboratory of Material Physics and Chemistry Under Extraordinary Conditions, School of Physical Science and Technology, Northwestern Polytechnical University, Xi'an, China

CHAPTER MENU

Within the microwave spectrum, the interaction of electromagnetic (EM) waves with absorptive materials facilitates the transformation of EM energy into thermal or alternative energy forms. This transformation is mediated by a spectrum of mechanisms, including but not limited to defect-induced polarization, interfacial polarization, conduction losses, and eddy current losses, along with exchange and natural resonances. The efficacy and characteristics of these energy dissipation pathways are closely tied to the materials' microscopic attributes, such as structural defects, interfacial properties, electrical conductivity, dipolar characteristics, and the degree of saturation magnetization. A focal point in this field is the establishment of a connective mechanism bridging the microscopic and macroscopic dimensions. This entails a detailed correlation between the microscopic attributes of materials and macroscopic electromagnetic wave (EMW) absorption parameters, including optimal thickness, effective absorption bandwidth (EAB), matching frequencies, and minimum reflection loss. This correlation aims to unravel the complex interplay between microscopic factors and EM loss mechanisms, paving the way for a novel physical model. The conceptualization of this bridge mechanism between micro and macro levels promises to enhance the comprehension of the underlying coupling processes. It seeks to delineate a sequence commencing with the regulation of microscopic factors, progressing through the operational mechanisms and culminating in optimized performance. Such insights are instrumental for the controlled modulation of EMW absorption characteristics, representing a considerable advancement in the domains of materials science and engineering.

Electromagnetic Wave Absorbing Materials: Fundamentals and Applications, First Edition.
Edited by Hongjing Wu, Jun Luo, and Meiyin Yang.

5.1 Introduction to Micro Factors

5.1.1 Defects

In the realm of materials science, the term "defects" is attributed to the deviations observed within the crystal lattice of a material from its ideal or flawless structural arrangement. These deviations, or imperfections, play a pivotal role in modulating a wide array of the material's physical and chemical attributes, encompassing electrical, magnetic, mechanical, and optical properties. The classification of these defects is broadly categorized based on their dimensional scale and intrinsic characteristics. First, point defects, which are the most diminutive form of imperfections, predominantly influence either a solitary atom or a minuscule cluster of atoms. These defects may emerge as atomic-scale vacancies, characterized by the absence of an atom at a lattice site, impurity atoms, which are foreign atoms situated at lattice points, or interstitial atoms, located in the interstices of the lattice points. Second, line defects, or dislocations, manifest as linear discontinuities within the lattice, interrupting the regular atomic arrangement of the crystal. These dislocations are further subcategorized into screw dislocations and edge dislocations. Each type imparts a substantial effect on the material's mechanical properties, notably in terms of strength and ductility. Lastly, planar defects or interfaces are identified at the junctures where two distinct crystal structures or phases converge. This category includes grain boundaries, which demarcate the segregation of crystal grains, and phase boundaries or surface defects, all indicative of discontinuities in the crystal structure.

In materials science, intrinsic imperfections such as vacancies, impurities, and dislocations can engender localized electric dipoles. As EMW traverse these materials, these dipoles undergo polarization under the influence of alternating EM fields. This polarization process involves the absorption of energy, thereby inducing losses in the traversing EMW. Defect engineering, a predominant focus in the field of EMW-absorbing materials, primarily encompasses point defects, including oxygen vacancies, sulfur vacancies, and the regulation of heteroatom doping. Oxygen vacancies, defined as the absence of oxygen atoms in a lattice structure, are a common phenomenon in both crystalline and amorphous materials, particularly in oxides. The proportion and presence of such vacancies can be discerned through deconvoluted high-resolution O 2p X-ray photoelectron spectroscopy (XPS) spectra. It is pertinent to note that XPS is a surface-sensitive technique. Given that EM wavelengths in the range of 2–18 GHz fall within the centimeter wave category, capable of penetrating absorbers, surface detection methods like XPS might not fully reveal the concentration of oxygen vacancies and the associated polarization intensity they induce. Integrating this approach with other prevalent techniques, such as high-resolution transmission electron microscopy (HRTEM), photoluminescence (PL) spectroscopy, scanning electron microscopy (SEM), X-ray absorption spectroscopy (XAS), positron annihilation spectroscopy (PAS), and electron paramagnetic resonance/electron spin resonance (EPR/ESR), could yield more precise results. Similarly, defect engineering involving sulfur vacancies and heteroatom doping can be investigated using analogous methodologies.

Defect-induced polarization, a phenomenon stemming from defect sites in absorptive materials, can be effectively augmented by introducing additional defect points. This is achieved through techniques such as thermal treatment, heteroatom doping (e.g. F, N, P, B doping), and reduction processes, which enhance the defect sites in absorptive materials, thereby increasing the polarization losses caused by these defects. For instance, Tong and coworkers [1] demonstrated the controlled introduction of Fe/Co/Ni atoms into layered $CeOHCO_3$ dendrites via heteroatom doping. This approach enabled regulation of lattice/oxygen vacancy content, thus modulating the sample's saturation magnetization and electrical conductivity to optimize EMW absorption performance (Figure 5.1a). Similarly, Ji and coworkers [2] optimized the EMW absorption capabilities by adjusting the polarization loss in semiconductors through oxygen vacancy regulation and modulating the electrical loss in carbon through nitrogen doping. This was accomplished by combining these two components (Figure 5.1b). Wu and coworkers [3] employed fluorine ion modulation engineering to simultaneously control multiple factors, including morphology, composition, interfaces, defects, and conductivity, thus optimizing EMW absorption performance (Figure 5.1c). While these studies substantiate that defect-induced polarization is an effective form of dissipating EMW, there remain certain challenges. For instance, defect-induced polarization is often equated with dipole polarization, and the role of defects at interfaces is generally attributed to mere interfacial polarization. Moreover, due to the intrinsic nature of the defects, their modulation also alters the material's conductivity, complicating the isolation of their sole contribution to EMW absorption. In other words, discerning the individual contributions of each mechanism to EMW absorption remains challenging.

Currently, elucidating the individual contributions of distinct loss mechanisms plays a crucial role in advancing the EMW absorption properties of materials. Wu and coworkers [4] provided compelling evidence regarding the paramount role of lattice defects in modulating the EMW absorption performance, utilizing a distinctive system comprising multilayer hollow spinel structures (Figure 5.1d). Furthermore, the current research on the EMW absorption by point defects primarily focuses on the effect of point defect content, with limited exploration of the impact of defect types and distributions. However, the type and distribution of defects also significantly influence the EMW absorption performance of materials. For instance, by controlling the types of anions and the formation of anionic and cationic vacancies (Figure 5.1e, f), Wu and coworkers [5, 6] have demonstrated the importance of defect types in modulating the EMW absorption performance. Future research should focus on controlling the types (point defects, line defects, surface defects), content, and distribution of defects and investigate their correlation with macroscopic parameters such as matching thickness, absorption bandwidth, matching frequency, and loss peak in order to establish a novel physical model for EMW absorption from the microscale to the macroscale.

5.1.2 Interfaces

In the realm of science and engineering, interfaces typically refer to the interfacial or contact surfaces between two or more different materials. These materials can be solids, liquids, or gases, and their interfaces serve as the sites for their interaction and exchange of energy, mass, charge, or other physical or chemical properties. Interfaces hold significant importance in the field of materials science, as they can influence the performance and behavior of materials. For instance, interfaces in solid materials can impact mechanical properties, electronic transport properties, chemical reactions, and interfacial diffusion. Interfaces between liquids and gases can affect surface tension, droplet formation, and evaporation phenomena. In the realm of EMW absorption, the modulation of a material's EMW absorption performance primarily relies on interfacial polarization.

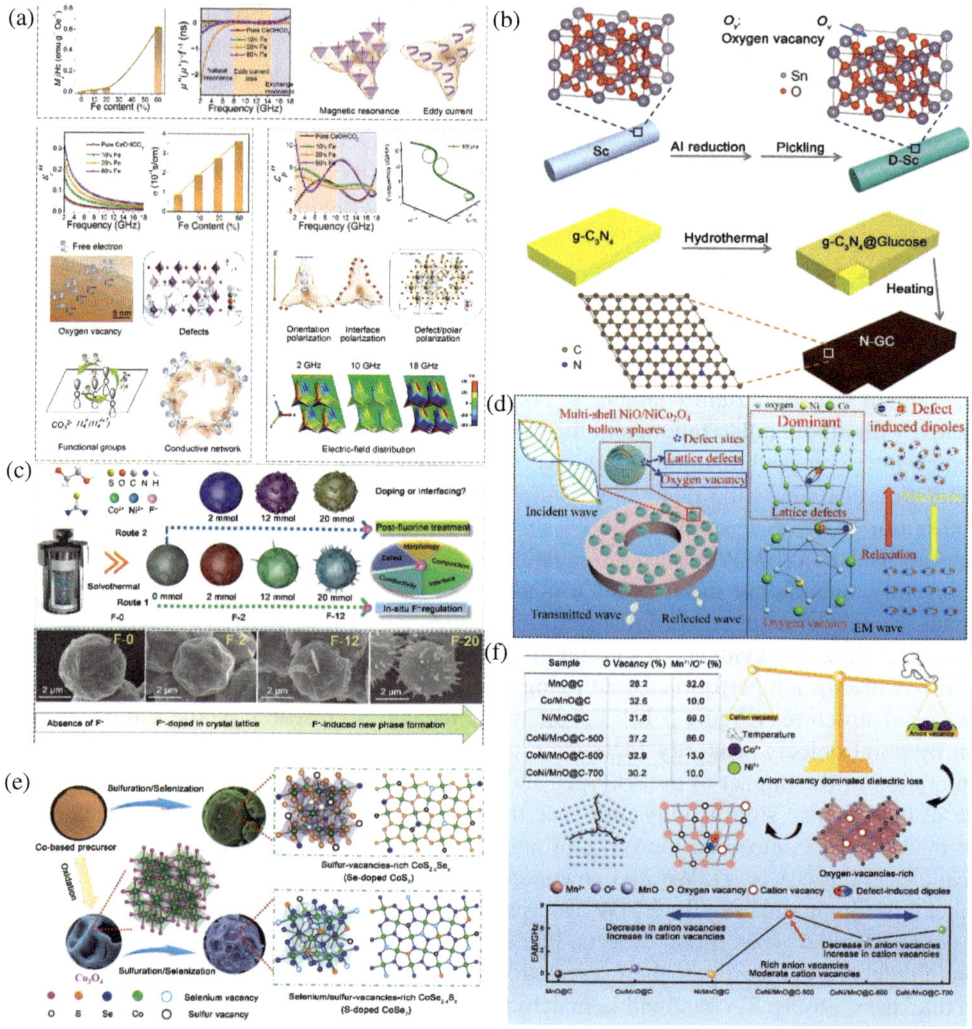

Figure 5.1 (a) EM loss mechanisms present in Fe-doped $CeOHCO_3$-layered dendritic crystal system. Source: Wang et al. [1]/with permission of Elsevier; schematic diagram of (b) the introduction of oxygen vacancies in tin oxide, and nitrogen doping in carbon. Source: Quan et al. [2]/with permission of John Wiley & Sons. (c) The F-ion modulation strategy. Source: Liu et al. [3]/with permission of John Wiley & Sons. (d) The lattice defect-induced polarization. Source: Qin et al. [4]/with permission of John Wiley & Sons. (e) The modulation of the anionic vacancies. Source: Liu et al. [5]/with permission of John Wiley & Sons. (f) The EM loss mechanism for the simultaneous modulation of anionic and cationic vacancies. Source: Liu et al. [6]/with permission of John Wiley & Sons.

Interfacial polarization refers to the phenomenon where the charge distribution at the interface between two different materials becomes asymmetric under the influence of an external electric field. This polarization is caused by the differences in the electrical properties of the materials and the redistribution of charges at the interface. To date, interfacial polarization in EMW-absorbing materials is mainly constructed in two forms, namely homogenous or heterogeneous interfaces. It is believed that the differences in dielectric properties (the ability to retain charge) and the conductivity of different components are the reasons for charge redistribution and subsequent interfacial polarization. The strength of this interfacial polarization is generally related to two factors: the number of interfaces and the conductivity differences of the materials constituting the interfaces. The difference in the number of interfaces is typically characterized by the intensity of TEM, SEM, XRD, EDS spectra, and so on. The conductivity differences of the materials constituting the interfaces can be evaluated through means such as work function and first-principles calculations. For example, Lv et al. [7] reported the homogeneous and heterogeneous interfaces in SnO and SnO_2 samples, as well as the corresponding polarization effects. The samples exhibited both uniform interfaces (SnO/SnO and SnO_2/SnO_2 interfaces) and nonuniform interfaces (SnO/SnO_2 interface). Based on the results, they found that compared to the heterogeneous interface, the homogeneous interface could not provide a strong interface polarization effect, as calculated by the work function in Figure 5.2a. The work functions of the SnO lattice and SnO_2 lattice in the homogeneous interface were 0.338 and 0.534 eV, respectively. The work function difference between SnO and the SnO_2 lattice was 1.986 eV, which is much higher than that of the homogeneous interface. Therefore, in most cases, the nonuniform interface and the resulting stronger interfacial polarization are of greater interest. Nonuniform interfaces can be easily obtained through the coupling of different materials. The study of composite EMW absorbers has been extensively reported, with composite materials composed of magnetic materials and highly conductive materials being typical representatives of such absorbers. The magnetic component can provide magnetic loss, while the highly conductive component will provide dielectric loss (conduction loss), combined with the interfacial polarization provided by the heterogeneous interface between them, resulting in enhanced EMW attenuation capabilities. For example, Che and coworkers [8] successfully synthesized the Fe/Fe_3O_4@C spindle-on-tube architecture utilizing a one-step copolymerization-etching strategy. This novel EM heterostructure showcases notable capabilities in inducing robust interfacial polarization. Additionally, the highly dispersed Fe/Fe_3O_4 nanoparticles configure a multidimensional magnetic network within the spindle morphology, thereby amplifying their interaction with incident microwaves and reinforcing their magnetic loss proficiency. Consequently, the fabricated hierarchical Fe/Fe_3O_4@C microtubes manifest outstanding EMW absorption performance, boasting a maximum reflection loss of −55.4 dB and the EAB reaching 4.2 GHz (Figure 5.2b).

Furthermore, another approach to interfacial polarization is the construction of porous or hollow structures. The introduction of bubbles creates interfaces between the material and air, and the different dielectric properties of the material and air medium may lead to the existence of interfacial polarization. However, this porous structure is more suitable for materials with high conductivity, such as sulfides and carbon materials. On one hand, these materials can achieve better impedance matching due to the inherent highly conductive components being replaced by air. On the other hand, the interaction between EMWs and pores may result in multiple scattering, prolonging the transmission path of EMWs. Moreover, the possible occurrence of interfacial polarization and reduced density are important factors in the construction of high-performance absorbers. Therefore, researchers have improved the EMW performance of high dielectric materials through the construction of foam structures [10], core-shell structures [11], and aerogel structures [12]. For instance, Kong and coworkers [9] utilized a self-sacrificed strategy followed by a subsequent hydrothermal method to synthesize yolk-shell structured magnetic multicomponent nanocomposites Co_2P/CoS_2@C@

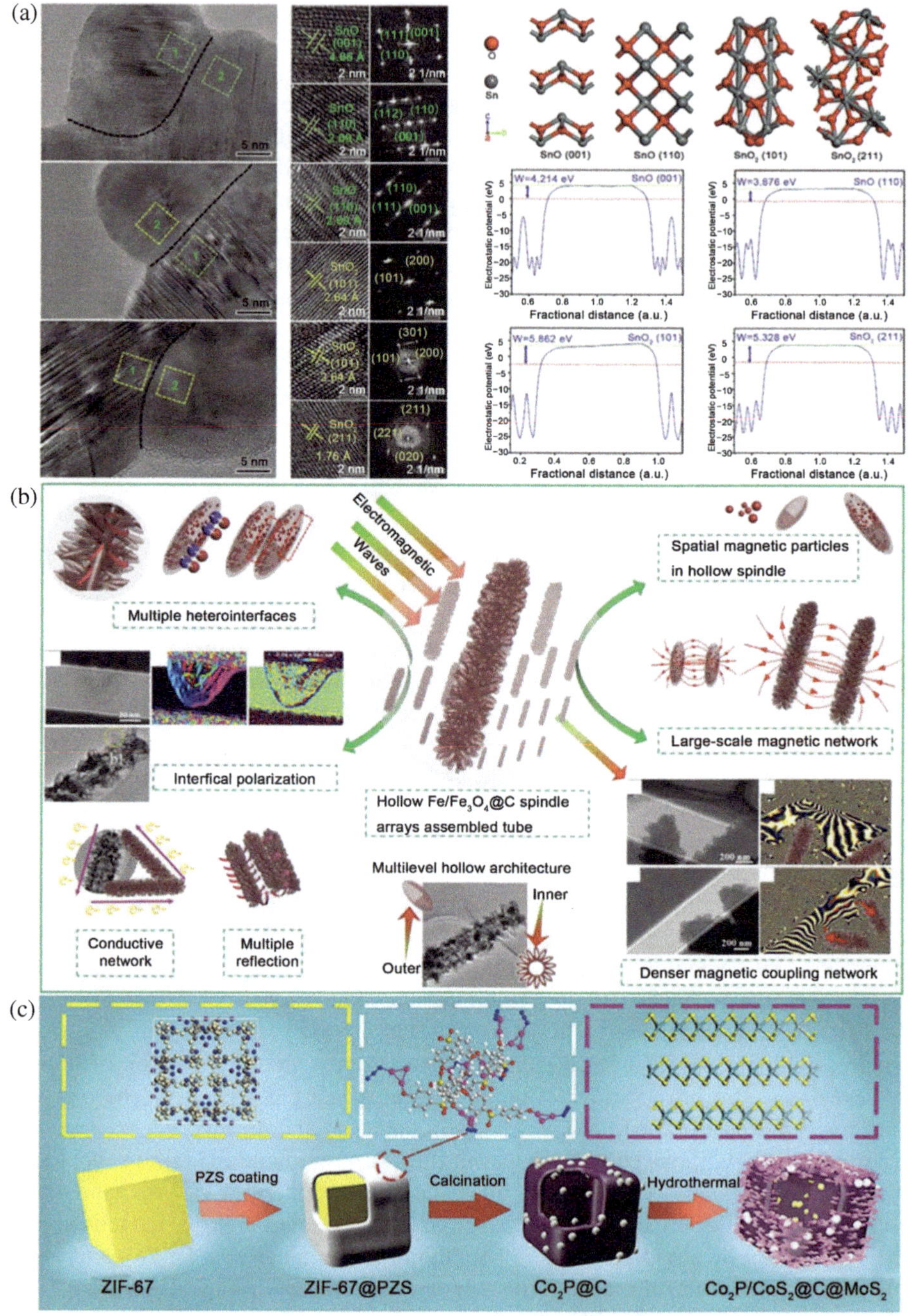

Figure 5.2 (a) The calculation of homogeneous and heterogeneous interfaces and their respective work functions in SnO and SnO_2 samples. Source: Lv et al. [7]/with permission of Elsevier. (b) A schematic representation illustrating the optimization of EMW absorption performance through interfacial polarization and magnetoelectric synergy. Source: Xu et al. [8]/with permission of Elsevier. (c) Schematic diagram for the formation of hierarchical $Co_2P/CoS_2@C@MoS_2$. Source: He et al. [9]/with permission of Springer Nature.

MoS_2, as depicted in Figure 5.2c. These nanocomposites, characterized by their distinctive structure and the synergistic interplay between outstanding dielectric and magnetic losses, showcase notable performance in EMW absorption, including strong absorption capacity, wide EAB, and thin matching thickness.

Although there has been extensive research on interfacial polarization in the field of EMW absorption, there are still some unresolved issues regarding the construction of interfaces and the understanding of interfacial polarization, for example, the optimal interface configuration design for different materials, the contribution of interfacial polarization to the process of EMW loss, and the construction of numerous heterogeneous interfaces, among others. Future research should focus on the correlation between interface types, quantities, distributions, and macroscopic parameters of material EMW absorption, such as matching thickness, absorption bandwidth, matching frequency, and peak loss. By establishing the correspondence between interface parameters and EMW absorption performance, a controllable adjustment of material EMW absorption performance can be achieved.

5.1.3 Conductivity

The conductivity of a material is derived from its ability to transport internal charge carriers and is used to describe the material's directionality toward current, measured in $S\ m^{-1}$. The conductivity of a material can be allocated to electronic conductivity and ion conductivity, representing the mobility of the two charge carriers: electrons and ions, respectively. Electronic conductivity mainly occurs in substances such as metals that have free electrons. In these substances, the valence electrons of atoms break free from the atomic nucleus and become free electrons, which can freely move within the crystal structure and actively participate in the conductive process. Ionic conductivity mainly occurs in substances such as electrolyte solutions, molten salts, and solid electrolytes. Among these substances, the dissociated positive/negative ions in electrolyte molecules or lattices can undergo directional movement under the action of an electric field, forming an electric current.

Conduction loss is a result of the conversion of incident EMWs into current within an absorber. The intrinsic resistance of the current in the absorber generates Joule heat, which contributes to the consumption of EMW energy. This type of loss is prominent in high-conductivity absorbers such as metal materials (alloy materials), carbon materials (graphene, carbon nanotubes), and conductive polymers (polyaniline, polypyrrole). It is important to note that while strong conductivity loss depends on the conductivity of the absorber, excessive conductivity can cause impedance mismatch, resulting in most of the EMWs being reflected on the material's surface rather than entering its interior. Therefore, a balance between conductivity loss and impedance matching should be maintained in the design of absorbers to achieve maximum attenuation of EMWs. Currently, two models have been established to study the conductivity loss of EMW absorbers. The first model is the electron transfer model, which focuses on the free motion behavior of electrons during transmission. For instance, when current is transmitted in one-dimensional carbon nanotubes or two-dimensional graphene generated under an external electric field, it flows along the radial direction of the nanotubes or graphene plane [13]. The second model is the electron-hopping model, which describes the electron transfer process between components, interfaces, and defects. This model requires sufficient filling of EMW absorbers to create a large conductive network. Insufficient filling results in a higher energy barrier that limits electron hopping.

However, when the conductive network is adequately filled, the energy barrier is lowered to a suitable range, allowing electrons to undergo hopping behavior and increasing microcurrents in the conductive network. In addition, the electron-hopping effect can be enhanced by raising the temperature [14].

The conductivity loss capability of absorbers can be assessed using the following equation: $\varepsilon_c'' = \sigma/2\pi f\varepsilon_0$, where σ, f, and ε_0 represent the conductivity of the material, EMW frequency, and vacuum permittivity, respectively [15]. It is important to note that, at a given EMW frequency, the contribution of the absorber's conductivity loss is directly proportional to its conductivity. Therefore, evaluating the conductivity of the material can provide an assessment of its conductivity loss. Generally, electronic conductivity can be characterized and measured through four-probe measurement methods, Hall effect measurement methods, and magnetic susceptibility measurement methods. The Hall effect measurement method is applicable to metal and semiconductor materials, whereas the magnetic susceptibility measurement method is only suitable for materials with lower conductivity. Ionic conductivity can be evaluated using electrochemical impedance spectroscopy (EIS), which is suitable for solutions or solid materials containing ions.

Increasing the content of conductive components is advantageous for the construction and optimization of conductive networks, thus enhancing the conductivity loss of absorbers. For instance, Cao and coworkers investigated the impact of reduced graphene oxide (RGO) content on the dielectric properties of RGO/SiO_2 composite materials (Figure 5.3a) [16]. When the RGO content is low, no conductive network is formed, and the imaginary permittivity is primarily determined by polarization loss. Conversely, at high RGO content levels, the contribution of conduction loss surpasses polarization loss and becomes the dominant mechanism for EMW loss. Consequently, the EMW absorption performance of the absorber can be controlled by adjusting the RGO content in the matrix. This approach resulted in effective absorption across the entire frequency range of the X-band, with a minimum reflection loss (RL_{min}) of −42 dB. Ensuring the uniform dispersion of conductive absorbers, such as RGO, at the micro-/macroscale is a prerequisite for achieving excellent conductivity loss. Pu and coworkers proposed a strategy to enhance the conductivity loss of absorbers by introducing nanopores into RGO to prevent restacking [18]. Due to the abundant conductive network and anti-stacking properties of nanoporous RGO, incident EMWs are effectively dissipated within the RGO, leading to remarkable EMW absorption performance (EAB of 5.0 GHz, RL_{min} at the frequency point of 13.2 GHz is −32.1 dB). Moreover, adjusting the microstructure of the absorber in a rational manner is beneficial for the material to possess both strong conductivity loss ability and impedance-matching characteristics. According to Clar's theory [19, 20], the presence of heteroatoms at the edges of graphene can induce localized localization of π electrons, thereby reducing the conductivity of graphene. Therefore, constructing edge structures can serve as a viable strategy for regulating the conductivity of graphene. Inspired by this, Li and coworkers achieved groundbreaking results by synthesizing edge-rich graphene (ERG) with excellent EMW absorption performance, as depicted in Figure 5.3b [17]. Theoretical calculations indicate that ERG possesses a nonzero bandgap, which is conducive to impedance matching. This unique structure contributes to strong conductivity loss, resulting in an RL_{min} of −26.7 dB for ERG and an EAB covering the entire X-band. The aforementioned findings strongly demonstrate the effectiveness of conductivity loss in regulating the EMW absorption performance of materials. However, the current research on conductivity loss has certain limitations, and future research directions should

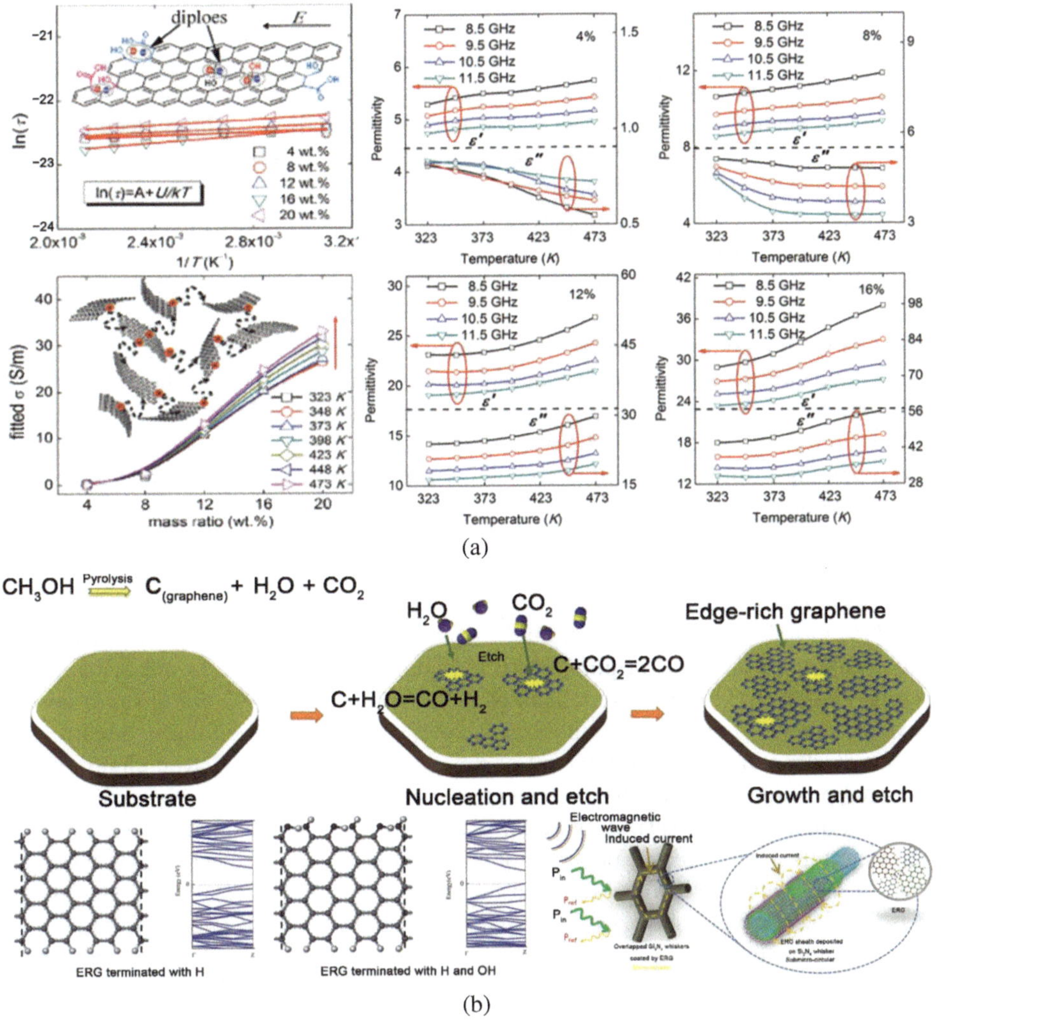

Figure 5.3 (a) The variation of relaxation time with mass ratio, alongside the alterations in conductivity and complex permittivity with temperature. Source: Wen et al. [16]/with permission of John Wiley & Sons. (b) Schematic diagram illustrating the synthesis of edge-rich graphene (ERG) and the mechanisms underlying EMW loss. Source: Ye et al. [17]/with permission of John Wiley & Sons.

focus on addressing the following aspects. First, while conductivity loss demonstrates excellent attenuation effects on high-frequency EMWs, its impact on low-frequency EMWs commonly utilized in electronic devices and detection instruments is not significant. Second, in the high-frequency range, there is a lack of quantitative theoretical models that can provide reliable guidance for the prediction and interpretation of the continuous multiple absorption peaks induced by conductivity loss [21]. Lastly, more precise synthesis strategies need to be explored to prepare higher-quality conductive absorbers, with the aim of optimizing EMW absorption performance.

5.1.4 Dipole

In a comprehensive physical system, a dipole moment typically consists of two equal and opposite charges. In material systems, a dipole moment can manifest as a molecule, wherein the positive and negative charge centers reside in different regions of the molecule. It can also manifest as an atom, where positive and negative charges are located in distinct parts of the atom. The presence of dipole moments gives rise to significant physical phenomena that influence the physicochemical properties of materials. For instance, dipoles can interact with other dipoles or external electric fields, resulting in various physical effects such as intermolecular forces, changes in the polarization of dielectric materials, and modifications in optical refractive indices.

Dipole polarization commonly refers to the relative displacement of positive and negative charges within a molecule under the influence of an external electric field. The potential difference established by the external field acts as a driving force, causing the positive charges to move toward the direction of the field and the negative charges to move in the opposite direction, resulting in polarization. Specifically, when an external electric field acts upon a molecule, the positive and negative charge centers within the molecule undergo relative displacement, leading to the creation of an electric dipole moment along the field direction. This dipole moment can be represented by a vector pointing toward the positive charge center, and its magnitude depends on the distance between the charges and their magnitudes. Dipole polarization is a temporary phenomenon, as the positive and negative charge centers of the molecule return to their original positions, and the dipole moment disappears once the external electric field is removed. This is because dipole polarization arises from the relative displacement of charges within the molecule, which is achieved through rearrangement of the electron cloud. Upon removal of the external field, the electron cloud returns to its original position, and the charge centers of the molecule restore to their initial configuration. Nowadays, the polarization effect of dipoles finds wide applications in various fields. In materials science, dipole polarization influences the optical, electrical, and thermal properties of materials. In chemical reactions, dipole polarization affects reaction rates and pathways. In biology, dipole polarization explains the interaction and structural stability of many biomolecules. In summary, dipole polarization has significant applications in numerous domains and plays a crucial role in understanding molecular structure and intermolecular interactions [22–26].

The phenomenon of dipole polarization within materials occurs when subjected to EM fields. However, due to the alternating nature of the electric field, rapid changes in the direction of the electric field may cause the dipoles to lag behind the external field during the reorientation process, leading to polarization relaxation. The polarization relaxation accompanying the frequency of the alternating electric field is effectively reflected in the complex dielectric constant of the material, and the relaxation phenomenon of dipole polarization can be further described using the Debye relaxation theory. The dispersion and absorption effects in a sufficient number of liquids and dielectrics can theoretically be expressed by the following formula [27]:

$$\varepsilon - \varepsilon_\infty = \frac{\varepsilon_0 - \varepsilon_\infty}{1 + (i\omega\tau_0)^{1-\alpha}} \tag{5.1}$$

The symbol ε represents the complex dielectric constant, where ε_0 and $\varepsilon\infty$ represent the dielectric constants at frequencies approaching zero and infinity, respectively. τ_0 denotes the generalized relaxation time, and $\omega = 2\pi f$ is the angular frequency, with α being a constant ranging from 0 to 1. Equation (5.1) requires the trajectory of the dielectric constant on the complex plane to be an arc, with its endpoints on the real axis and its center below the real axis. According to the Debye relaxation theory, the orientation of polar molecules in alternating currents is influenced by both particle thermal motion and molecular interactions, with the latter being represented by the concept of viscous damping. The molecules are considered spherical entities within a macroscopically viscous continuous medium, and when α approaches 0, the equation can be expressed as [27]:

$$\varepsilon' - \varepsilon_\infty = \frac{\varepsilon_0 - \varepsilon_\infty}{1 + (\omega\tau_0)^2} \tag{5.2}$$

$$\varepsilon''_{\mathrm{p}} = \frac{(\varepsilon_0 - \varepsilon_\infty)\omega\tau_0}{(\omega\tau_0)^2} \tag{5.3}$$

The symbol $\varepsilon''_{\mathrm{p}}$ represents the variation of the imaginary part of the dielectric constant caused by polarization. From the aforementioned Eqs. (5.2) and (5.3), it can be observed that the real part of the dielectric constant decreases with increasing frequency, while the imaginary part exhibits a prominent relaxation peak at $\omega\tau_0 = 1$. This process enables effective dissipation of EMWs by the material, converting EM energy directly into internal energy. Therefore, materials that absorb EMWs, including those containing defects, interfaces, single atoms, polar molecules, or polar groups, undergo dipole polarization and relaxation processes under the influence of alternating EM fields, consuming a significant amount of EMW energy and playing a crucial role in EMW absorption.

Due to the unique dielectric response characteristics and strong EMW absorption capability of dipole polarization, controlling the types and quantities of dipoles within the material has become an effective approach to enhance the EMW absorption capacity of materials. Yang et al. [28] utilized a high-temperature carbonization method to modify the microstructure of $C@MnO_2$ composite material shells, and the degree of graphitization of carbon cores, thereby controlling the number of dipole centers at the interface between the graphitized region and the disordered region, as well as the dipole centers induced by defects in the disordered region. This enabled the adjustment of the polarization relaxation intensity of the composite material, further regulating its EM parameters, optimizing impedance-matching characteristics, and achieving high-performance wave absorption capabilities. Song et al. [29] confirmed that the introduction of oxygen vacancies in $SiC@SiO_2$ nanofibers during the oxidation process, as well as the introduction of dislocations during the fiber formation process, led to a nonuniform distribution of charges, inducing dipole polarization losses, and enhancing the EMW absorption performance of the material (Figure 5.4a). Zhang et al. [15] introduced polar groups, polar surface functional groups (such as N-doping agents, C—O, etc.), and defects into N-doped C anchored with single Mn atoms, constructing various forms of dipoles. This novel approach enabled the creation of diverse dipole configurations, resulting in enhanced dipole polarization losses and improved EMW absorption capabilities of the material (Figure 5.4b).

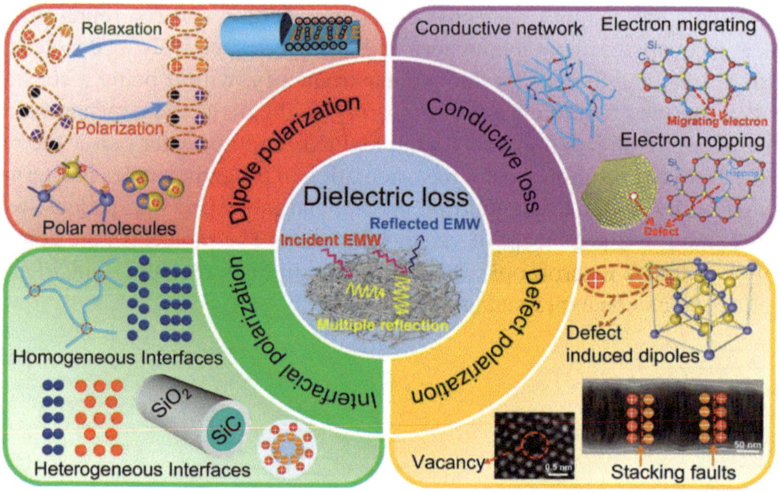

(a)

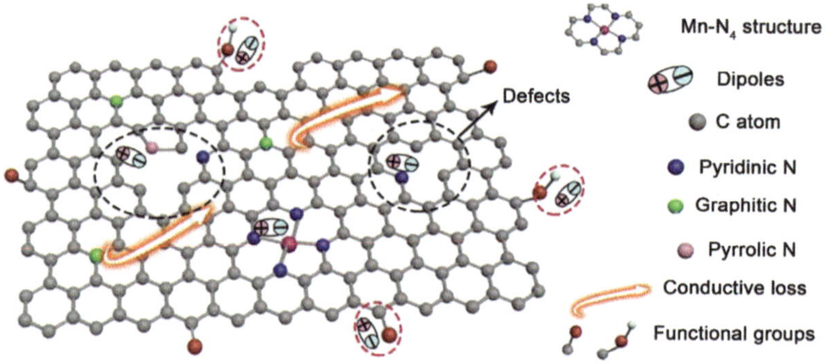

(b)

Figure 5.4 (a) Schematic of the EMW absorption mechanisms for the SiC@SiO_2. Source: Song et al. [29]/Springer Nature/CC BY 4.0. (b) Schematic illustration of the EMW absorption mechanism of 3D Mn-NC. Source: Zhang et al. [15]/Springer Nature/CC BY 4.0.

5.1.5 Saturation Magnetization

The saturation magnetization intensity represents the maximum achievable magnetization intensity in a material when subjected to an external magnetic field. The magnetization of a material primarily arises from the magnetic moments associated with its magnetic atoms or ions. Materials with larger magnetic moments and well-ordered arrangements exhibit higher saturation magnetization. Furthermore, magnetic domains, which are microscopic magnetic structural units within a material, and the formation and energy of magnetic domain walls can influence the magnetization process and the saturation magnetization strength of the material. In general, higher saturation magnetization corresponds to superior magnetic properties of the material.

The resonance phenomenon resulting from the equivalent anisotropic field within magnetic materials is referred to as natural resonance, whereas exchange resonance pertains to the heightened energy dissipation brought about by the displacement of magnetic domain walls and the reordering of magnetic domains when the frequency of the applied alternating magnetic field approaches the exchange frequency of the material. These two resonance phenomena are commonly regarded as the primary mechanisms responsible for low-frequency microwave losses. The frequencies at which natural resonance (f_r) and exchange resonance (f_e) occur can be expressed as follows [30, 31]:

$$K = \frac{1}{2}\mu_0 M_S H_C \tag{5.4}$$

$$H_a = \frac{4|K|}{3\mu_0 M_S} \tag{5.5}$$

$$f_r = \frac{\gamma H_a}{2\pi} \tag{5.6}$$

$$f_e = \frac{\gamma}{2\pi}\left(\frac{C\mu_{kn}{}^2}{r^2 M_s}\right) \tag{5.7}$$

In these formulas, K represents the crystal anisotropy performance, H_a represents the equivalent magnetic crystal anisotropy field, μ_0 denotes the vacuum magnetic permeability, γ represents the spin magnetic ratio, while M_S and H_C represent the saturation magnetization and coercivity of the magnetic material, C represents the exchange constant of the material, and r represents the radius of the magnetic particle. μ_{Kn} represents the root of a spherical Bessel function. It can be observed from the aforementioned equation that magnetic crystal anisotropy, coercivity, and anisotropic energy collectively influence the frequencies of natural resonance and exchange resonance. By increasing the saturation magnetization intensity, the strength of the equivalent magnetic field arising from magnetic anisotropy can be enhanced, thereby augmenting magnetic loss. Typically, the resonance peak resulting from magnetic loss is found within the low-frequency EMW range, and resonance occurs due to the mutual interaction of the magnetic exchange field and external alternating magnetism, resulting in the dissipation of EMW energy [32]. The saturation magnetization strength of magnetic materials can be accurately measured using Hall effect measuring instrument (HEMIs), superconducting quantum interferometers (SQUIDs), and high-precision vibrating sample magnetometers (VSMs).

A higher saturation magnetization can enhance the magnetic permeability of absorbers, thereby improving the dynamic magnetic loss performance of materials in alternating magnetic fields. For instance, Che and coworkers devised a composite EM-absorbing material by integrating FeCoNi medium-entropy alloy into a one-dimensional carbon-based framework using an improved electrospinning technique, as shown in Figure 5.5a [33]. Micromagnetic simulation has demonstrated that magnetic medium-entropy alloys possess one-dimensional anisotropy and augmented saturation magnetization (Figure 5.5b). Axial electron holography technology reveals that the prepared one-dimensional FeCoNi alloy composite material exhibits a high-density continuous magnetic network (Figure 5.5c), indicating its favorable magnetic loss capability when subjected to an external EM field and a significant attenuation effect on low-frequency (2–6 GHz) microwaves. Duan and coworkers investigated the EMW absorption capability of the imaginary part of the relative complex permeability within the frequency range of 2–17 GHz by controlling the fabrication of soft magnetic nickel nanocone arrays with sharp geometrical shapes, large aspect ratios, uniform sizes, and parallel arrangements [34]. The present investigation partially elucidates the significant contribution of magnetic loss to EMW absorption. However, the assessment of magnetic

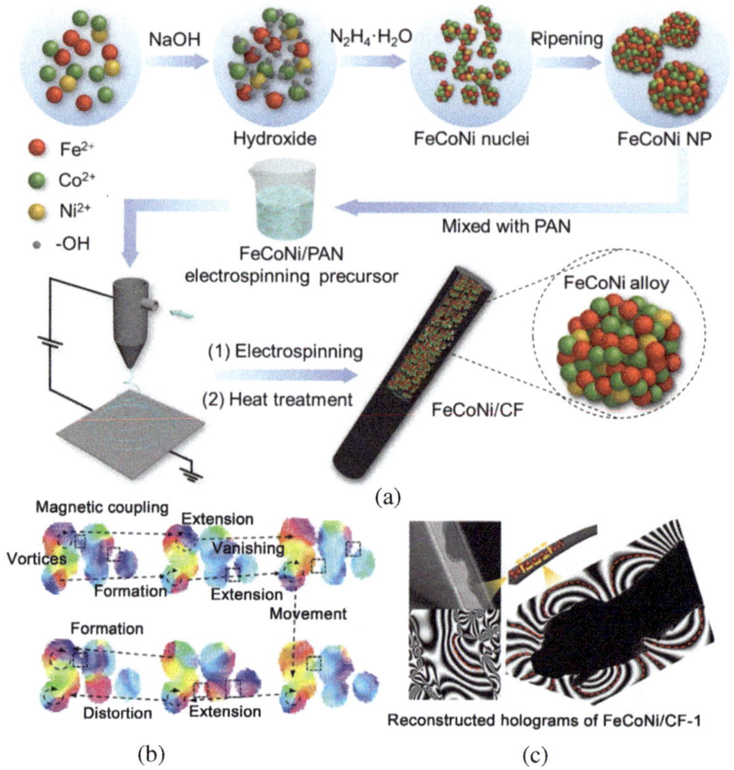

Figure 5.5 (a) Schematic representation delineating the synthesis strategy for FeCoNi/CF composites. Source: Yang et al. [33]/Springer Nature/CC BY 4.0. (b) Micromagnetic simulation of 1D axially distributed FeCoNi NP (the external magnetic field frequency is 6 GHz, and the diameter of the microsphere is 200 nm). Source: Yang et al. [33]/Springer Nature/CC BY 4.0. (c) Electron holograms and corresponding stray magnetic field distributions of FeCoNi/CF samples. Source: Yang et al. [33]/Springer Nature/CC BY 4.0.

loss remains primarily qualitative or comparative. The forthcoming direction of magnetic loss research is expected to emphasize the quantification of diverse microscopic parameters utilizing theoretical approaches and subsequently verifying them through advanced characterization techniques. Moreover, the examination of magnetic losses unavoidably entails the involvement of dielectric losses, thereby impeding the ability to disentangle and analyze the distinct influence of magnetic losses independently. Consequently, there is a pressing need for the advancement of more comprehensive theoretical research in this domain.

5.2 Regulation of Microscopic Attributes

5.2.1 Conventional Regulation

In the field of EMW absorption, material composites and material doping strategies are the primary means of microscale manipulation. Material composites refer to the combination of two or more components with different properties to form a composite material with new characteristics and performance, such as improved thermal and electrical conductivity, enhanced strength and stiffness, increased wear resistance and corrosion resistance, and reduced weight [35–37]. Generally, these composite materials often possess the advantages of two or more materials, thereby exhibiting EMW absorption properties. In the field of EMW absorption, magnetic loss materials such as Fe, Co, Ni, and corresponding ferrites have demonstrated good reflection loss and broadband EMW absorption performance. However, their development in the field of microwave absorption is constrained by factors such as high density and poor corrosion resistance. On the other hand, pure dielectric loss materials exhibit a significant difference in complex permittivity and complex permeability, resulting in impedance mismatch and reflection of EMWs on the material surface, making it difficult for them to undergo internal dissipation. Therefore, researchers often combine magnetic loss materials with dielectric loss materials to prepare EMW absorption materials with good impedance-matching characteristics and excellent attenuation properties. Compared with other EM composite absorption materials, the magnetic loss, dielectric loss, and impedance matching of metal–organic frameworks (MOFs) can be adjusted by in situ generated metal/metal oxide nanoparticles or clusters, thus promoting EMW absorption [38, 39]. Lu and coworkers [40] employed a one-step hydrothermal method followed by subsequent thermal decomposition processes to fabricate hierarchical flower-like structured ZnO/FeNi composite materials. Due to the strong synergistic loss mechanism between dielectric and magnetic components alongside favorable structural characteristics, the layered flower-like ZnO/FeNi specimens exhibit optimal EMW absorption performance, with a maximum reflection loss of −52.08 dB and the widest EAB reaching 6.56 GHz (Figure 5.6a). Yin and coworkers [41] utilized ZIF-8 as a sacrificial template to enhance the EMW absorption performance of hollow-derived LDH composite materials through temperature control and composition design. During the temperature-control process, chemical reactions were adjusted to fine-tune the composition of the products, resulting in the construction of a refined structure and the optimization of LDHs' conduction loss. Moreover, the modulation of different phase interfaces, generated by controlling the components, enhanced interfacial polarization and optimized impedance matching. Consequently, the prepared NCZ ($Ni_3ZnC_{0.7}/Co_3ZnC@C$) exhibited a minimum reflection loss of −58.92 dB and a maximum EAB of 7.36 GHz at a thickness of 2.4 mm (Figure 5.6b).

Material doping refers to the process of introducing other elements or compounds into a material to alter its properties or enhance its functionalities. Doping can be achieved through various methods, such as solid-state reactions, solution methods, and gas-phase methods. The objective of material doping can be to modify the material's electrical, magnetic, optical, or thermal properties, or to increase its conductivity, thermal conductivity, and mechanical strength, among other functionalities. Commonly used doping elements include metals, nonmetals, and impurity ions. From the perspective of ion doping, it can be further classified into anion doping and cation doping.

Anion doping refers to the process of introducing negatively charged ions into a material to alter its properties or enhance its functionalities. The objectives of anion doping can include modifying the material's conductivity, optical properties, magnetic properties, chemical stability, and enhancing its functionalities such as catalysis and adsorption. Commonly used anion dopants include oxygen ions, nitrogen ions, and sulfur ions, among others. It is well known that oxygen vacancies can serve as shallow donors to increase the electron conductivity of metal oxides [42]. Additionally, experimental evidence has shown that sulfur

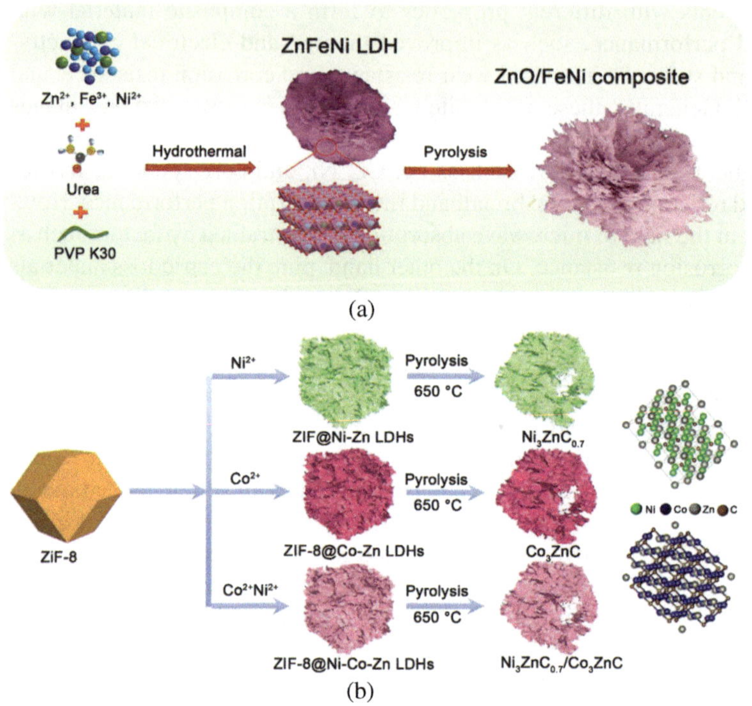

Figure 5.6 (a) Schematic illustration of the synthesis of hierarchical ZnO/FeNi flowers. Source: Gan et al. [40]/with permission of Elsevier. (b) Schematic illustration of the preparation of composites. Source: Zhou et al. [41]/with permission of Elsevier.

vacancy-enriched $CoNi_2S_4$ exhibits significantly enhanced conductivity compared to pure $CoNi_2S_4$ due to improved hybridization of Ni and Co d states and Sp states close to the Fermi level [43]. The formation of band structures by introducing states within the bandgap is advantageous for improving electronic conductivity [44, 45]. Liu et al. [5] discovered that anion vacancy engineering can have significant impacts on dielectric loss, and the introduction of appropriate concentrations of different anion vacancies in transition metal chalcogenides (TMCs) can provide a greater degree of control to enhance conduction/polarization loss. Anion vacancy doping thus represents a promising choice for designing and developing high-performance EMW-absorbing materials.

In comparison to anion doping, cation doping exhibits distinct positions and modes of action within the material. Commonly used cation dopants include metal ions such as iron, cobalt, and manganese, among others. Defects, edges, and impurity atom doping can induce polarization loss, which is advantageous for enhancing the EMW absorption performance of absorbers. Defects disrupt the symmetry of the crystal structure, resulting in the surrounding atoms at the defect site possessing more or fewer electrons, thereby enhancing the polarization loss of the absorber. Common cation doping leads to charge redistribution and induces electric dipoles, contributing to the improvement of polarization. It is worth noting that cation doping typically generates defects and edges within the host material, making it an extremely effective strategy for enhancing the EMW absorption performance of absorbers [46]. For instance, during the preparation of cobalt-doped graphene aerogels, numerous mesopores are formed, leading to a substantial presence of defects and edge sites within the aerogel. The synergistic effect of cobalt ion doping endows the cobalt-doped graphene aerogel with exceptional EMW absorption performance (37,220 dB cm^{-2} g^{-1}), significantly surpassing that of the pristine aerogel [47].

Doping and material composites are commonly employed strategies for manipulating microscopic attributes in materials; however, they also possess certain limitations. For instance, the distribution and concentration of doped elements or composite materials may be challenging to precisely control, resulting in material performance heterogeneity. The precise and independent manipulation of microscopic attributes through doping and material composites is difficult, making it challenging to elucidate the interplay between microscopic attributes and their correlation with the EMW absorption properties of materials, thus elucidating the bridging mechanisms between micro- and macroscales. Therefore, when optimizing EMW absorption performance through doping or material composites, it is necessary to comprehensively consider these limitations and undertake rational material selection and process optimization to achieve the desired variations in microscopic attributes and corresponding EMW absorption properties.

5.2.2 Physical External Fields Regulation

Physical external fields refer to a type of field with specific physical characteristics that act on materials, enabling the microscopic manipulation of materials through the application of external physical forces or energies. Physical external fields can primarily be categorized as electric fields, acoustic fields, magnetic fields, plasma fields, and optical fields, all of which exhibit spatial distributions determined by specific physical quantities [48–50]. Generally, physical external fields can impact the microstructure and properties of materials through various means.

Among them, electric fields serve as an important means of manipulating physical external fields, as they can influence the crystal structure, crystal defects, and phase transitions, and have been widely applied in the fields of EM absorption, capacitors, sensors, and optoelectronic converters [51–53]. Specifically, the following can be observed: (1) When an external electric field acts on a crystal, it can induce the displacement and rearrangement of ions and electrons within the crystal, leading to lattice deformation and distortion, thereby altering the structural stability and morphology of the crystal. (2) In materials with polar crystal structures, an applied electric field can cause polarization of electrons and ions within the crystal, resulting in the generation of a polarized charge distribution, which modifies the dielectric properties of the crystal and affects its dielectric constant and capacitance. (3) Electric fields can alter the charge distribution and the local electric field surrounding defects, thereby influencing the migration rate and direction of defects (such as vacancies and interstitials) within the crystal, impacting the material's electrical conductivity, mechanical strength, thermal conduction, and magneto-optical properties. (4) Electric fields can effectively regulate the electronic band structure and interaction energy within the material, inducing temperature changes that lead to phase transitions, thereby affecting the material's phase transition behavior and properties. (5) Electric fields can modify the potential energy barriers, strength, and direction of atomic/molecular interactions within the material, inducing the emergence of new phases under the influence of the electric field or altering the stability and transition temperature of existing phases. In recent years, electric fields have found extensive applications in material synthesis and the manipulation of microstructural properties.

For instance, Xu et al. [54] achieved the direct growth of graphene-like carbon nanosheets (GCNs) in zirconia ceramics by applying a direct current electric field during the hot-press sintering process, leading to the in situ fabrication of GCNs/zirconia nanocomposites with EM shielding efficiency (>40 dB) and mechanical strength (>1000 MPa). The study revealed that the application of a direct current electric field significantly enhanced carbon infiltration during sintering, resulting in the in situ formation of GCNs at zirconia grain boundaries. By adjusting the current density, the quantity of in situ formed GCNs could be controlled, thereby achieving tunable conductivity. Additionally, Evans et al. [55] employed an electric field to induce nanoscale precision dislocations in Er(Mn,Ti)O_3, facilitating modulation of its local structure and electronic response (Figure 5.7a). Utilizing high-resolution imaging techniques and density functional theory, they obtained direct images of the dislocations and investigated their impact on local electrical transport behavior. Ultrasound field refers to a three-dimensional acoustic field formed during the propagation of sound waves in a

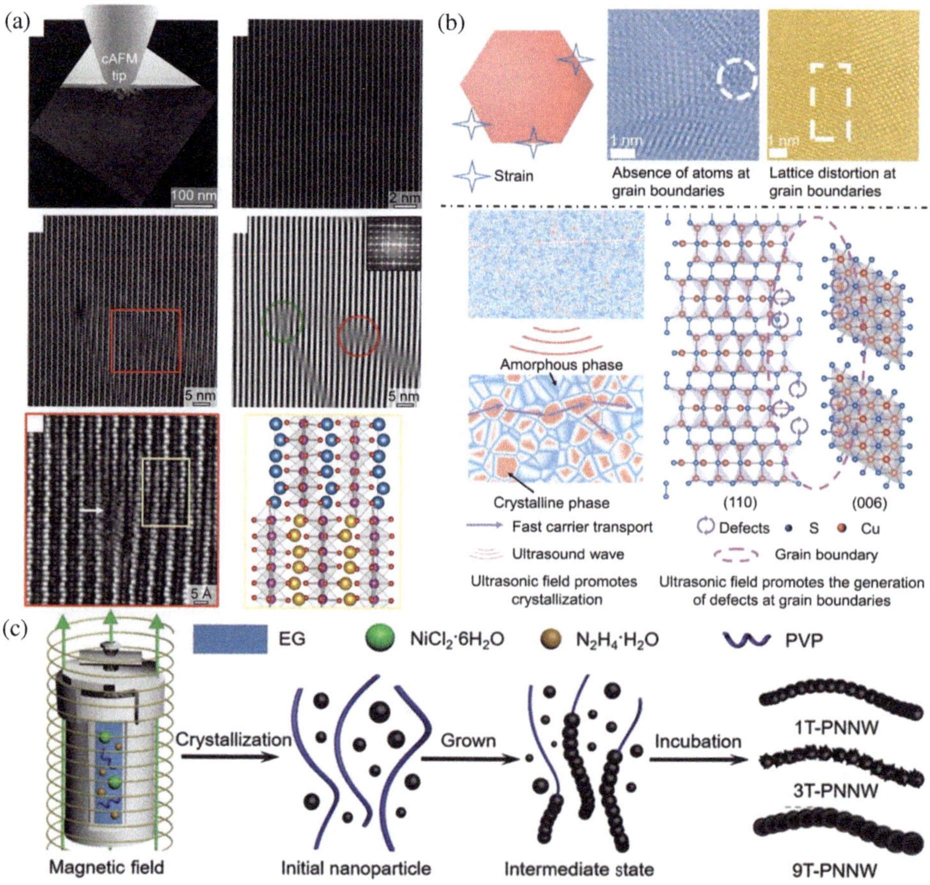

Figure 5.7 (a) The generation of dislocations under the action of an electric field. Source: Evans et al. [55]/American Chemical Society/CC BY 4.0. (b) Schematic diagram of the mechanism of ultrasonic field. Source: Chen et al. [56]/with permission of John Wiley & Sons. (c) Schematic illustration of the synthesis of PNNWs assisted by magnetic field. Source: Qian et al. [57]/John Wiley & Sons/CC BY 4.0.

medium, determined by the propagation characteristics of ultrasound waves and the properties of the medium. In an ultrasound field, parameters such as the propagation velocity, frequency, and amplitude of the sound waves vary with the properties of the medium, resulting in different distributions of the acoustic field. Recently, ultrasound field has emerged as a means to investigate the structure of media and manipulate the properties of materials. Compared to other manipulation methods, ultrasound field offers the following advantages: (1) Noninvasiveness: Ultrasound field manipulation is a noninvasive approach that does not require direct contact with the object being manipulated, thus avoiding damage to the object. (2) Precision: Ultrasound field manipulation enables precise control of parameters such as the direction of sound wave propagation, frequency, and amplitude, allowing for accurate manipulation of the object. (3) Real-time capability: Ultrasound field manipulation allows for real-time monitoring of the response of the manipulated object, enabling timely adjustment of acoustic field parameters to achieve real-time manipulation effects. (4) Tunability: Ultrasound field manipulation can achieve different manipulation effects on the object by adjusting parameters such as frequency and amplitude of the sound waves, demonstrating high tunability. (5) Non-polluting: Ultrasound field manipulation is a novel, clean, and environmentally friendly technology that does not require the use of chemical reagents or other pollutants, resulting in no environmental contamination and compliance with environmental requirements.

Ultrasonic fields have emerged as a versatile synthesis method for nanomaterials, offering advantages such as high production yield, low energy consumption, and short reaction time [58–60]. Ultrasonic cavitation induces the formation, growth, and collapse of bubbles in the solution, resulting in drastic changes in the local field. The generated hotspots can reach temperatures of approximately 5000 K and pressures of approximately 20 MPa, with heating and cooling rates of approximately 1000 K s^{-1}. These unique conditions lead to unconventional chemical reactions, thereby displaying distinctive characteristics in the preparation of nanomaterials. In addition, ultrasonic fields can induce transient changes in temperature, pressure, and heat transfer, providing energy for particle crystallization and growth, as well as facilitating nucleation. The mechanical effects of ultrasonic fields can drive grains to high-stress conditions, leading to atomic losses and lattice distortions. For instance, ultrasonic fields can increase the strain in $BiFeO_3/Ti_3C_2$, enhancing the surface's imbalanced charge state and releasing more space charge [61]. Ultrasonic fields have also found applications in various fields of polymer science, including polymer synthesis, polymer processing, and functionalization for targeted applications. The chemical and mechanical effects generated by ultrasound in liquid media have been actively utilized for chemical reactions. Ultrasound serves as an external stimulus that initiates free radical polymerization (FRP) through solvent homolysis, thereby generating free radicals. Unlike traditional FRP, ultrasonic-induced polymerization does not involve any external toxic initiators, providing excellent control over the polymerization process in terms of time, molecular weight, and microstructure [59]. Furthermore, ultrasonic field regulation can enhance reaction rates and increase polymerization degrees during the polymerization process. By breaking the mass transfer limitations in the polymerization reaction, ultrasound can improve reaction efficiency and polymer yield. Additionally, ultrasonic field regulation can alter parameters such as temperature, concentration, and molecular structure in the polymerization reaction, thereby controlling the molecular weight and distribution of the polymer. The influence of ultrasonic fields extends beyond these examples; for instance, Chen et al. [56] employed ultrasonic fields to modulate the inherent properties of EMW-absorbing materials. The ultrasonic-induced cavitation effect significantly promotes the crystallization of CuS and leads to the accumulation of S defects at grain boundaries. The degree of CuS crystallization and the content of grain boundary defects are regulated by ultrasonic fields and various ions. The enrichment of S defects at grain boundaries and in the crystallinity significantly improves the relaxation and conduction losses of CuS (Figure 5.7b).

The magnetic field is a vector field that describes the influence of the magnetic field on moving charges, currents, and magnetic materials. Moving charges in a magnetic field experience a force perpendicular to their own velocity and the magnetic field, while a varying magnetic field exerts forces on a range of nonmagnetic materials through its impact on the motion of their external atomic electrons. The magnetic field is a non-contact, environmentally friendly, high-energy, and multi-effect physical field that can alter the thermodynamic and kinetic conditions of material preparation processes, transferring high-intensity energy to the atomic scale of matter without direct contact. This, in turn, modifies the microstructure and properties of materials, directly influencing the migration, matching, and arrangement of atoms, molecules, ions, or grains. Consequently, magnetic fields have become important technological tools for developing new materials and optimizing material performance, with magnetic field manipulation methods widely applied in scientific research and industrial production [62–64]. In metallic materials, magnetic field manipulation can significantly optimize alloy performance, promote dislocation motion, reduce residual stresses, refine grains, facilitate second-phase precipitation, and influence alloy texture [65]. In the fields of biology and medicine, magnetic field manipulation also holds great potential for a wide range of applications, including drug control and release, drug delivery, cancer treatment, and magnetic hyperthermia. It allows for the alteration of the structure and functional properties of magnetic materials under static or alternating magnetic fields, enabling remote magnetic field control and manipulation [66]. Likewise, in the field of EMW absorption, magnetic field manipulation also holds extensive application prospects. First, by adjusting the magnetic properties and magnetic field response capability of materials, effective absorption of EMWs within different frequency ranges can be achieved. Second, magnetic field manipulation can realize absorption of multiple frequency bands of EMWs. By adjusting the intensity and direction of the magnetic field, the EMW absorption capability of materials can be controlled, enabling absorption of EMWs within different frequency bands. Lastly, dynamic modulation of the magnetic properties and magnetic field response capability of materials can be achieved by changing the intensity and direction of the magnetic field, thereby achieving dynamic control over EMW absorption performance. Qian et al. [57] synthesized PVP-directed nickel nanowires (PNNWs) using a magnetic solvent thermal method, directly controlling the aspect ratio, surface roughness, and magnetic properties of the nickel nanowires by varying the applied magnetic field intensity (Figure 5.7c). The study showed that high magnetic field intensity resulted in PNNWs with high aspect ratios, large specific surface areas, and excellent magnetic properties, thereby significantly enhancing their EMW performance. At a thickness of 4.5 mm and a frequency of 4.08 GHz, the minimum reflection loss reached −29.82 GHz, while at a thickness of 1.5 mm, the EAB extended from 14.4 to 18.0 GHz. Cai et al. [67] prepared Co-MOF 74 derivative/polyacrylamide dual-functional gels using magnetic field-assisted techniques and in situ FRP. The optimal gel exhibited a minimum reflection loss of −51.6 dB and a maximum EAB of 5.73 GHz at a low loading of 5 wt%. The magnetic nanoparticles in the MOF derivatives, under the influence of the magnetic field, oriented the fillers in the polymer matrix, forming ordered heat conduction pathways that improved thermal conduction perpendicular to the plane's direction, thereby facilitating rapid heat transfer.

The term "plasma field" refers to the EM field generated by a plasma, which consists of positively and negatively charged ions and free electrons. The plasma field possesses EM properties and can be described through the propagation and interaction of EMWs. The characteristics and behavior of the plasma field are influenced by factors such as the properties, temperature, density, charge distribution, and external EM fields of the plasma [68–70]. In the field of EMW absorption, the manipulation of plasma fields holds great potential for various applications. First, by adjusting factors such as the density, energy-level structure, and interaction of ions and electrons within the plasma, the frequency response and absorption intensity of EMW-absorbing materials can be controlled. Second, plasma manipulation can be utilized to reduce EMW noise and interference. By manipulating the charge distribution and EM response of the plasma, absorption and scattering of EMWs can be achieved, thereby reducing their propagation and interference. Lastly, adjusting plasma parameters such as density, temperature, and electric field enables the dynamic control of the EM characteristics of the plasma absorption layer, thereby achieving dynamic control of EMW absorption performance. To address the thickness dependence of plasma-absorbed EMWs and further reduce the target's radar cross-section (RCS), Hao et al. [71] designed a novel hybrid structure of metasurfaces and plasma. By designing the equivalent circuit of an EM resonance-based metamaterial absorber, a metasurface with three absorption peaks was created. Numerical and experimental verifications were conducted on the reflective and absorptive properties of the hybrid structure. The research demonstrates that compared to plasma or metasurfaces alone, this hybrid structure exhibits significantly improved absorption capabilities and is more practical due to its lower plasma thickness requirement. Deng et al. [72] discovered that an absorption device composed of low-pressure gas discharge tubes and radar-absorbing materials (RAMs) can provide a new solution for broadband EM stealth. The combination of plasma and RAM was found to significantly broaden the absorption bandwidth and enhance absorption performance, resulting in excellent synergistic stealth capabilities. Experimental and simulation results indicate that through adjusting the spatial layout of the combined plasma layer and the type of RAM, broadband and wide-angle adjustable EMW absorption and RCS reduction can be achieved, creating opportunities for selective stealth in microwave transmission and equipment. Therefore, the integration of plasma arrays and RAM for wave manipulation provides valuable insights for the development of numerous applications, such as radar antenna stealth, spatial filters, and high-power microwave shielding.

5.3 The Current State and Future Potential of Bridge Mechanism Between Micro and Macro Levels

The relationship between microscopic attributes and macroscopic parameters in EMW-absorbing materials is fundamental for the controllable modulation of their EMW absorption performance and holds significant implications for the design and application of such materials. By gaining a comprehensive understanding of the relationship between microscopic attributes and macroscopic parameters, novel EMW-absorbing materials with exceptional performance can be developed to meet the requirements of various fields.

Current status and prospects for the development of a bridge mechanism between micro and macro levels are primarily associated with the regulation of microscopic attributes, characterization of microscopic attributes, and elucidation of micro–macro correlations. In terms of regulating microscopic attributes, conventional control measures have been employed to modulate microscopic attributes related to material EMW absorption performance. For instance, annealing or doping techniques have been used to control the generation of point defects. Compositional composites or porous designs have been employed to construct interfaces, and the introduction of highly conductive components has enhanced conductivity. These methods enable a certain degree of regulation over microscopic attributes but with limited precision and the potential introduction of confounding factors. As a result, only qualitative relationships between EMW absorption and microlevel factors can be established, making it challenging to conduct semi-quantitative studies. Furthermore, conventional control approaches also possess limitations, such as the ability to solely control defect content while struggling to manage defect types and distributions, as well as interface types and strengths. In terms of characterizing microlevel factors, current reliance is primarily on conventional characterization techniques such as XRD, SEM, XPS, TEM, PL, Raman spectroscopy, and EIS. These techniques can only qualitatively determine defect content, interface composition, conductivity values, and the presence of dipoles, without quantitatively assessing the contributions of various microlevel factors. Regarding the elucidation of micro-macro correlations, current research primarily focuses on the regulation of EMW absorption performance through the combined effects of multiple factors, lacking control over individual loss mechanisms and investigation into EM correlations. Additionally, there is a lack of in-depth understanding regarding the correlation between control measures and variations in microscopic attributes.

Looking ahead, the establishment of a bridge mechanism between micro and macro levels in EMW absorption materials can be achieved by incorporating the latest advances in materials science. Novel means of control, such as physical fields, play a crucial role in manipulating microscopic attributes of materials. By subjecting materials to external electric, magnetic, optical, acoustic, and plasma fields, the microstructure, electron states, and lattice vibrations can be altered, enabling control over microscopic attributes. For instance, electric fields can modulate conductivity, electronic structure, and surface charge distribution, while magnetic fields can control magnetism, permeability, and magnetic domain structures. Light fields can influence band structures, electronic states, and optical properties, while acoustic fields can regulate lattice vibrations, phonon states, and acoustic properties, among others. Introducing multiple physical fields during material synthesis or post-processing holds the promise of achieving precise control over microscopic attributes. Therefore, future control of microscopic attributes can be realized by introducing multiple physical fields during material synthesis or post-processing, enabling controllable modulation of individual factors and their interactions.

Furthermore, with the continuous development of materials science and characterization techniques, novel methods for characterizing microscopic attributes have emerged, offering more accurate insights into the microstructure and properties of materials. The application of these innovative techniques can better verify changes in microscopic attributes and establish their correlation with control measures. In the future, the characterization of microscopic attributes in EMW absorption materials can employ techniques such as off-axis electron holography, in situ XRD, in situ XPS, in situ Raman spectroscopy, ESR/EPR, XAS, PAS, ultraviolet-visible diffuse reflectance spectroscopy (UV-Vis-DRS), and aberration-corrected transmission electron microscopy (AC-TEM), among others.

To elucidate the correlation between micro and macro levels, advanced characterization techniques can be employed to examine changes in microscopic attributes, enabling the semi-quantitative detection of these factors and establishing their correlation with control measures. By linking variations in microscopic attributes with macroscopic parameters such as matching thickness, absorption bandwidth, matching frequency, and peak loss, the relationship between microscopic attributes and EMW absorption mechanisms can be explored, leading to the development of novel physical models that bridge the gap between micro and macro levels in EMW absorption.

References

1 Wang, X., You, F., Wen, X. et al. (2022). Doping $Ce(OH)CO_3$ laminated dendrites with Fe, Co and Ni for defect steered wide-frequency microwave absorption. *Chem. Eng. J.* 445: 136431.

2 Quan, B., Shi, W., Ong, S. et al. (2019). Defect engineering in two common types of dielectric materials for electromagnetic absorption applications. *Adv. Funct. Mater.* 29: 1901236.

3 Liu, J., Zhang, L., and Wu, H. (2022). Enhancing the low/middle-frequency electromagnetic wave absorption of metal sulfides through F^- regulation engineering. *Adv. Funct. Mater.* 32: 2110496.

4 Qin, M., Zhang, L., Zhao, X., and Wu, H. (2021). Defect induced polarization loss in multi-shelled spinel hollow spheres for electromagnetic wave absorption application. *Adv. Sci.* 8: 2004640.

5 Liu, J., Zhang, L., and Wu, H. (2022). Anion-doping-induced vacancy engineering of cobalt sulfoselenide for boosting electromagnetic wave absorption. *Adv. Funct. Mater.* 32: 2200544.

6 Liu, Y., Zhou, X., Jia, Z. et al. (2022). Oxygen vacancy-induced dielectric polarization prevails in the electromagnetic wave-absorbing mechanism for Mn-based MOFs-derived composites. *Adv. Funct. Mater.* 32: 2204499.

7 Lv, H., Wu, C., Tang, J. et al. (2021). Two-dimensional SnO/SnO_2 heterojunctions for electromagnetic wave absorption. *Chem. Eng. J.* 411: 128445.

8 Xu, C., Liu, P., Wu, Z. et al. (2022). Customizing heterointerfaces in multilevel hollow architecture constructed by magnetic spindle arrays using the polymerizing-etching strategy for boosting microwave absorption. *Adv. Sci.* 9: 2200804.

9 He, Z., Xu, H., Shi, L. et al. (2024). Hierarchical Co_2P/CoS_2@C@MoS_2 composites with hollow cavity and multiple phases toward wideband electromagnetic wave absorption. *Small* 20: 2306253.

10 Qin, M., Zhang, L., Zhao, X., and Wu, H. (2021). Lightweight Ni foam-based ultra-broadband electromagnetic wave absorber. *Adv. Funct. Mater.* 31: 2103436.

11 Zhao, B., Yan, Z., Liu, L. et al. (2024). A liquid-metal-assisted competitive galvanic reaction strategy toward Indium/oxide core–shell nanoparticles with enhanced microwave absorption. *Adv. Funct. Mater.* 34: 2314008.

12 Tian, Y., Zhi, D., Li, T. et al. (2023). Graphene-based aerogel microspheres with annual ring-like structures for broadband electromagnetic attenuation. *Chem. Eng. J.* 464: 142644.

13 Wen, B., Cao, M., Hou, Z. et al. (2013). Temperature dependent microwave attenuation behavior for carbon-nanotube/silica composites. *Carbon* 65: 124–139.

14 Song, W., Cao, M., Hou, Z. et al. (2009). High dielectric loss and its monotonic dependence of conducting-dominated multiwalled carbon nanotubes/silica nanocomposite on temperature ranging from 373 to 873 K in X-band. *Appl. Phys. Lett.* 94: 233110.

15 Zhang, X., Shi, Y., Xu, J. et al. (2022). Identification of the intrinsic dielectric properties of metal single atoms for electromagnetic wave absorption. *Nano Micro Lett.* 14: 27.

16 Wen, B., Cao, M., Lu, M. et al. (2014). Reduced graphene oxides: light-weight and high-efficiency electromagnetic interference shielding at elevated temperatures. *Adv. Mater.* 26: 3484.

17 Ye, F., Song, Q., Zhang, Z. et al. (2018). Direct growth of edge-rich graphene with tunable dielectric properties in porous Si_3N_4 ceramic for broadband high-performance microwave absorption. *Adv. Funct. Mater.* 28: 1707205.

18 Chen, C., Pu, N., Liu, Y. et al. (2018). Microwave absorption properties of holey graphene/silicone rubber composites. *Compos. Part B Eng.* 135: 119–128.

19 Fujii, S. and Enoki, T. (2012). Clar's aromatic sextet and π-electron distribution in nanographene. *Angew. Chem. Int. Ed.* 51: 7236–7241.

20 Song, Q., Yan, H., Liu, K. et al. (2018). Vertically grown edge-rich graphene nanosheets for spatial control of Li nucleation. *Adv. Energy Mater.* 8: 1800564.

21 Liu, X., Nie, X., Yu, R., and Feng, H. (2018). Design of dual-frequency electromagnetic wave absorption by interface modulation strategy. *Chem. Eng. J.* 334: 153–161.

22 Ma, S., Xu, H., Wang, M. et al. (2019). Polarization effects of transition dipoles on photoluminescence and photocurrent in organic-inorganic hybrid perovskites. *Nano Energy* 65: 104004.

23 Lai, L., Li, B., Tian, S. et al. (2023). Giant electrostrain in lead-free textured piezoceramics by defect dipole design. *Adv. Mater.* 35: 2300519.

24 Liu, X., Cheng, P., Zhang, X. et al. (2021). Enhanced solar-to-hydrogen efficiency for photocatalytic water splitting based on a polarized heterostructure: the role of intrinsic dipoles in heterostructures. *J. Mater. Chem. A* 9: 14515–14523.

25 Martin, L. (2021). Whirls and swirls of polarization. *Science* 371: 992–993.

26 Hsu, H., Huang, B., Chin, S. et al. (2019). Photodriven dipole reordering: key to carrier separation in metalorganic halide perovskites. *ACS Nano* 13: 4402–4409.

27 Cole, K. and Cole, R. (1941). Dispersion and absorption in dielectrics I. alternating current characteristics. *J. Chem. Phys.* 9: 341–351.

28 Yang, L., Lv, H., Li, M. et al. (2020). Multiple polarization effect of shell evolution on hierarchical hollow C@MnO_2 composites and their wideband electromagnetic wave absorption properties. *Chem. Eng. J.* 392: 123666.

29 Song, L., Zhang, F., Chen, Y. et al. (2022). Multifunctional SiC@SiO_2 nanofiber aerogel with ultrabroadband electromagnetic wave absorption. *Nano Micro Lett.* 14: 152.

30 Zhang, Z., Zhao, Y., Li, Z. et al. (2021). Synthesis of carbon/SiO_2 core-sheath nanofibers with Co-Fe nanoparticles embedded in via electrospinning for high-performance microwave absorption. *Adv. Compos. Hybrid Mater.* 5: 513–524.

31 Zong, M., Huang, Y., Zhao, Y. et al. (2013). Facile preparation, high microwave absorption and microwave absorbing mechanism of RGO–Fe_3O_4 composites. *RSC Adv.* 3: 23638–23648.

32 Wang, L., Huang, M., Qian, X. et al. (2021). Confined magnetic-dielectric balance boosted electromagnetic wave absorption. *Small* 17: 2100970.

33 Yang, B., Fang, J., Xu, C. et al. (2022). One-dimensional magnetic FeCoNi alloy toward low-frequency electromagnetic wave absorption. *Nano Micro Lett.* 14: 170.

34 Zhang, W., Zhang, J., Wu, P. et al. (2020). Parallel aligned nickel nanocone arrays for multiband microwave absorption. *ACS Appl. Mater. Interfaces* 12: 23340–23346.

35 Yan, J., Huang, Y., Liu, X. et al. (2021). Polypyrrole-based composite materials for electromagnetic wave absorption. *Polym. Rev.* 61 (3): 646–687.

36 Guo, Y., Ruan, K., Wang, G., and Gu, J. (2023). Advances and mechanisms in polymer composites toward thermal conduction and electromagnetic wave absorption. *Sci. Bull.* 68: 1195–1212.

37 Elhassan, A., Abdalla, I., Yu, J. et al. (2020). Microwave-assisted fabrication of sea cucumber-like hollow structured composite for high-performance electromagnetic wave absorption. *Chem. Eng. J.* 392: 123646.

38 Rehman, S., Xu, S., Li, Z. et al. (2023). Hierarchical-bioinspired MOFs enhanced electromagnetic wave absorption. *Small* 2306466.

39 Lin, J., Wu, Q., Qiao, J. et al. (2023). A review on composite strategy of MOF derivatives for improving electromagnetic wave absorption. *iScience* 26: 107132.

40 Gan, F., Rao, Q., Deng, J. et al. (2023). Controllable architecture of ZnO/FeNi composites derived from trimetallic ZnFeNi layered double hydroxides for high-performance electromagnetic wave absorbers. *Small* 19: 2300257.

41 Zhou, Z., Zhou, X., Lan, D. et al. (2024). Modulation engineering of electromagnetic wave absorption performance of layered double hydroxides derived hollow metal carbides integrating corrosion protection. *Small* 20: 2305849.

42 Ji, D., Fan, L., Tao, L. et al. (2019). The kirkendall effect for engineering oxygen vacancy of hollow Co_3O_4 nanoparticles toward high-performance portable zinc–air batteries. *Angew. Chem. Int. Ed.* 58: 13840.

43 Li, Z., Zhao, D., Xu, C. et al. (2018). Reduced $CoNi_2S_4$ nanosheets with enhanced conductivity for high-performance supercapacitors. *Electrochim. Acta* 278: 33.

44 Li, G., Blake, G., and Palstra, T. (2017). Vacancies in functional materials for clean energy storage and harvesting: the perfect imperfection. *Chem. Soc. Rev.* 46: 1693.

45 Fang, Z., Peng, L., Qian, Y. et al. (2018). Dual tuning of Ni–Co–A (A = P, Se, O) nanosheets by anion substitution and holey engineering for efficient hydrogen evolution. *J. Am. Chem. Soc.* 140: 5241.

46 Zhong, W., Li, B., Ma, Z. et al. (2023). Double salt-template strategy for the growth of N, S-codoped graphitic carbon nanoframes on the graphene toward high-performance electromagnetic wave absorption. *Carbon* 202: 235–243.

47 Xu, J., Liu, M., Zhang, X. et al. (2022). Atomically dispersed cobalt anchored on N-doped graphene aerogels for efficient electromagnetic wave absorption with an ultralow filler ratio. *Appl. Phys. Rev.* 1: 011402.

48 Zhou, Y., Li, J., Wu, Y., and Luo, Z. (2020). Role of external field in polymerization: mechanism and kinetics. *Chem. Rev.* 120: 2950–3048.

49 Li, J., Yin, W., Pan, J. et al. (2023). External field assisted hydrogen evolution reaction. *Nano Res.* 16 (7): 8638–8654.

50 Hu, C., Tu, S., Tian, N. et al. (2021). Photocatalysis enhanced by external fields. *Angew. Chem. Int. Ed.* 60: 16309–16328.

51 Chen, Y., Lee, H., Verba, R. et al. (2017). Parametric resonance of magnetization excited by electric field. *Nano Lett.* 17: 572–577.

52 Wu, Y., Zhang, X., Xu, X. et al. (2020). A review on the effect of external fields on solidification, melting and heat transfer enhancement of phase change materials. *J. Energy Storage* 31: 101567.

53 Akamatsu, M., Sakai, N., and Matile, S. (2017). Electric-field-assisted anion–π catalysis. *J. Am. Chem. Soc.* 139: 6558–6561.

54 Xu, J., Liu, Z., He, S. et al. (2022). Electric field-assisted in situ fabrication of carbon/zirconia nanocomposites with tunable conductivity for electromagnetic interference shielding applications. *Compos. Part B* 246: 110254.

55 Evans, D., Småbråten, D., Holstad, T. et al. (2021). Observation of electric-field-induced structural dislocations in a ferroelectric oxide. *Nano Lett.* 21: 3386–3392.

56 Chen, G., Liang, H., Yun, J. et al. (2023). Ultrasonic field induces better crystallinity and abundant defects at grain boundaries to develop CuS electromagnetic wave absorber. *Adv. Mater.* 35: 2305586.

57 Qian, Y., Meng, X., Liu, H. et al. (2023). Magnetic field-induced synthesis of one-dimensional nickel nanowires for enhanced microwave absorption. *Adv. Mater. Interfaces* 10: 2201604.

58 Tan, G., She, L., Liu, T. et al. (2017). Ultrasonic chemical synthesis of hybrid mpg-C_3N_4/$BiPO_4$ heterostructured photocatalysts with improved visible light photocatalytic activity. *Appl. Catal. B* 207: 120.

59 Kumar, A., Padmakumar, A., Kalita, U. et al. (2023). Ultrasonics in polymer science: applications and challenges. *Prog. Mater. Sci.* 136: 101113.

60 Shi, C., Liu, Y., Qi, R. et al. (2021). Hierarchical N-doped carbon spheres anchored with cobalt nanocrystals and single atoms for oxygen reduction reaction. *Nano Energy* 87: 106153.

61 Li, J., Liu, X., Zheng, Y. et al. (2023). Achieving fast charge separation by ferroelectric ultrasonic interfacial engineering for rapid sonotherapy of bacteria-infected osteomyelitis. *Adv. Mater.* 35: 2210296.

62 Yang, P., Cai, R., Ruan, H. et al. (2023). In situ synthesis of heterostructured Fe-Ni nanowires with tunable electromagnetic wave absorption capabilities. *ACS Appl. Nano Mater.* 6: 14322–14331.

63 Ai, J., Cheng, S., Miao, Y. et al. (2023). Graphene/electrospun carbon nanofiber sponge composites induced by magnetic particles for multi-functional pressure sensor. *Carbon* 205: 454–462.

64 Luo, Y., Chen, Z., Wen, S. et al. (2022). Magnetic regulation of the luminescence of hybrid lanthanide-doped nanoparticles. *Coord. Chem. Rev.* 469: 214653.

65 Hu, Y., Zhao, H., Yu, X. et al. (2022). Research progress of magnetic field regulated mechanical property of solid metal materials. *Metals* 12: 1988.

66 Yao, J., Yao, C., Zhang, A. et al. (2022). Magnetomechanical force: an emerging paradigm for therapeutic applications. *J. Mater. Chem. B* 10: 7136.

67 Cai, J., Zhao, H., Liu, H. et al. (2024). Magnetic field vertically aligned Co-MOF-74 derivatives/polyacrylamide hydrogels with bifunction of excellent electromagnetic wave absorption and efficient thermal conduction performances. *Compos. Part A* 176: 107832.

68 Lin, L., Peng, X., Wang, M. et al. (2016). Light-directed reversible assembly of plasmonic nanoparticles using plasmon-enhanced thermophoresis. *ACS Nano* 10: 9659–9668.

69 Aupiais, I., Grasset, R., Guo, T. et al. (2023). Ultrasmall and tunable TeraHertz surface plasmon cavities at the ultimate plasmonic limit. *Nat. Commun.* 14: 7645.

70 Mertens, J., Kleemann, M., Chikkaraddy, R. et al. (2017). How light is emitted by plasmonic metals. *Nano Lett.* 17: 2568–2574.

71 Hao, Z., Li, J., Xu, B. et al. (2023). Composite wave-absorbing structure combining thin plasma and metasurface. *Plasma Sci. Technol.* 25: 045504.

72 Deng, X., Ding, C., Wang, Y. et al. (2022). Investigation into wideband electromagnetic stealth device based on plasma array and radar-absorbing materials. *Plasma Sci. Technol.* 24: 114006.

6

New Dielectric Physical Models for Electromagnetic Wave Absorption

Hongsheng Liang

MOE Key Laboratory of Material Physics and Chemistry Under Extraordinary Conditions, School of Physical Science and Technology, Northwestern Polytechnical University, Xi'an, China

CHAPTER MENU

It is well known that the enhancement of electromagnetic (EM) wave absorption performance is closely related to the loss mechanism. Currently, the study of electromagnetic wave (EMW) absorption mechanisms under the microscopic perspective faces a series of challenges, including problems such as unclear understanding of the physical mechanisms of EM losses. In order to deeply investigate the mechanism of EMW absorption, it is especially crucial to establish clear physical models. The models are not only a key tool to reveal the complexity of the loss mechanisms leading to wave absorption properties but also play a transformative role in enhancing the overall performance of EMW-absorbing materials. In this chapter, we summarize the advanced research on dielectric physical modeling of EMW-absorbing materials. This is mainly divided into dielectric physical modeling from a microscopic perspective, dielectric physical modeling in terms of structural design, and novel dielectric physical modeling spawned by new application scenarios.

6.1 Dielectric Microphysical Model

6.1.1 Dimension Distribution-Induced Interfacial Polarization Model

Recently, various absorptive materials have flourished, including carbon materials, transition metal oxides/sulfides, and conductive polymers [1, 2]. Most of these materials belong to

Electromagnetic Wave Absorbing Materials: Fundamentals and Applications, First Edition.
Edited by Hongjing Wu, Jun Luo, and Meiyin Yang.

dielectric materials, primarily relying on conduction loss and polarization loss to attenuate incident EMWs. Given the significance of dielectric loss, decades of experimental research have accumulated a wealth of knowledge about dielectric materials with different microstructures and their EMW absorption characteristics. One prominent dielectric material is a supporting structure composed of surface nanoscale particles (NPs) and a substrate [3]. Due to the diffusion of nonmetallic elements from NPs into the substrate and metallic elements from the substrate into NPs, a strong interaction may exist between NPs and the substrate. Consequently, the composition, distribution, and morphology of NPs may undergo changes due to the strong interaction between NPs and the substrate. Undoubtedly, the composition and distribution of these NPs are likely to have a significant impact on the dielectric performance of the material, but there are few dedicated studies on this. In the literature, these heterostructures may lead to symmetry breaking and result in spatial charge distribution; however, the detailed mechanism of interface polarization remains largely lacking. Until now, there has been no investigation into the primary contributions of dielectric loss in the supporting material [4]. Supplementing the fundamental understanding of the impact of surface NPs on dielectric loss holds profound significance for designing efficient EMW absorbers based on dielectric loss.

Zhao et al. [5] prepared a series of carbon-based composites using Cu-MOF-74 as a precursor and decorated them with a large number of NPs of different compositions and distributions (Figure 6.1a). The well-ordered and voluminous bouquet-like structure of the precursor offered plenty of growth sites for the NPs. Sulfur powder (300/600 mg) positioned upstream of the tube furnace served as the sulfur source for partial/complete sulfidation of both NPs and Cu atoms in the substrate. The quantity of sulfur powder played a pivotal role in determining the surface morphology of NPs, resulting in varied forms: irregular and dispersed (0 mg), uniform and clustered (300 mg), and compact and diminutive (600 mg). EM parameters revealed that clustered NPs facilitated interfacial polarization, sulfur elements induced the formation of —C=S—C— dipoles in the interaction between NPs and the carbon substrate, and the introduction of sulfides augmented conduction loss, thereby significantly amplifying the material's dielectric loss. Moreover, the sample with an optimal particle distribution exhibited the highest complex permittivity. It demonstrated robust EMW absorption with a dissipation capacity of −33.5 dB and exceptional broadband absorption (7.6 GHz) at a thin thickness of 2.3 mm (Figure 6.1b).

In summary, the advantages of MOF-74-S1 can be summarized as follows: (1) The outside surfaces of NPs are different from each other due to the complex composition. Therefore, the interfacial charges at NPs/NPs interfaces will accumulate and form strong interfacial polarization, which is also different from the sulfide layer in MOF-74-S2 because there are no obvious NPs in this sample. (2) The aggregated NPs form many local electric fields, which can promote the interfacial polarization. (3) These pentad composites (C/Cu/Cu_2O/$Cu_{2-x}S$) are nonhomogeneous, and the charges tend to accumulate at the multiphase boundaries, which can also result in substantial interfacial polarization and relaxation. Therefore, the evident interfacial polarization of MOF-74-S1 should be responsible for its strongest dielectric loss, considering that the dipole polarization of MOF-74-S1 is weaker than that of MOF-74-S2, and the almost identical conduction loss for both samples. The aforementioned mechanism is finely depicted in Figure 6.1c.

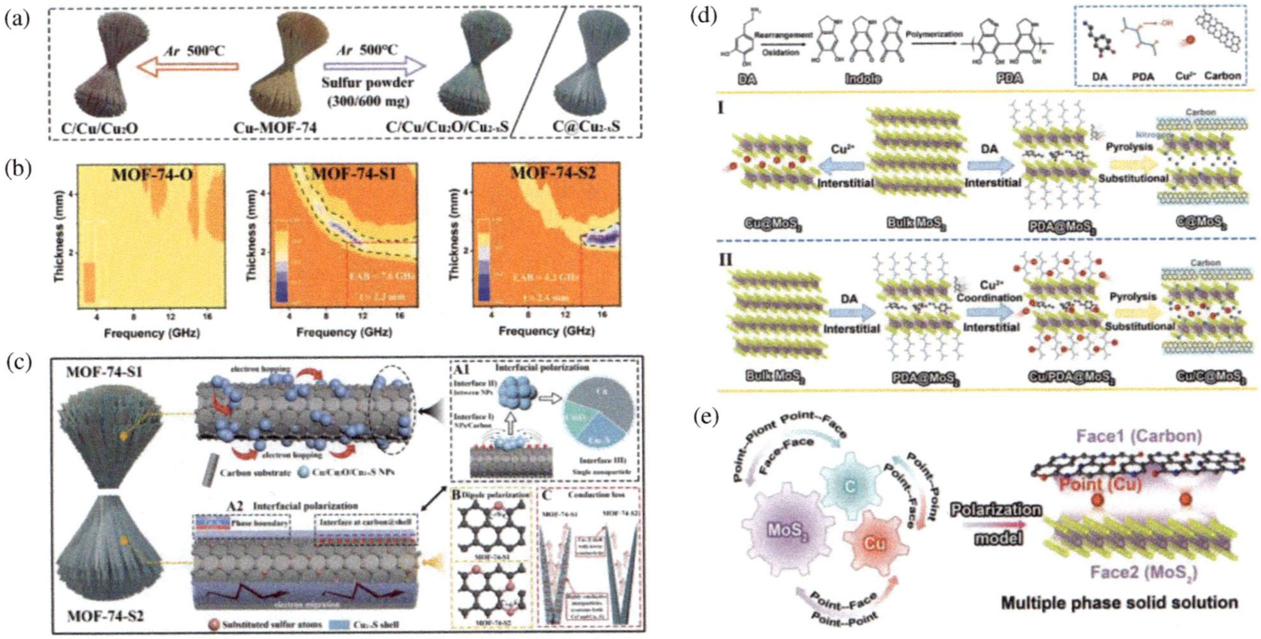

Figure 6.1 (a) Schematic illustration of the synthesis process. (b) The electromagnetic (EM) wave absorption performance of the sample. (c) Schematic diagram of the absorption mechanism in MOF-74-S1 and MOF-74-S2. (d) Schematic illustration of the preparation of MoS_2-based multiple-phase solid solution. (e) Schematic representation of polarization models in multiple-phase MoS_2 solid solution.

6.1.2 Three Types of Polarization Site Models

Typically, polarization manifests itself through the processes of "orientation–deorientation" of polarization sites or the "aggregation–dispersion" of charges within diverse heterostructures [6, 7]. Subsequently, the repetitive and periodic response of polarization centers to incident EM fields can lead to the transformation of energy forms and the attenuation of energy [8, 9]. Consequently, EMWs undergo attenuation during the progression of polarization relaxation. In the context of EMW absorption, polarization processes not only transpire at face–face (heterointerface) and point–point (dipole) interactions but also involve heterogeneous point–face sites. The polarization occurring at face–face and point–point sites is conventionally categorized as interfacial polarization and dipole polarization, respectively. Additionally, interactions between point and face can initiate polarization relaxation [10]. However, there is currently a lack of comprehensive exploration regarding the simultaneous and distinct modulation of polarization characteristics across various polarization sites to achieve controllable polarization loss. Given the considerations outlined previously, it is imperative to devise a series of medium materials for research with tunable structures of polarization sites. This approach is crucial for elucidating the mechanism of polarization loss and addressing the gap in understanding the concurrent manipulation of diverse polarization characteristics.

In a pioneering study, Gao et al. [11] introduced a novel metal–organic synergetic method to fabricate a series of MoS_2-based solid solutions with adjustable interstitial, substitutional, and vacancy structures (Figure 6.1d). This innovative approach allowed for the analysis of the polarization loss mechanism. The study systematically examined the synergistic effects of point–point, point–face, and face–face polarizations on EMW absorption. The EM properties of defected MoS2, Cu-intercalated MoS_2 (Cu@MoS_2), and carbon-modified MoS_2 (C@MoS_2) were initially investigated to understand point–point, point–face, and face–face polarization losses, respectively. Subsequently, the metal–organic cooperative interactions between Cu and polydopamine (PDA) facilitated the construction of Cu-intercalated MoS_2 with carbon modification (Cu/C@MoS_2). This composite material achieved synergistic manipulation of all three types of polarization. Finally, leveraging the charge migration performance of semiconductors through electrochemical analysis, the research established polarization models for MoS_2 and even solid solution structural EM materials. This study not only serves as a reliable case for the fabrication of EMAs with tunable polarization sites but also provides valuable insights for theoretical analyses of the polarization loss mechanism.

The synergistic effects of the three types of polarization are analyzed, and the contributions of vacancies, gaps, substitutions, and heterointerfaces to the polarization loss, and even the EMW absorption, are analyzed based on structural and electrochemical studies. Vacancies cannot generate effective polarization loss relying on their own contribution, since the minimum RL value of pure MoS_2 with an S vacancy was only −5.55 dB, which verified the invalid effect of the point–point hybrid structure. The modification of Cu was the main source of polarization loss, which led to the minimum RL value of Cu@MoS_2 reaching −30.30 dB. In other words, the point–face structure formed by introducing point hybrid outside the polar surface will promote effective polarization loss. Comparing the EMW absorption performance of C@MoS_2 and Cu/C@MoS_2, it can be inferred that the heterogeneous interface between carbon matrices and MoS2 films also cannot induce energy loss through the polarization relaxation, which is in line with the classical dielectric polarization theory. Additionally, similar to the role of a vacancy, the point–point interaction formed by N substitution cannot optimize EMW absorption, indicating that the dipoles polarization usually cannot lose EM energy alone. Fortunately, the regulation of carbon dramatically improved the interaction between Cu and MoS_2, while a greater charge delocalization and migration can be formed to cause energy loss. The ε' and ε'' values of Cu/C@MoS_2 increased exponentially with the enhancement of carbon, especially the ε' and ε'' values of Cu/C@MoS_2 were 1.36 and 4.16 times higher than MoS_2, while those of C@MoS_2 remained almost unchanged. In this case, it can be concluded that quadra-tunable heterogeneous structures in Cu/C@MoS_2 solid solution formed a great synergistic effect among vacancy, interstitial, substitutional, and heterointerface structures. Finally, as shown in Figure 6.1e, a polarization loss model of Cu/C@MoS_2 can be built, in which the synergistic polarization was induced among point–point, point–face, and face–face heterogeneous sites constructed by S vacancy, interstitial Cu, substitutional N, and the heterointerface between carbon and MoS_2, thereby leading to excellent polarization loss.

6.1.3 Electron Hopping and Electron Migrating Model

Conductive loss occurs when EMWs propagate through absorbers, leading to the conversion of their energy into electric current. As the electric current travels through EMW absorbers, their inherent resistance generates Joule heat, thereby consuming EMW energy [12, 13]. This phenomenon is observed in absorbers with high conductivity, such as carbon materials (graphene, carbon fibers (CFs), and carbon nanotubes) and conducting polymers (polypyrrole and polyaniline). However, excessively high conductivity should be avoided, as it can result in impedance mismatch and poor EMW dissipation capacity. Conductive loss is directly related to the conductivity of EMW-absorbing materials. Therefore, it can be effectively promoted by increasing the conductivity of absorbing materials. The introduction of highly conductive components into EMW-absorbing materials serves to enhance conductive loss. Additionally, strategies like increasing the filling ratio can also contribute to elevating conductive loss.

Cao and coworkers [14] presented the shadow of multi-walled carbon nanotube (MWCNT) concentration on the loss tangent and microwave attenuation of MWCNTs/SiO_2 composites at high temperatures. The results show that temperature has a great influence on the microwave attenuation properties of MWCNT-based composites. Interestingly, they proposed a new dielectric model. The resistivity-temperature relationship ($\sigma(T)$) of the composites is determined by electronic transport. The layers of the MWCNTs are conductive and exhibit a nonparallel and imperfect layer configuration. Two modes of electronic transport are considered in the MWCNT system: migrating and hopping electrons (Figure 6.2a,b). When the EMW propagates into the composite, electrons within the MWCNTs respond directionally to the EMW. The electrons can migrate within one graphene layer or jump across defects or interfaces between layers, representing migrating and hopping electrons, respectively. Due to the high aspect ratio of the MWCNTs, they can easily form a conductive network in the silica matrix. Silica, being an insulator, increases the energy barrier between MWCNTs (Figure 6.2c,d). The electronic transport modes in the MWCNT network are depicted in Figure 6.2e,f. Hopping electrons can jump across the interface between MWCNTs, enhancing the micro-current in the MWCNT network when the filling concentration is sufficient. The potential conditions for hopping electrons are the appropriate energy barrier between MWCNTs. Higher MWCNT content results in a denser micro-current network by decreasing the energy barrier. In Figure 6.2f, electron transport in the MWCNT network is simplified as a series

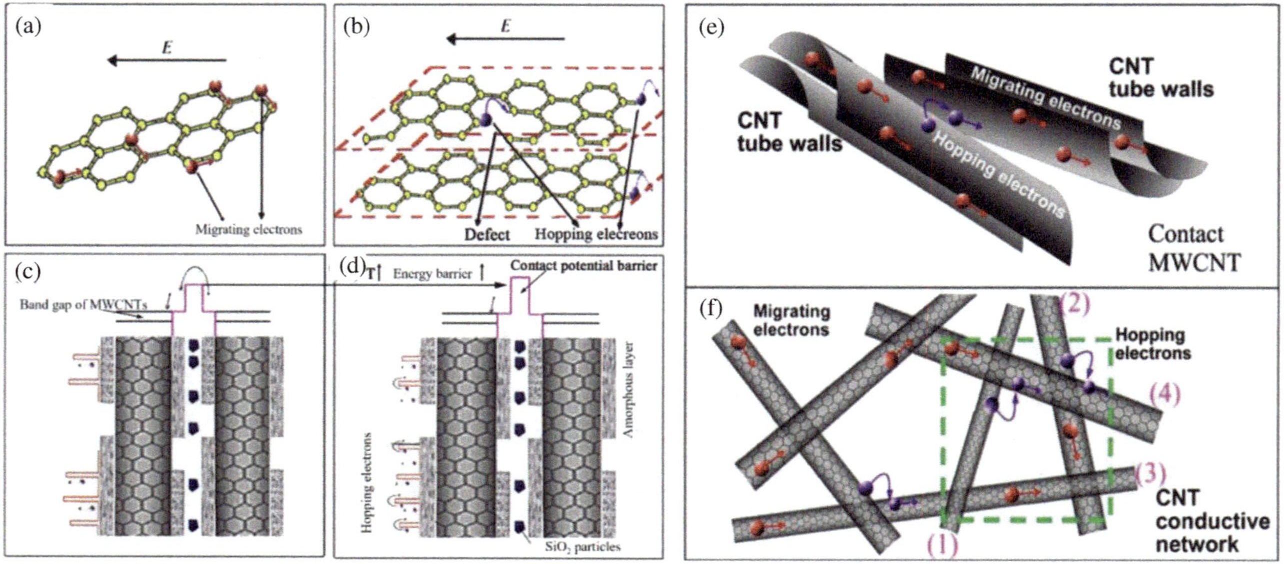

Figure 6.2 (a) Migrating electrons in the graphitic layer. (b) Hopping electron transport in the outmost MWCNT layers. (c) The characterization of hopping electron transport in composite. (d) Hopping electron transport at high temperature. Illustrations of electron transport of MWCNTs: (e) Transport in contact MWCNTs. (f) Transport in MWCNT network.

of circuit modes, including the parallel connection of migrating and hopping electronic transport in the MWCNTs. The hopping electron transport between the MWCNTs. The electronic transport in the MWCNT network is influenced by the MWCNT content and temperature.

6.1.4 Three-Dimensional Conductive Network Model in Foam

Porous melamine foam, due to its attractive features such as mechanical stability, lightweight nature, and porous structure, has gradually emerged as a potential structural support for EMA materials. Recently, melamine-based hybrid foams combined with EMW dissipative materials like MXene, graphene, Ni, and Ag nanowires have been developed for EMA or electromagnetic interference (EMI) applications. Moreover, this simple synthesis process has shown potential for large-scale applications [15]. However, the current dissipative units in melamine-based EMA materials are singular, and the effective absorption bandwidth (EAB) needs further expansion. Integrating various EMW dissipative materials into melamine foam can enhance the EMW absorption capacity through nonuniform interface polarization [16]. Among numerous EMA materials, graphene and MWCNTs are two promising candidates with strong EMW loss capabilities, high stability, and a large surface area [17, 18]. Additionally, they exhibit excellent performance in flame retardancy, thermal insulation, and hydrophobicity. Therefore, assembling graphene and MWCNT into melamine foam with appropriate micro/nanostructures and compositions is an effective approach to broaden the bandwidth of EMA materials, providing multifunctionality such as self-cleaning, thermal insulation, infrared stealth, and flame retardancy.

Ma et al. [19] reported an effective method for the preparation of multifunctional continuous melamine-based hybrid foam through a process involving impregnation, coating, and hydrazine reduction (Figure 6.3a). This foam consists of a three-dimensional (3D) porous melamine framework, a two-dimensional (2D) continuous graphene network, and a one-dimensional (1D) MWCNT interwoven network (Figure 6.3b). The 3D porous structure of the continuous graphene and MWCNT melamine foam (CGMF) exhibits excellent impedance matching characteristics, ensuring efficient transmission and dissipation of EMW through the continuous graphene and MWCNT networks. The synthesized CGMFs not only possess lightweight properties (~0.013 $g\,cm-^{3}$) but also demonstrate an ultra-wide EAB of 10.8 GHz and a minimum RL_{min} of – 50.43 dB at a thickness of 4.6 mm. Furthermore, these CGMFs exhibit good mechanical stability and hydrophobicity, with a contact angle of 126.5°, enabling effective self-cleaning functionality. Additionally, due to the multilayered porous structure of graphene and MWCNT, as well as their nonflammable properties, flame-retardant CGMFs exhibit excellent thermal insulation and infrared stealth capabilities. Figure 6.3c illustrates the potential absorption mechanism of the 3D porous CGMF1.0 (1:2). First, the 3D porous structure contributes to the impedance matching of EMW between air and the interface of the absorption materials, reducing the reflection ratio of EMW on the surface. Subsequently, multiple inner multi-reflections and multi-scatterings increase the propagation path, further enhancing the exhaustion of EMW. Second, with the introduction of electrically conductive MWCNT and reduced graphene oxide (RGO), abundant dipole and interfacial polarization can form under the alternating EM field on the heterogeneous interface of MWCNT and RGO, the melamine foam (MF) skeleton, and RGO. Additionally, due to the bi-continuous conductive network constructed by interweaving MWCNTs

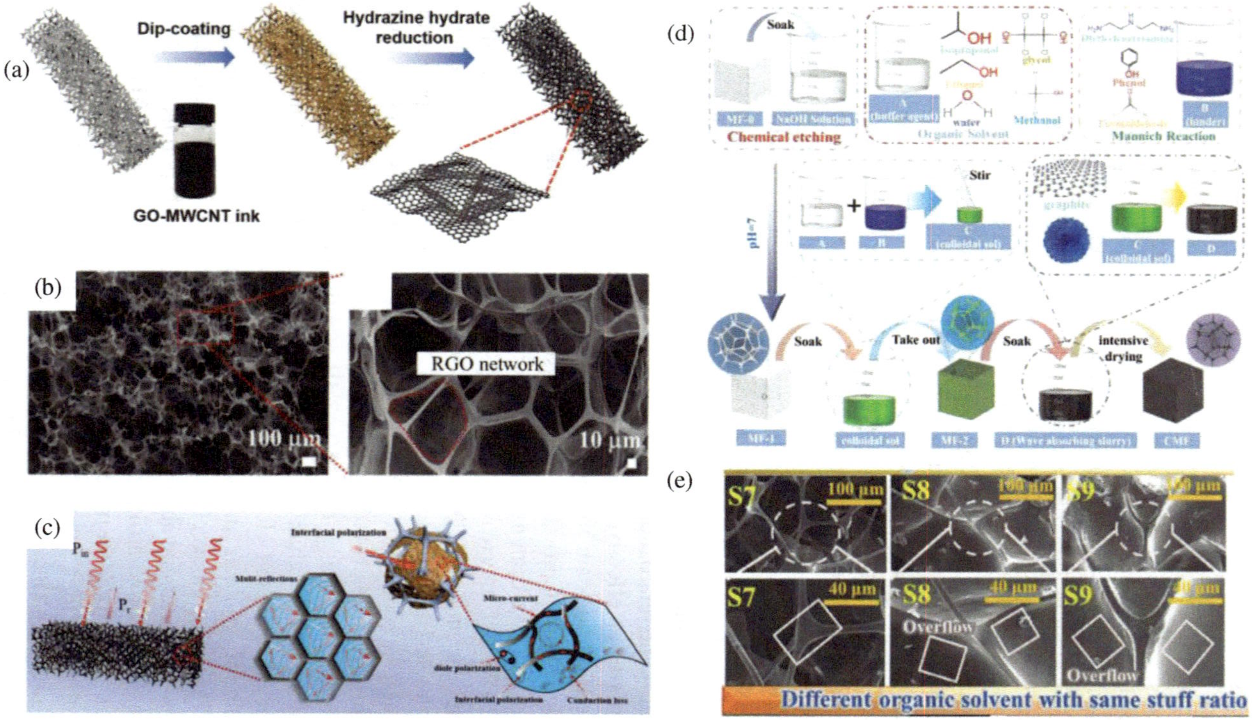

Figure 6.3 (a) Schematic illustration of the fabrication process of CGMFs. (b) SEM image of the sample at different magnifications. (c) Schematic diagram of the possible EMA mechanism of CGMFs. (d) The preparation process of CMFs via an organic polymer assisted in direct adsorption strategy. (e) SEM image of the sample at different magnifications (30% stuff ratio).

and lamellar RGO sheets, the conductive loss verified by Cole–Cole semicircles serves as another dielectric dissipation mechanism for CGMF1.0 (1:2). This mechanism generates induced currents when irradiated, and these long-range-induced currents rapidly attenuate in the resistive network, converting into thermal energy.

Li et al. [20] developed a room-temperature synthesized foam-based microwave-absorbing device carbonized melamine foam (CMF) using melamine and phenolic aldehyde amine as the 3D foam skeleton and cementing compounds, respectively. They incorporated graphite as the dielectric loss component and thermal conductivity bridge (Figure 6.3d). The improvement process involved the use of alkaline solutions (NaOH and KOH) for etching, enhancing the roughness and active sites of the MF to facilitate better attachment. Subsequently, the phenolic aldehyde amine, synthesized through the Mannich Reaction using phenol, formaldehyde, and diethylenetriamine, was dissolved in organic solvents to create a colloidal sol. The solubility of phenolic aldehyde amine in the organic solvent directly determined the loading capacity for graphite powder. The etched MF was immersed in the colloidal solution for the initial attachment, allowing the solution to thoroughly penetrate the foam network and enhance the stability of the final foam. This step played a crucial role in minimizing the risk of powder detachment caused by reduced adhesion after the powder was mixed into the gel solution. Subsequently, the EMW absorption slurry (graphite organic colloidal sol) was prepared through sonication treatment to promote the dispersion of bulk graphite powders (Figure 6.3e). The CMF demonstrates unique versatility attributed to the porous structure of the melamine foam and the synergistic effect between graphite and organic binder compound components. The porous structure not only imparts lightweight characteristics to CMF but also endows it with anechoic potential due to the tortuous nature of the sound transmission path. Furthermore, the mesoscopic network formed by graphite and phenolic aldehyde amine significantly enhances the EMW absorption performance and thermal conductivity of CMF, given graphite's strong electrical and thermal conductivity. Notably, CMF achieves outstanding EWA performance with an ultra-low RL_{min} of − 41.7 dB and an ultra-wide EAB of 9.8 GHz at a thickness of only 2.60 mm. The fabrication of CMFs avoids energy-intensive methods such as freeze-drying, hydrothermal, and high-temperature calcination, leading to reduced manufacturing costs and improved scalability.

6.1.5 Disordered Structure on the Atomic Scale–High-Entropy Models

High-entropy materials (HEM), defined as a composition consisting of oxygen and five or more metal cations with equimolar or near equimolar ratios in the 5–35% atomic concentration range, are characterized by concentrated solid solutions. They have garnered significant attention in various functional material applications, including catalysis and energy storage [21, 22]. In general, the crystal structure and properties of HEM are influenced by four core factors: sluggish diffusion, configurational entropy, lattice distortion, and cocktail effects. While some reports exist on high-entropy absorption materials, the high-entropy effect in binary high-entropy oxides (HEO) is not clearly evident [23–25]. Additionally, the internal details of the local microstructure in HEO, such as their short-range element distribution, lattice distortions, oxygen defects, and shear strains, remain to be elucidated. Furthermore, a comprehensive understanding of the local atomic features is crucial for optimizing the properties of these materials. Therefore, it is necessary to delve deeper into the microstructural evolution with increased entropy to comprehend the anomalies and unprecedented characteristics of HEO.

Zhao et al. [26] have designed and developed an effective high-entropy strategy to enhance microwave attenuation performance. Starting with the perovskite $BaTiO_3$, they designed a high-entropy $(Ca_{0.2}Sr_{0.2}Ba_{0.2}La_{0.2}Pb_{0.2})TiO_3$ compound by introducing equimolar ratios of Ca, Sr, La, and Pb at the Ba site (Figure 6.4a). The concept of a stable high-entropy structure allows establishing a connection between local element distribution, atomic structure, defects, and dielectric properties. The results indicate that the synthesized high-entropy $(Ca_{0.2}Sr_{0.2}Ba_{0.2}La_{0.2}Pb_{0.2})TiO_3$ perovskite exhibits an ultra-high density of grain boundaries, distorted lattice, and shear strain, leading to strong interfaces, dipoles, and defect-induced polarization. The large strain generated by intense planar distortion is favorable for electron-carrier transport. Furthermore, electron holography was employed to assess the charge distribution induced by strain in various perovskites (Figure 6.4b–f). When compared to the $BaTiO_3$ sample, the high-entropy-stabilized $(Ca_{0.2}Sr_{0.2}Ba_{0.2}La_{0.2}Pb_{0.2})TiO_3$ exhibited a significant charge distribution around grain boundaries and locations with concentrated strain. The quantitative charge density distribution line profiles (Figure 6.4d) reveal that the charge density of high-entropy $(Ca_{0.2}Sr_{0.2}Ba_{0.2}La_{0.2}Pb_{0.2})TiO_3$ is more than 10 times higher than that of $BaTiO_3$. The accumulation of positive/negative charges around numerous boundaries and strong strain locations is beneficial for enhancing dielectric polarization loss. First-principles computations were further employed to replicate the electrical characteristics for a better understanding of the lattice distortion caused by the high-entropy effect. The differential charge density maps between the oxygen and A-site atoms of several ceramic oxides are presented in Figure 6.4g. Clearly, as the high-entropy effect gradually increases, the polarized region around the O atom considerably expands, indicating the presence of substantial electron enrichment effects, particularly in the $(Ca_{0.2}Sr_{0.2}Ba_{0.2}La_{0.2}Pb_{0.2})TiO_3$ sample. This is attributed to the existence of metal elements with different valences within the oxide, leading to the formation of abundant oxygen vacancies and inducing structural relaxation of the ions around these vacancies. When an oxygen atom abruptly disappears, forming an oxygen vacancy, electrons become highly bound around the A-site atom, resulting in electron enrichment. It can be hypothesized that due to electron enrichment effects, the abundance of local electrons in the metal atoms contributes to stronger dipole polarization, thereby promoting microwave losses. It is noteworthy that the effective bandwidth of high-entropy $(Ca_{0.2}Sr_{0.2}Ba_{0.2}La_{0.2}Pb_{0.2})TiO_3$ is almost twice that of $BaTiO_3$. This demonstrates that the designed high-entropy strategy can be an attractive approach for developing promising EM absorption materials.

However, as the study progresses, more features of HEMs are revealed, which pass through a series of unfavorable consequences caused by high-entropy strategies. For HEMs containing large radius atoms, the electron spin channel fights against strong disordered scattering, leading to reduced electron transport [27, 28]. In addition, high group entropy may promote the formation of single-phase solid solutions rather than multiphase compounds, resulting in a weak response to the interfacial structure of HEMs. In order to address the aforementioned challenges, the following research framework should be constructed in depth: (1) An effective design guideline must be adopted to generate abundant heterogeneous interfaces within the HEM. (2) In addition to the modulation of the interfaces, it is equally important to maintain a large number of defect positions in the HEM to obtain a significant density of polarization centers. (3) Although the design of heterogeneous interfaces has been mentioned previously, due to its single-phase solid solution nature, it is impossible to maintain an abundance of heterogeneous interfaces. Inspired by conventional interface engineering, a new multiphase interface engineering provides a possible approach to realize the research framework mentioned earlier.

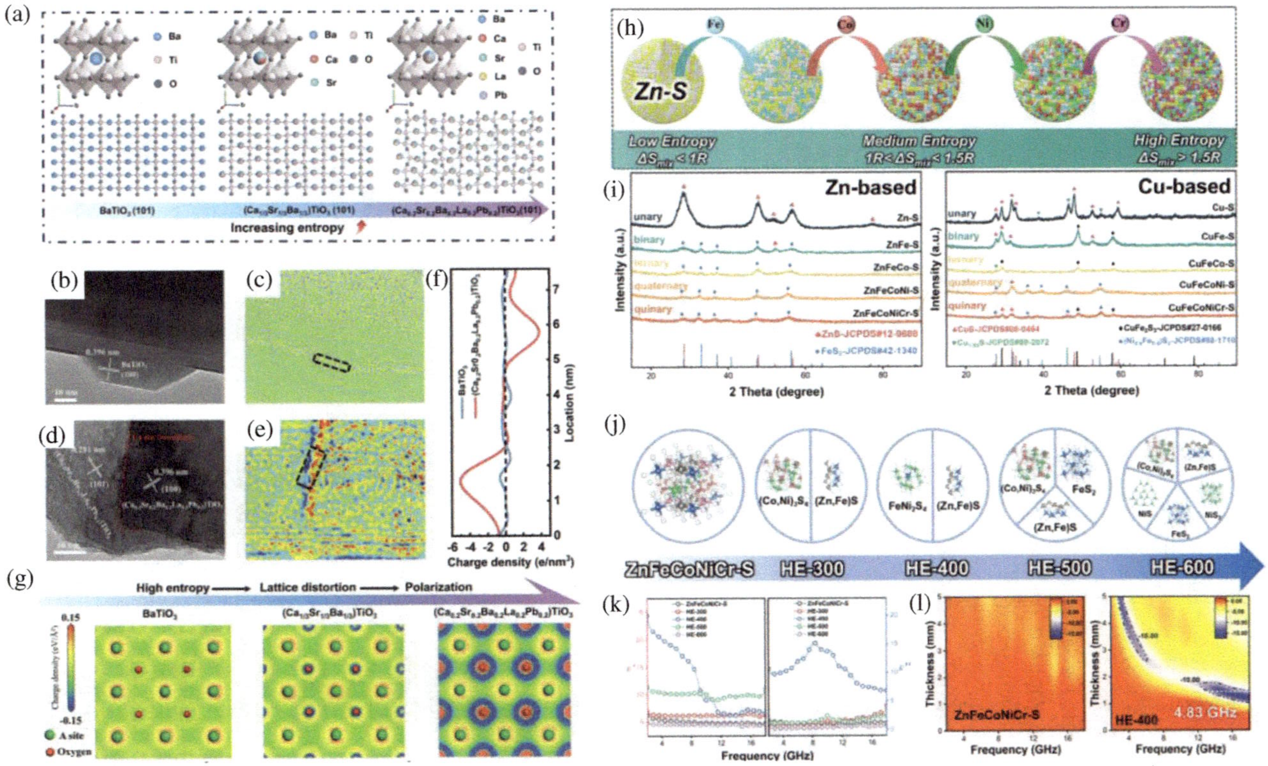

Figure 6.4 (a) Schematic diagram of the distorted titanate perovskite lattices with increasing entropy. (b) HRTEM image and (c) the reconstructed charge density distribution map of $BaTiO_3$, in which the color reflects the magnitudes of different charge density. (d) HRTEM image and (e) the reconstructed charge density distribution map of high-entropy $(Ca_{0.2}Sr_{0.2}Ba_{0.2}La_{0.2}Pd_{0.2})TiO_3$. (f) The quantitative charge density distribution line profiles in the region of the black dashed frames for the $BaTiO_3$ and $(Ca_{0.2}Sr_{0.2}Ba_{0.2}La_{0.2}Pd_{0.2})TiO_3$. (g) Simulated charge mappings of the three $BaTiO_3$-based perovskites, suggesting the presence of considerable polarization with increasing entropy. (h) Increased mixing entropy driving Zn-based sulfides to form a high-entropy-stabilized structure. Powder XRD pattern of (i) Zn-based sulfides and Cu-based sulfides. (j) The XRD pattern of HE-300, HE-400, HE-500, and HE-600. (k) The EM parameters of HE-300, HE-400, HE-500, and HE-600. (l) 2D-RL plot of HE-400 sample.

Hui et al. [29] proposed a novel multiphase interface engineering to mitigate the detrimental effects of weak heterogeneous interfaces on the EMW absorption properties of high-entropy sulfides. Stable high-entropy zinc-based sulfides and phase-separated copper-based sulfides were synthesized using a solvothermal method (Figure 6.4h,i). The results indicate that in high-entropy zinc-based sulfides, the high configurational entropy leads to a reduction in charge density at Co atomic sites and a weaker electron exchange effect between Co/Ni and Fe atoms, resulting in decreased EM parameters. As expected, the high-entropy zinc-based sulfide (ZnFeCoNiCr-S) failed to effectively absorb EMW due to its limited interface effects and low complex dielectric constant. In contrast, the multiphase copper-based sulfide (CuFeCoNiCr-S) exhibited satisfactory EMW absorption performance (i.e. at 2.00 mm, effective absorption bandwidth: 6.70 GHz), attributed to multiphase-induced interface polarization losses, particularly induced by the specific interface of CuS (105)/$CuFe_2S_3$ (200). Furthermore, the single-phase ZnFeCoNiCr-S transformed into a phase-separated structure through thermal decomposition at 400 °C (Figure 6.4j). The EMW absorption performance was significantly enhanced (1.45 mm, 4.83 GHz) due to strong interface polarization induced by the $FeNi_2S_4$ (311)/(Zn, Fe)S (111) phase interface (Figures 6.4k,l). This study reveals the limitations of EMW absorption properties in high-entropy sulfides and proposes a multiphase interface engineering strategy to enhance interface charge accumulation and polarization, achieving optimized EMW absorption characteristics.

6.2 Physical Models Related to Structural Design

6.2.1 Hierarchical Structural Models for Improved Impedance Matching

Microcosmic 3D hierarchical structural design has proved to be an effective strategy to obtain high-performance microwave absorbers. Based on current understanding, materials related to 1D MA leverage anisotropy and a high aspect ratio. This tendency facilitates the formation of a carrier transport path in the axial direction under EM fields, leading to energy consumption of EMWs through conducting loss [30, 31]. In contrast, 2D MA-related materials contribute to EMW attenuation primarily through polarization enhancement, attributed to their high specific surface area and the multiple reflections occurring between layers [32, 33]. On the other hand, 3D materials enhance impedance matching by introducing pores or gaps, thereby improving the MA properties [34]. Notably, recent studies have indicated that hierarchical 3D structures containing 1D or 2D cells demonstrate remarkable microwave absorption properties. Recently, a new class of materials with 2D layered structures, known as MXene (metal carbides and nitrides), has been incorporated into the realm of novel MA materials due to its high specific surface area, excellent conductivity, and quasi-metallic properties. However, most research tends to directly composite 2D $Ti_3C_2T_x$ MXene with other components, inevitably leading to aggregation and compromising its inherent superiority. Additionally, in the fabrication of these MXene-based MA materials, the optimization of impedance matching should be considered, as their reflection of EMW is influenced by high electrical conductivity.

Therefore, a hierarchical ternary core–sheath composite with MXene as the intermediate dielectric layer can be designed to achieve high-performance MA. Specifically, the inner core needs to adopt some kind of 1D template with excellent mechanical properties and easy surface modification to firmly support the MXene, thus forming a 1D structure. On the other hand, the outermost material should have the ability to improve the impedance matching and to protect the MXene from oxidation. Wang et al. [35] prepared a layered CF@MXene@MoS_2 (CMM) core–sheath structure with a 1D MXene sheath and an edge MoS_2 sheath using CF as a 1D template and MoS_2 as the outer layer encapsulation. The fabrication process of the hierarchical CMM absorber is depicted in Figure 6.5a. Carbon fiber was rendered positively charged through cetyltrimethylammonium bromide (CTAB) modification and immersed in a multilayered $Ti_3C_2T_x$ suspension. Due to its abundant functional groups (—OH, —F, —O), $Ti_3C_2T_x$ flakes acquired a negative charge and electrostatically adhered to the surface of CTAB-modified CF (CF-CTAB). Consequently, the carbon microfibers were entirely covered by $Ti_3C_2T_x$ MXene, giving rise to a distinctive 1D MXene sheath. Next, 2D MoS_2 nanosheets were grown on CF@MXene (CM) using a hydrothermal method to modify the surface of MXene. MXene flakes closely stacked along the axis direction of the CF, forming a hollow cylindrical sheath. The outermost MoS_2 nanosheets were vertically aligned on the surface of the MXene sheath, constructing an additional sheath. In summary, the hierarchical ternary microrods were established, comprising an inner CF core, an intermediate MXene sheath, and an outer MoS_2 sheath (Figure 6.5b1–d2). As expected, tunable and efficient MA between the whole X-band and Ku-band can be achieved by regulating the loads of MoS_2 (Figure 6.5e).

The mechanism of EM losses induced by the hierarchical ternary model is described (Figure 6.5f). First, MoS_2 nanosheets with a stacked edge-on structure contribute to the absorption of an incident wave, minimizing undesired reflection. Microrods with a high aspect ratio easily come into contact with each other, forming a 3D conductive network within the wax matrix. In the alternating EM field, the excited electrons in CF and $Ti_3C_2T_x$ migrate along the axial direction or hop to other contacted microrods through the conductive network, converting EM energy into thermal energy and thus consuming the EMW. Moreover, the hierarchical ternary composite provides numerous heterointerfaces, enhancing interfacial polarization, as discussed earlier. Dipolar polarization caused by residual cetyltrimethylammonium bromide (CTA^+), hydroxyl groups (—OH) on CF, abundant functional groups of $Ti_3C_2T_x$, and defects in MoS_2 also contributes to dielectric loss. Additionally, layered $Ti_3C_2T_x$ and MoS_2 induce multiple reflections and scattering of the EMW, prolonging the propagation path and reinforcing EMW dissipation. Consequently, the unique 3D ternary composite fully leverages the advantages of its low-dimensional components, resulting in unprecedented MA performance. The optimal RL curves for the four types of prepared samples are extracted in Figure 6.5e, demonstrating the highly efficient MA of the CMM composites. Differing from spherical core–shell structures that typically employ magnetic particles as the core, the CMM microrods utilize the 3D conductive network for conductive loss instead of magnetic loss in consuming EMW energy. Even when compared with other 1D core–shell structures, the CMM microrods exhibit strong stability due to the close combination of the sheath and the completely wrapped structure.

6.2.2 Core–Shell Structure Model

In the pursuit of broadband and high-performance microwave absorbers, a robust magnetic loss is indispensable for achieving balanced EM parameters and a wider frequency response. Numerous studies have validated that the combination of ferromagnetic particles and dielectric materials can yield MA materials with absorption in multiband regions. Recently, metal–organic framework (MOF) derivatives have demonstrated unique functional

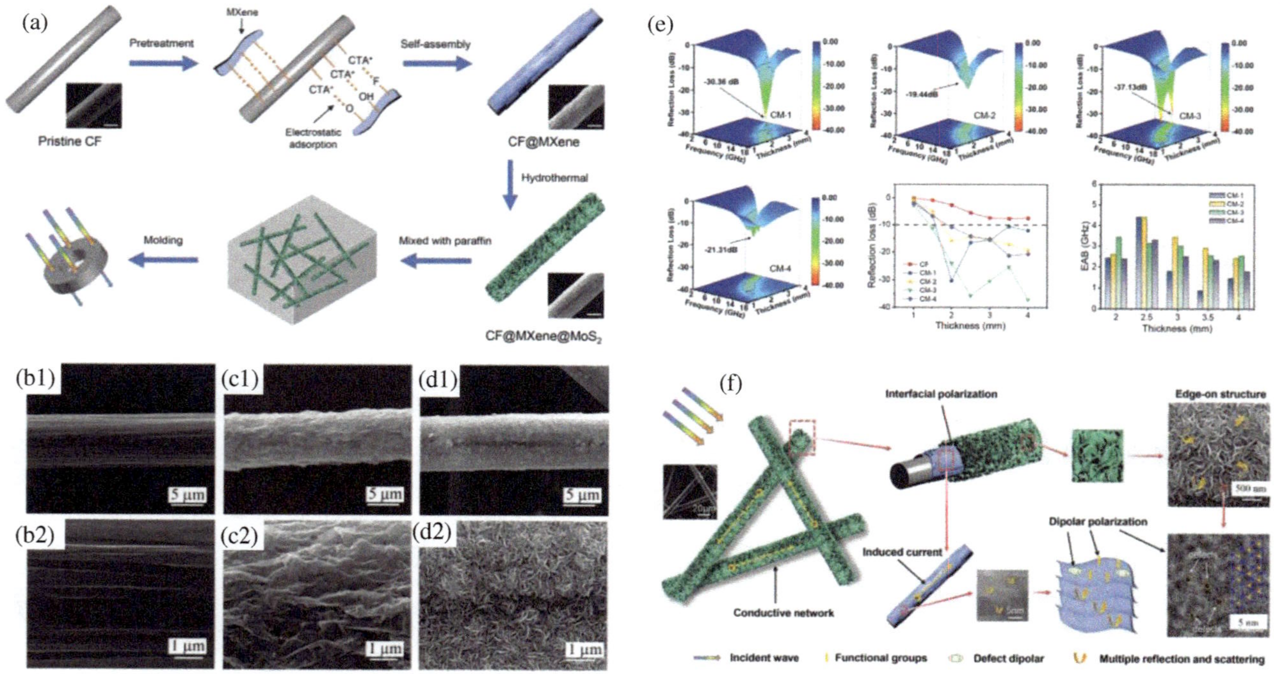

Figure 6.5 (a) Schematic illustration of the synthesis process of the CMM absorber. SEM images of as-obtained materials. (b1, b2) CF-CTAB. (c1, c2) CM. (d1, d2) CMM. (e) The EMW absorption performance of the sample. (f) Schematic illustration of microwave absorption mechanisms for the CMM system.

advantages [36, 37]. Due to their adjustable microstructure, intrinsic heteroatom doping, and reduced magnetic metal particles, MOF-derived MA composites not only exhibit magnetic-dielectric synergy but also possess favorable impedance matching, exemplified by Ni@C, Co@C, Fe@C, and others [38, 39]. Additionally, single atoms find wide applications in the field of energy due to their high atom utilization. Doping single atoms (such as W, Co, and Ru) represents an effective pathway to modify the electronic state of the metal element and adjust electrical conductivity, leading to improvements in catalytic properties. To further enhance dielectric loss and expand absorption frequency, the combination of a strong magnetic core and MOF-derived dielectric carbon shell with single atoms is a promising avenue worth exploring.

Huang et al. [40] successfully synthesized single-atom-doped Fe_3O_4@Zn-N-carbon (FZNC) composites as magnetic media synergistic microwave absorbers using a faceted method. Through skillful annealing treatment, Zn and N are transformed at the atomic level and homogeneously anchored in the carbon matrix to form stacked heteroatom-doped pod-like dielectric shells (Figure 6.6a). Leveraging the magnetic Fe_3O_4 core, FZNC composites demonstrate outstanding microwave absorption performance, characterized by both reflection loss capability and broadband response characteristics. To achieve maximum polarization, the complex permittivity of the Fe_3O_4@Zn-N carbon composites can be adjusted by tuning the pod-like Zn-N-doped carbon shells through repeated ZIF-8 growth cycles. The obtained FZNC composites exhibit remarkable absorption, with RL_{min} values reaching − 61.9 dB at 2.5 mm and effective absorption bandwidths extending as high as 15.2 GHz. The monoatomic Zn-doped dielectric carbon shells tightly adhere to the Fe_3O_4 inner core, creating a unique energy absorption system with several advantages. The doping of individual Zn and N atoms with different electronegativity results in high-density polarization, maintaining maximum dissipative behavior, while the magnetic Fe_3O_4 core experiences stronger magnetic loss due to magnetic coupling. The magnetic-dielectric synergy of the core–shell Fe_3O_4@Zn-N-carbon microspheres provides an ultra-wideband microwave response.

The core–shell Fe_3O_4@Zn-N-carbon microsphere is composed of a Fe_3O_4 sphere and a pod-like ZIF-derived carbon shell, creating abundant interfaces and contact regions. The reduced FeO particles, occupying the original position of Fe_3O_4, establish unique FeO-Fe_3O_4 heterojunction interfaces, providing advantages in trapping electron/hole carriers and forming interfacial energy barriers. The pod-like Zn-N-carbon dodecahedral framework contains numerous squeezed nested interfaces, including those between adjacent dielectric dodecahedra and the magnetic Fe_3O_4 core, facilitating high-density interfacial polarization (Figure 6.6b). To explore the intrinsic magnetic properties, the magnetic flux line distribution of the synthesized FZNC composites is visualized via Lorenz electron microscope (Figure 6.6c). These images show that the magnetic flux lines of a single FZNC-1 microsphere perfectly surround its surfaces, forming a closed loop with two focused points. The radiated magnetic flux lines of FZNC-1 microsphere exceed its size and extend to the outside space. FZNC-2 microsphere, obtained under similar conditions, exhibits a comparable magnetic lines distribution, benefiting from the magnetic enhancement of adjacent microspheres and a higher density distribution of magnetic flux lines in the same space. Additionally, the enhanced magnetic responding ability is confirmed by the unique coupling phenomenon among Fe_3O_4@Zn-N-carbon microspheres. In Figure 6.6c, three closely touching FZNC-3 microspheres show an integral magnetic line distribution instead of disarray distributions, revealing compulsory assembly magnetic interaction. The magnetic flux lines radiate out from the first microsphere, step over the second microsphere, and finally focus on

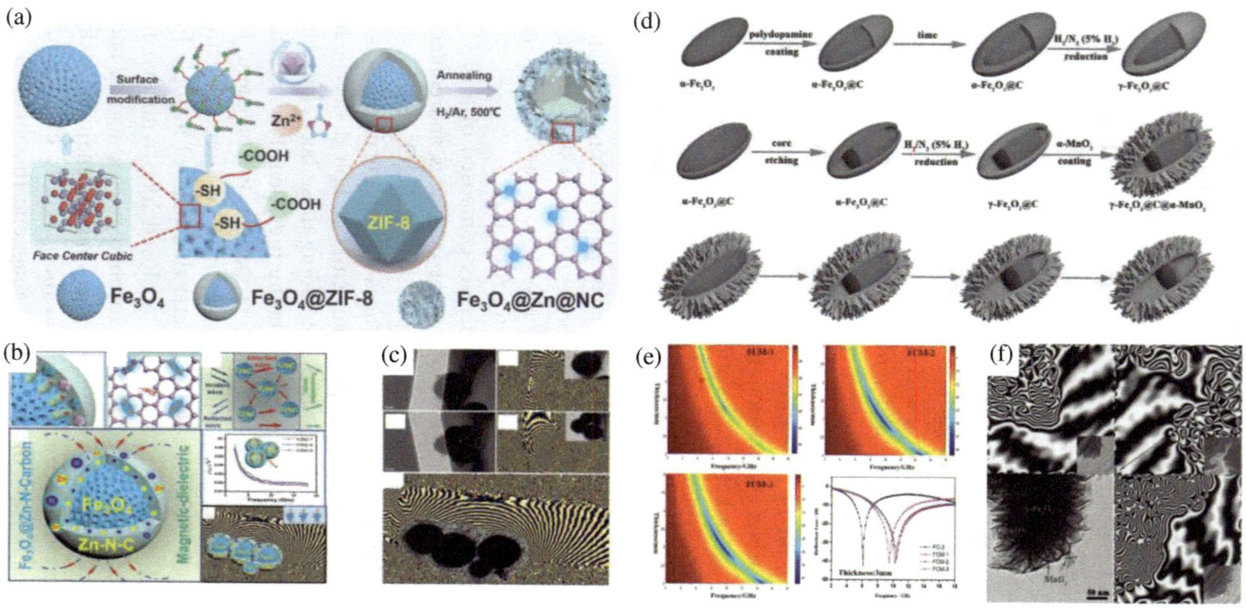

Figure 6.6 (a) Schematic preparation process of Fe_3O_4@Zn-N-Carbon composites. (b) Schematic illustration of microwave absorption mechanisms for the FZNC composites. (c) The electron holography images and the corresponding stray magnetic field distribution for the samples. (d) Schematic illustrations of the synthetic process for hierarchical ellipsoidal shape γ-Fe_2O_3@C and γ-Fe_2O_3@C@α-MnO_2. (e) The EMW absorption performance of the sample. (f) Off-axis electron holography of the samples.

the third microsphere, constructing a unique magnetic coupling effect. The special magnetic interaction and coupling behaviors of Fe_3O_4@Zn-N-carbon microspheres greatly enhance the potential for intrinsic storage and loss ability. Large-scale magnetic coupling among Fe_3O_4@Zn-N-carbon microspheres promotes the overall magnetic properties and enhances the magnetic loss capacity.

6.2.3 Double-Shell Structure Model

Hierarchical nanostructures, consisting of a magnetic core and dielectric shells, have garnered significant attention in research due to their unique ability to combine the advantages of multiple components. γ-Fe_2O_3, known for its morphology versatility and chemical stability in comparison to magnetite, has been widely utilized. Various shapes of iron oxide@C core–shell structures, including rods, rings, spheres, and dendrites, have been fabricated and employed in catalysts, electrochemistry, and microwave absorption applications. More recently, there has been extensive research on the synthesis and application of double-shell materials due to their tunable properties [41, 42], such as enhanced electro-responsiveness in double-shell SiO_2@TiO_2 [43] and wideband microwave absorption in CoNi@SiO_2@TiO_2 [34]. Consequently, it is anticipated that a core–shell structure comprising a γ-Fe_2O_3 core and a carbon/MnO_2 shell can exhibit broadband and tunable microwave absorption performance. However, there is a dearth of research on novel types of microwave absorbers constructed using γ-Fe_2O_3, carbon, and α-MnO_2.

Che and coworkers [44] proposed a facile and effective method for the preparation of multicomponent γ-Fe_2O_3@C (FC) and γ-Fe_2O_3@C@α-MnO_2 (FCM) core–shell nanospindles. The synthesis of γ-Fe_2O_3@C and γ-Fe_2O_3@C@α-MnO_2 is illustrated in Figure 6.6d. First, nanospinners of α-Fe_2O_3 were prepared through a hydrothermal method. Second, a PDA layer was uniformly coated onto the surface of elliptical α-Fe_2O_3 particles at room temperature. Due to the encapsulation property of dopamine condensation, the polymer layer took on a typical elliptical morphology. Third, α-Fe_2O_3@dopamine was reduced in a furnace under an H_2/N_2 (5% H_2) atmosphere to obtain γ-Fe_2O_3@C (Figure 6.6d). Finally, γ-Fe_2O_3@C@α-MnO_2 with a special cavity structure, was successfully prepared by reducing $KMnO_4$ and chemically bipolar etching. Specifically, the different thicknesses of the carbonized layers were regulated by the reaction time of dopamine polymerization. The γ-Fe_2O_3@C nanospindle exhibits a strong RL of $-$ 45.0 dB and a broad effective bandwidth of 3.89 GHz at a 2 mm thickness of the carbon shell (Figure 6.6e). The effective bandwidth of the γ-Fe_2O_3@C@α-MnO_2 nanospindle is approximately four times wider (9.2 GHz) than that of the γ-Fe_2O_3@C nanospindle FC-3 (2.26 GHz, 3 mm).

The following outlines novel microwave absorption models by exploring the dependence of RL results in FC and FCM nanospindles on carbon coating, α-MnO_2, and the core etching process, indicating multifaceted effects in these material systems. The primary associated mechanisms are detailed later: (1) In the hierarchical absorber system comprising a magnetic core and dielectric shell, magnetic absorption consistently emerges as a dominant factor. Off-axis electron holography was conducted to analyze the magnetic stray field distributed from the spindle (Figure 6.6f). Reconstruction graphs reveal high-density magnetic flux lines radiating from the γ-Fe_2O_3 core, indicating its intrinsic ferromagnetism. Both the carbon shell and α-MnO_2 shell are unable to impede the strong magnetic flux stray lines emanating from the γ-Fe_2O_3 core. The stray magnetic flux lines can penetrate through the nonmagnetic shells, affirming the ferromagnetic nature of the entire spindle. Moreover, strong coupling interference of magnetic flux lines between neighboring nanospindles is confirmed by electron holography analysis. This robust magnetic field, induced by the γ-Fe_2O_3 core, significantly contributes to the strong RL observed in FC and FCM samples. (2) The ε'' overall performance of the FCM series in the 2–18 GHz range consistently aligns well with their RL values. Conversely, μ'' values exhibit an opposite trend, indicating that the microwave absorption enhancement primarily arises from dielectric loss in the MnO_2 shells rather than from γ-Fe_2O_3. With an increase in frequency, ε' decreases, attributed to dielectric relaxation. Additionally, the wrapped carbon shell exhibits better conductivity compared to the loosely stacked α-MnO_2 shell. The incorporation of α-MnO_2 leads to a noticeable reduction in ε', resulting in a better match between ε' and μ'. Furthermore, the tan $\delta\varepsilon$ values of FCM, defined as the ratio between ε'' and ε' of complex permittivity, significantly increase due to this reduction. These factors are likely the main contributors to the broader effective absorption bandwidth observed in FCM samples, indicative of a better impedance balance between FCM spindles and air. (3) The core–shell structure, comprising ferromagnetic and dielectric materials, introduces both magnetic loss and dielectric dissipation. Additionally, the loosely stacked α-MnO_2 layer on FC contributes to multiple scattering of the incident microwave. The olive-shaped particle induces anisotropic dipole dielectric polarization. The self-assembled ultrathin lamellar α-MnO_2 nanosheets, along with varied volumes of dipolar-distribution voids between the core and shells, provide ample sites for repeated scattering and intrinsic reflection. Microwave energy propagating through the structure can be dissipated by multiple interfacial polarization events. The dipolar cavity accumulates abundant free space charge, ensuring the penetration of the incident microwave into the magnetic cores with minimal obstruction, enhancing the microwave absorption properties of the samples through both dielectric loss and improved impedance matching with air. Furthermore, interface polarization, electronic polarization, ion polarization, resonance, and relaxation occurring between the core and shell contribute to absorption. The synergistic effects result from a combination of these multiple factors.

6.3 Intelligent Off/On Switchable Model

In practical applications of EMW technology, the need for intelligent microwave absorption materials has become more critical due to changing surroundings and situations. There is an urgent demand for materials capable of adjusting their MA performance according to real-time requirements [6, 45, 46]. For instance, in the civil field, industrial robots are replacing traditional workers, and MA materials are integrated into these robots to prevent EM signal leakage. However, the MA performance needs to switch to the off-mode when the robots receive real-time wireless instructions from remote workers. Similarly, in the military field, stealth fighters use significant amounts of MA materials to reduce their EM signature. However, during daily training, the MA performance should switch to the off-mode to avoid safety issues arising from dense flying. In both civil and military scenarios, the demand for intelligent MA materials with switchable off/on MA performance is highly desirable, as they offer greater adaptability to evolving needs. Traditional MA materials mostly fall under the category of passive stealth materials, and once their fabrication is complete, their MA performance remains fixed. To address this limitation, researchers have proposed strategies for preparing real-time tunable MA materials. Under external stimuli such as voltage, heat, and light, certain materials undergo inherent changes in their EM properties, leading to real-time variable MA performance. For instance, Lv et al. [47] demonstrated that Sn/SnS/SnO_2/C hybrid materials exhibit tunable MA performance with excellent dielectric response at low voltage. However, these reports still face some crucial issues that might hinder their potential applications. First, most reports struggle to achieve a well-defined "off-mode" for their MA performance. Specifically, there are still some frequency bands where effective absorption occurs in their "off-mode," potentially strongly impacting the reception and transmission of wireless communication signals. Therefore, research on intelligent MA materials is still in its early stages, and developing smart MA materials with real-time and significant changes in MA performance remains a substantial challenge.

Cheng et al. [48] first developed a composite aerogel of VO_2/reduced graphene oxide (rGO) with excellent switchable microwave absorption MA performance. Initially, VO_2 nanoparticles were prepared using a hydrothermal method. Subsequently, surface modification was performed to introduce NH_2 functional groups, facilitating nanocomposite formation with graphene oxide (GO) (Figure 6.7a). The resulting VO_2 derivative is denoted as VA. Third, VA nanoparticles were dispersed in an ethanol solution and mixed with a GO solution. The mixture underwent a simple solvent-thermal reaction to form a hydrogel, wherein amidation occurred between NH_2 of VA and COOH of GO. The obtained hydrogel was freeze-dried to obtain the final composite aerogel composed of rGO and VA. The mass ratio of GO to VA in the initial solution was set at 1:1 and 1:3, and the corresponding composite aerogels are denoted as G/VA-100 and G/VA-300. For comparison, a pure rGO aerogel was prepared and named GF. To further validate the results, an in situ XRD test was conducted for VA nanoparticles, G/VA-300, and pure GF. The in situ heating/cooling process revealed several changes in G/VA-300 (Figure 6.7b) and VA, which were not observed in pure GF. In the case of G/VA-300, four typical peaks of monoclinic VO_2 (M phase) were evident when the temperature was set at 60 °C. As the temperature increased to 80 °C, these peaks shifted to different angles, indicating the transformation of VA in G/VA-300 to the rutile phase (R phase). Upon decreasing the temperature from 80 to 40 °C, these peaks reverted to the monoclinic phase, consistent with the XRD spectra change of pure VA. This process unequivocally demonstrated the phase change behavior of VA in the composite aerogel during the temperature-increasing procedure, contributing to the electrical properties. This significant contrast is attributed to the phase change behavior of VA in the composite aerogel. As a result, with the temperature increased, the VO_2-based material exhibited real-time phase transition behavior. This further resulted in changes in the conductivity of the composite aerogel, leading to switchable MA performance. Specifically, two different switch modes ("on-off" and "off-on") can be achieved based on the different starting materials. Compared to other reports, the switchable effect is significant, with the highest EAB and RL values being 7.27 GHz and − 49 dB, respectively.

Pan et al. [49] constructed a V_2C-VO_2(M) heterostructure. The preparation of the V_2C-VO_2(M) heterostructure involved a solution-based oxidation technique followed by a high-temperature phase transition process, as depicted in Figure 6.7a. In summary, multilayer V_2C MXene was synthesized by selectively removing Al layers from the MAX(V_2AlC) precursor using aqueous HF acid. The V_2C-VO_2(B) heterostructure was produced through a subsequent hydrothermal reaction carried out over a 20-hour period at 180 °C in the presence of multilayer V_2C and hydrogen peroxide (H_2O_2). By exploring a range of hydrothermal temperatures (160 °C, 200 °C), V_2C-VO_2(B) heterostructures with varying degrees of oxidation

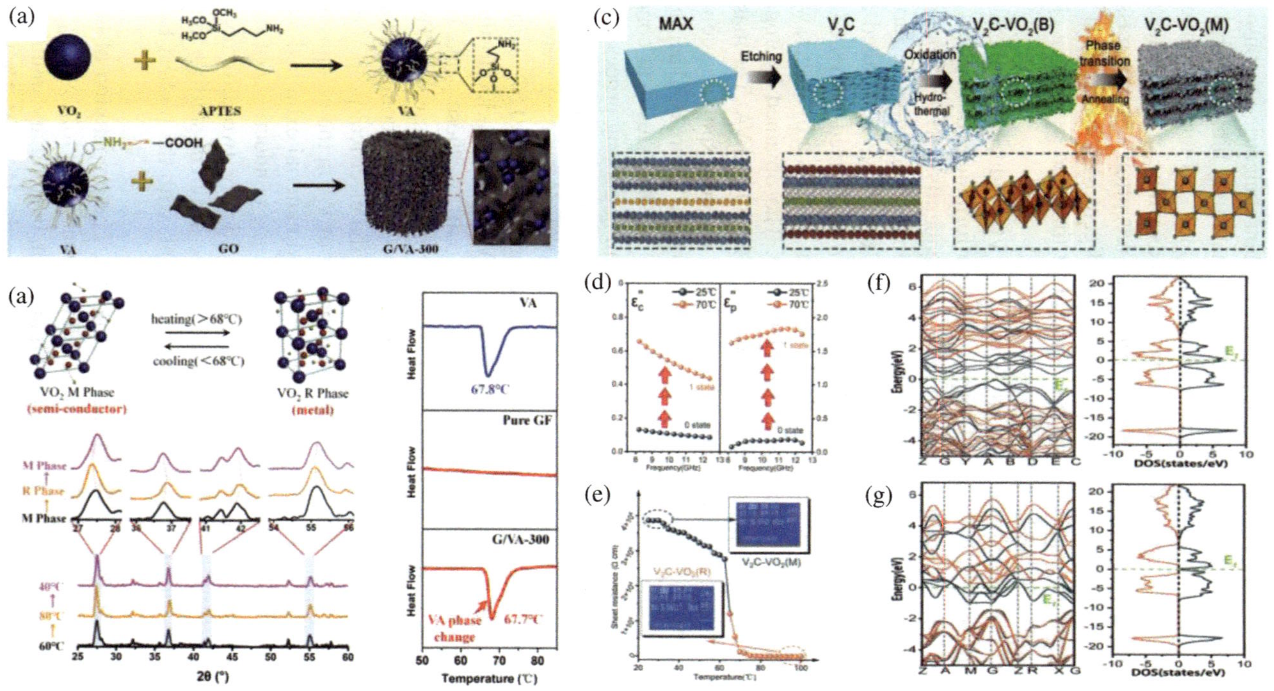

Figure 6.7 (a) Surface decoration of VO_2 by 3-aminopropyltriethoxysilane (APTES), the assembly of VA and GO solution for composite aerogel by solvothermal reaction. (b) Phase change behavior of VO_2 from M phase to R phase; differential scanning calorimetry (DSC) curves of VA, GF, and G/VA-300 and in situ XRD spectra of G/VA-300 at different temperatures. (c) Schematic illustration of the fabrication of V_2C-VO_2(M) heterostructures via a solution-based oxidation method and subsequent high-temperature phase transition. (d) The conductive loss and polarization loss of the OV_2C-2 at 25 °C and 70 °C. (e) Square resistance versus temperature curve of the OV_2C-2 from 25 °C to 100 °C. Band structure and corresponding density of state of (f) VO_2(M) and (g) VO_2(R).

were obtained (Figure 6.7c). At room temperature, VO_2 exists in multiple crystal structures (A, B, D, M, and P), with VO_2(M) being the only form that undergoes a metal–insulator transition at ≈68 °C. Given the sustained stability of VO_2(M) at high temperatures, vacuum heat treatment is employed in the conversion of V_2C-VO_2(B) heterostructures to V_2C-VO_2(M) counterparts.

The "quantized" dielectric loss is required for intelligent switching between wave absorption and wave transmission. Moderate conductive/polarization loss ("1 state") is necessary for efficient wave absorption, and wave transmission requires poor conductive/polarization loss ("0 state"). It can be generalized that the dielectric loss is quantitatively calculated into conductive loss and polarization loss according to the Debye theory (Figure 6.7d). The intrinsic conductivity and on-off heterostructure contribute to conductive loss in the "1/0 state." The quantized transition of conductivity loss between "0 states" and "1 state" is attributed to two factors: (1) the intrinsic conductivity of the adsorbent; (2) the conductive network established by the V_2C-VO_2 adsorbent within the PDMS binder. The square resistance versus temperature curve (Figure 6.7e) shows a continuous decline in the square resistance of the absorber with increasing temperature, experiencing a sudden alteration near the phase transition temperature (62 °C). The phase transition of VO_2 (M and R phases) is considered the primary factor leading to different conductivities, resulting in distinguishable charge transport ability in "0 states" and "1 state." Temperature fluctuations affect the overlap of Hubbard bands and distance of atoms, resulting in changes in electron concentration. Modifications in the crystal structure before and after the phase transition lead to alterations in the Fermi level, valence band, and conduction band. The band and electronic structures of VO_2(M) and VO_2(R) are investigated through first-principles density functional theory (Figure 6.7f,g). In the monoclinic VO_2(M) phase, a zig-zag chain comprising V atoms along the a_m axis is formed, causing the d_1 orbitals to split into antibonding and bonding configurations. This results in the Fermi-level residing within the bandgap. Conversely, the uniform distribution of V atoms in the VO_2(R) phase maintains degenerate d_1 orbitals, placing the Fermi level in the conduction band. The observed difference in the location of the two Fermi levels leads to a sudden change in insulation-metal properties during the phase transition. As the absorber is a biphasic system composed of the absorber and PMDS, the conductive network is another important factor determining conductive loss. The contact between the multilayer absorber creates a local conductive channel or network within PDMS. Over the phase transition temperature, VO_2(M) transforms into VO_2(R), exhibiting metallic properties, leading to a connected state of the local conductive network. However, at room temperature, the insulating characteristics of VO_2(M) curtail the formation of a local conductive network, inhibiting conductive loss. Therefore, due to the different contributions of intrinsic conductivity and local conductive network at room/high temperature, the conductivity loss of the absorber exhibits a "quantized" characteristic and shifts between "1 state" and "0 state."

6.4 Conclusion and Outlook

In conclusion, this chapter provides an overview of the physical models related to dielectric correlation that have been developed in recent years. Some concepts of each model and the corresponding data support are presented. This provides insights into the design, analysis, and improvement of future EMW -absorbing materials. Finally, we have looked forward to the future development trends of physical models for absorptive materials. This is primarily manifested in two aspects. One is that we will devote ourselves to the in-depth exploration of more microscopic systems with a view to achieving breakthroughs in performance, such as the physical modeling of monatomic materials. On the other hand, interdisciplinary research and practical complex application scenarios put forward higher requirements for multisystem-coupled physical modeling. For example, cross-modeling of wave-absorbing materials with tumor therapy, microsensors, and other fields requires more comprehensive consideration and research.

References

1 Wang, S., Li, D., Zhou, Y., and Jiang, L. (2020). Hierarchical $Ti_3C_2T_x$ MXene/Ni chain/ZnO array hybrid nanostructures on cotton fabric for durable self-cleaning and enhanced microwave absorption. *ACS Nano* 14: 8634–8645.

2 Wang, G., Gao, Z., Tang, S. et al. (2021). Microwave absorption properties of carbon nano coils coated with highly controlled magnetic materials by atomic layer deposition. *ACS Nano* 6: 11009–11017.

3 Dong, S., Zhang, X., Li, X. et al. (2020). SiC whiskers-reduced graphene oxide composites decorated with MnO nanoparticles for tunable microwave absorption. *Chem. Eng. J.* 392: 123817.

4 Liang, L., Song, G., Liu, Z. et al. (2020). Constructing $Ni_{12}P_5/Ni_2P$ heterostructures to boost interfacial polarization for enhanced microwave absorption performance. *ACS Appl. Mater. Interfaces* 12: 52208–52220.

5 Zhao, Z., Kou, K., Zhang, L., and Wu, H. (2022). Optimal particle distribution induced interfacial polarization in bouquet-like hierarchical composites for electromagnetic wave absorption. *Carbon* 186: 323–332.

6 Wang, X., Cao, W., Cao, M., and Yuan, J. (2020). Assembling nano-microarchitecture for electromagnetic absorbers and smart devices. *Adv. Mater.* 32: 2002112.

7 Wu, H., Wu, G., Ren, Y. et al. (2015). Co^{2+}/Co^{3+} ratio dependence of electromagnetic wave absorption in hierarchical $NiCo_2O_4$-$CoNiO_2$ hybrids. *J. Mater. Chem. C* 3: 7677.

8 Liang, X., Wang, G., Gu, W., and Ji, G. (2021). Prussian blue analogue derived carbon-based composites toward lightweight microwave absorption. *Carbon* 177: 97.

9 Liu, M., Yang, X., Shao, W. et al. (2021). Superior microwave absorbing properties of O, S, N codoped carbon planar helixes via carbonization of polypyrrole spiral nanowires. *Carbon* 174: 625.

10 Liang, L., Li, Q., Yan, X. et al. (2021). Multifunctional magnetic $Ti_3C_2T_x$ MXene/Graphene aerogel with superior electromagnetic wave absorption performance. *ACS Nano* 15: 6622.

11 Gao, Z., Ma, Z., Lan, D. et al. (2022). Synergistic polarization loss of MoS_2-based multiphase solid solution for electromagnetic wave absorption. *Adv. Funct. Mater.* 32: 2112294.

12 Cao, M., Song, W., Hou, Z. et al. (2010). The effects of temperature and frequency on the dielectric properties, electromagnetic interference shielding and microwave-absorption of short carbon fiber/silica composites. *Carbon* 48: 788–796.

13 Song, W., Cao, M., Hou, Z. et al. (2009). High dielectric loss and its monotonic dependence of conducting-dominated multiwalled carbon nanotubes/silica nanocomposite on temperature ranging from 373 to 873 K in K-band. *Appl. Phys. Lett.* 94: 233110.

14 Wen, B., Cao, M., Hou, Z. et al. (2013). Temperature dependent microwave attenuation behavior for carbon-nanotube/silica composites. *Carbon* 65: 124–139.

15 Qin, M.M., Xu, Y.X., Cao, R. et al. (2018). Efficiently controlling the 3D thermal conductivity of a polymer nanocomposite via a hyperelastic double-continuous network of graphene and sponge. *Adv. Funct. Mater.* 28: 1805053.

16 Gu, W.H., Tan, J.W., Chen, J.B. et al. (2020). Multifunctional bulk hybrid foam for infrared stealth, thermal insulation, and microwave absorption. *ACS Appl. Mater. Interfaces* 12: 28727–28737.

17 Qian, X., Zhang, Y.H., Wu, Z.C. et al. (2021). Multi-path electron transfer in 1D double-shelled Sn@Mo_2C/C tubes with enhanced dielectric loss for boosting microwave absorption performance. *Small* 17: 2100283.

18 Dong, L.Y., Hu, C.G., Song, L. et al. (2016). A large-area, flexible, and flame-retardant graphene paper. *Adv. Funct. Mater.* 26: 1470–1476.

19 Ma, W., Liu, X., Qiu, Z. et al. (2022). Hydrophobic and flame-retardant multifunctional foam for enhanced thermal insulation and broadband microwave absorption via a triple-continuous network of RGO/MWCNT-melamine composite. *Carbon* 196: 913–922.

20 Li, Z., Zhang, L., and Wu, H. (2023). A regulable polyporous graphite/melamine foam for heat conduction, sound absorption and electromagnetic wave absorption. *Small* 2305120.

21 Tomboc, G.M., Zhang, X., Choi, S. et al. (2022). Stabilization, characterization, and electrochemical applications of high-entropy oxides: critical assessment of crystal phase-properties relationship. *Adv. Funct. Mater.* 32: 2205142.

22 Xiang, H., Xing, Y., Dai, F.Z. et al. (2021). High-entropy ceramics: present status, challenges, and a look forward. *J. Adv. Ceram.* 10: 385–441.

23 Bérardan, D., Franger, S., Meena, A.K., and Dragoe, N. (2016). Room temperature lithium superionic conductivity in high entropy oxides. *J. Mater. Chem. A* 4: 9536–9541.

24 Pu, Y., Zhang, Q., Li, R. et al. (2019). Dielectric properties and electrocaloric effect of high-entropy $(Na_{0.2}Bi_{0.2}Ba_{0.2}Sr_{0.2}Ca_{0.2})TiO_3$ ceramic. *Appl. Phys. Lett.* 115: 223901.

25 Zhou, S., Pu, Y., Zhang, Q. et al. (2020). Microstructure and dielectric properties of high entropy $Ba(Zr_{0.2}Ti_{0.2}Sn_{0.2}Hf_{0.2}Me_{0.2})O_3$ perovskite oxides. *Ceram. Int.* 46: 7430–7437.

26 Zhao, B., Yan, Z., Du, Y. et al. (2023). High-entropy enhanced microwave attenuation in titanate perovskites. *Adv. Mater.* 35: 2210243.

27 Mu, S., Samolyuk, G., Wimmer, S. et al. (2019). Uncovering electron scattering mechanisms in NiFeCoCrMn derived concentrated solid solution and high entropy alloys. *npj Comput. Mater.* 5: 1.

28 Zhao, B., Du, Y., Yan, Z. et al. (2023). Structural defects in phase-regulated high-entropy oxides toward superior microwave absorption properties. *Adv. Funct. Mater.* 33: 2209924.

29 Hui, S., Zhou, X., Zhang, L., and Wu, H. (2023). Constructing multiphase-induced interfacial polarization to surpass defect-induced polarization in multielement sulfide absorbers. *Adv. Sci.* 11: 2307649.

30 Qiao, J., Zhang, X., Xu, D. et al. (2020). Design and synthesis of TiO_2/Co/carbon nanofibers with tunable and efficient electromagnetic absorption. *Chem. Eng. J.* 380: 122591.

31 Ding, J., Wang, L., Zhao, Y. et al. (2019). Boosted interfacial polarization from multishell TiO_2@Fe_3O_4@PPy Heterojunction for enhanced microwave absorption. *Small* 15: 1902885.

32 Zhang, D., Liu, T., Cheng, J. et al. (2019). Lightweight and high-performance microwave absorber based on 2D WS_2-RGO heterostructures. *Nano-Micro Lett.* 11: 38.

33 Meng, F., Wang, H., Huang, F. et al. (2018). Graphene-based microwave absorbing composites: a review and prospective. *Compos. Part B* 137: 260.

34 Liu, Q., Cao, Q., Bi, H. et al. (2016). CoNi@SiO_2@TiO_2 and CoNi@Air@TiO_2 microspheres with strong wideband microwave absorption. *Adv. Mater.* 28: 486.

35 Wang, J., Liu, L., Jiao, S. et al. (2020). Hierarchical carbon fiber@MXene@MoS_2 core-sheath synergistic microstructure for tunable and efficient microwave absorption. *Adv. Funct. Mater.* 30: 2002595.

36 Lu, Y., Wang, Y., Li, H. et al. (2015). MOF-derived porous Co/C nanocomposites with excellent electromagnetic wave absorption properties. *ACS Appl. Mater. Interfaces* 7: 13604.

37 Yan, W., Li, S., Yang, T. et al. (2020). Molecular vises for precisely positioning ligands near catalytic metal centers in metal-organic frameworks. *J. Am. Chem. Soc.* 142: 16182.

38 Shang, L., Yu, H., Huang, X. et al. (2016). Well-dispersed ZIF-derived Co, N-Co-doped carbon nanoframes through mesoporous-silica-protected calcination as efficient oxygen reduction electrocatalysts. *Adv. Mater.* 28: 1668.

39 Pan, Y., Sun, K., Liu, S. et al. (2018). Core–shell ZIF-8@ZIF-67-derived CoP nanoparticle-embedded N-doped carbon nanotube hollow polyhedron for efficient overall water splitting. *J. Am. Chem. Soc.* 140: 2610.

40 Huang, M., Wang, L., You, W., and Che, R. (2021). Single zinc atoms anchored on MOF-derived N-doped carbon shell cooperated with magnetic core as an ultrawideband microwave absorber. *Small* 17: 2101416.

41 Bing, Y., Zeng, Y., Liu, C. et al. (2015). Synthesis of double-shelled SnO_2 nano-polyhedra and their improved gas sensing properties. *Nanoscale* 7: 3276.

42 Kim, S., Jeong, J., Mkhoyan, K., and Jang, H. (2016). Direct observation of the core/double-shell architecture of intense dual-mode luminescent tetragonal bipyramidal nanophosphors. *Nanoscale* 8: 10049.

43 Lee, S., Lee, J., Hwang, S. et al. (2015). Enhanced electroresponsive performance of double-shell SiO_2/TiO_2 hollow nanoparticles. *ACS Nano* 9: 4939.

44 You, W., Bi, H., She, W. et al. (2017). Dipolar-distribution cavity γ-Fe_2O_3@C@α-MnO_2 nanospindle with broadened microwave absorption bandwidth by chemically etching. *Small* 13: 1602779.

45 Li, L., Zhao, S., Luo, X.-J. et al. (2021). Smart MXene-based Janus films with multi-responsive actuation capability and high electromagnetic interference shielding performances. *Carbon* 175: 594.

46 Li, D.-Y., Liu, L.-X., Wang, Q.-W. et al. (2022). Functional polyaniline/MXene/cotton fabrics with acid/alkali-responsive and tunable electromagnetic interference shielding performances. *ACS Appl. Mater. Interf.* 14: 12703.

47 Lv, H., Yang, Z., Wang, P.L. et al. (2018). A voltage-boosting strategy enabling a low-frequency, flexible electromagnetic wave absorption device. *Adv. Mater.* 30: 1706343.

48 Cheng, Z., Wang, R., Cao, Y. et al. (2022). Intelligent Off/On switchable microwave absorption performance of reduced graphene oxide/VO_2 composite aerogel. *Adv. Funct. Mater.* 32: 2205160.

49 Pan, F., Pei, K., Chen, G. et al. (2023). Integrated electromagnetic device with On-Off heterointerface for intelligent switching between wave-absorption and wave-transmission. *Adv. Funct. Mater.* 33: 2306599.

7

Integrated Foam-Type Electromagnetic Wave Absorption Materials

Qing Chang

MOE Key Laboratory of Material Physics and Chemistry Under Extraordinary Conditions, School of Physical Science and Technology, Northwestern Polytechnical University, Xi'an, China

CHAPTER MENU

Nowadays, the rapid advancement of information technology in electronic equipment in military and civil fields stimulates the demand for portable and wearable electronic devices. Therefore, lightweight and high-efficiency electromagnetic wave (EMW) absorption materials are intensely demanded. Multifunctional integrated foam-type EMW-absorbing materials such as foams, aerogels, and others can greatly enhance the EMW absorption capacity owing to the synergistic effect of components and structures. This chapter briefly introduces the advances in integrated foam-type EMW-absorbing materials, including carbon-based foam, ferrite-based foam, silicon carbide (SiC)-based foam, and conductive polymer composite (CPC) foam. The content covers sample preparation, electromagnetic parameter regulation, EMW absorption performance, and dissipation mechanism.

7.1 Carbon-Based Foam for EMW Absorption

Numerous EMW absorption materials have been widely studied to address the electromagnetic pollution caused by military and civil equipment in recent years. Among them, carbon-based materials are widely used because of their environment-friendly characteristics, strong physical and chemical stability, low density, excellent electrical conductivity, and adjustable dielectric properties. The three-dimensional (3D) porous structure of carbon materials has a variety of internal reflections and scattering, further trapping and attenuating EMW. In recent years, the manufacture of both pure carbon-based foams and carbon-based composite

Electromagnetic Wave Absorbing Materials: Fundamentals and Applications, First Edition.
Edited by Hongjing Wu, Jun Luo, and Meiyin Yang.

foams has greatly advanced. Carbon-based composite foams include heteroatoms doping carbon foam (CF) and composite foams formed by different carbon materials, and carbon material with ceramics, magnetic metals and their alloys, metal oxides, MXene, and so on by adjusting dielectric loss or magnetization loss to improve impedance matching and enhance absorption capacity.

7.1.1 Pure Carbon-Based Foams

Structural regulation strategy is the primary pathway to obtain multifunctional, pure carbon-based foam or aerogel with adjustable dielectric properties and EMW absorption performance. Shao's group successfully designed ultralight 3D-reduced graphene oxide (rGO) aerogels with tunable pore structures via a freeze–thaw assembly method, which involved chemical pre-reduction, freeze–thaw, further reduction, and freeze-dry processes [1]. By adjusting the pre-reduction time, the ultralight ($5.83\,\mathrm{mg\,cm^{-3}}$) and highly porous (94.9%) rGO-60 aerogel developed a uniform cellular structure that maximizes multiple reflections and scattering of EMW among microcellular free spaces, which achieved a minimum reflection loss (RL_{min}) of −61.63 dB with an effective absorption bandwidth (EAB) of 7.8 GHz for an ultralow filling ratio (0.74 wt%). The 3D porous conductive network with a uniform micropore structure formed by rGO-60 aerogel effectively promoted the conductivity loss. In addition, multiple random reflections of EMW occurred repeatedly at the internal interface of the cell wall. The conduction loss of rGO-60 aerogel and the synergistic effect of multiple reflections and scattering promoted the balance between impedance matching and attenuation, giving rGO-60 aerogel excellent EMW absorption performance. This investigation provides a new paradigm for the development of pure dielectric loss graphene aerogel-based absorbing materials by taking advantage of pore structure engineering and sheds light on the design of ultralight and multifunctional absorbing materials in the future. Similarly, Ji's team developed two kinds of chitosan-derived carbon aerogels (CAs) by adjusting the amount of chitosan and changing the pre-freezing method (Figure 7.1a) [2]. Two morphologies of bamboo tube structure (BTS) and isolated island structure (IIS) were controlled by the growth speed and direction of ice crystals in a refrigerator and freeze dryer differently. The EAB of the sample CA3.5-60 (the amounts of chitosan were 3.5 wt%, and 60 is the freeze dryer freezing temperature) reached 5.15 GHz at 1.4 mm, which was wider than that of the BTS aerogel, and almost covered the entire Ku band when the thickness was 1.0–1.4 mm. It cannot be ignored that the IIS has a multilayer sheet structure, and there are multiple reflections of EMW between the sheets. In addition, the loss by the absorber is predominantly derived from interface polarization loss between paraffin wax and the absorbing material, dipole polarization, and conduction loss; particularly, the defects in the microstructure of the absorber can easily develop the polarization point. Conductive loss mainly depends on the highly graphitized carbon network, which is beneficial for the jumping and transmission of electrons in materials. Therefore, this work provides a new reference for the preparation and application of ecofriendly and renewable carbon aerogels. Guang's group synthesized ultra-lightweight N-doped graphene foams (NGF) with open reticular structures using a self-assembly method and a subsequent freeze-drying process [4]. The N bonding configurations, electrical conductivity, open reticular structures, bulk density, and dielectric loss characteristics can be accurately tuned by adjusting the substitutional N atoms and annealing temperature. When the annealing temperature was 800 °C, the RL_{min} of NGF was below −40 dB, and the corresponding effective bandwidths exceeding −10 dB were wider than 12 GHz in

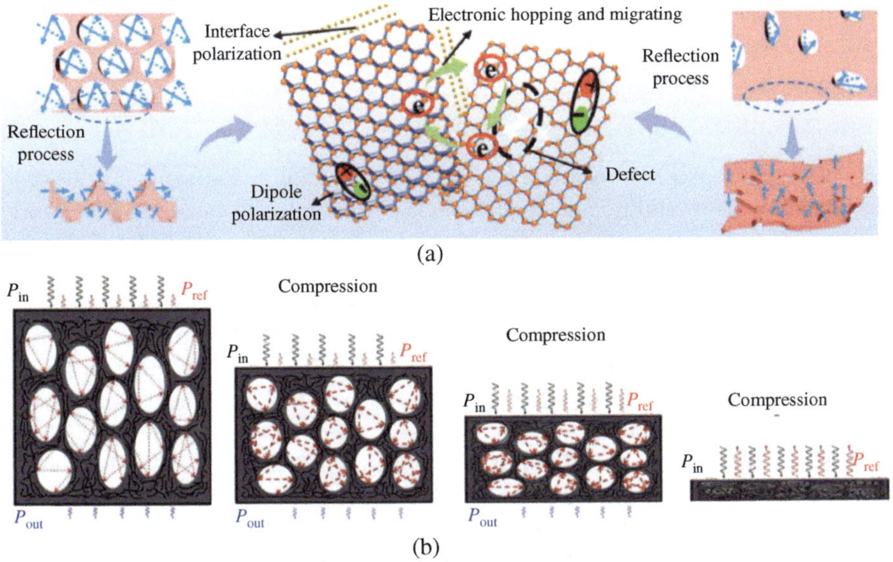

Figure 7.1 (a) EMA mechanism of BTS and IIS aerogels. Source: Chen [2]/with permission of Royal Society of Chemistry. (b) Schematic representation of the microwave absorption mechanism for the graphene aerogel under different compression strains. Source: Wang [3]/with permission of Elsevier.

the range of 2–5 mm, which can be used as lightweight and high-efficiency microwave-absorbing materials (MWAMs). The layered structure with a 3D interconnecting conductive network enables the absorbent to have a high conduction loss. After chemical doping, many heteroatoms and defects were created in the graphene lattice, which acted as polarization centers and induced dipolar polarization. Additionally, NGF with a large specific surface area and numerous open reticular pores will form larger solid-air interfaces. More space for electrons will accumulate at the interfaces due to the different permittivity and conductivity between solid graphene and interior air, leading to enhanced interfacial polarization. This study not only provides a fundamental point in understanding the intrinsic EMW absorption mechanism of mono-component NGF but also offers a facile method for rationally designing and constructing lightweight, high-efficiency absorbing materials. Liu's group also successfully fabricated the macroscopic 3D free-standing porous all-graphene aerogel with ultralight density and high compressibility through a mild in situ self-assembly and thermal annealing processes (Figure 7.1b) [3]. The microwave absorption performance of the graphene aerogel was effectively self-adjusted via a simple mechanical compression. The optimal absorbing value was up to 61.09 dB with a broad qualified bandwidth of 6.30 GHz at the thickness of 4.81 mm when the compression strain ratio of the sample was controlled to be 30%. Due to the existence of closed microcellular structures, random multiple reflections of EMW occurred repeatedly at the internal interface of the cell wall. The increased random multiple reflection and conductive dissipation will work together to promote the effective absorption of incident EMW. As the compressive strain increases (up to 50%), the further reduction of the internal free space greatly weakens the random multiple reflections. At the same time, the further enhancement of backscattering will cause more incident EMW to be reflected out of the aerogel, thus affecting the effective absorption of incident EMW. When the sample is fully compressed (75%), the free space of the aerogel is completely squeezed out. Based on these results, the microwave absorption properties of graphene aerogel can be effectively adjusted by a simple mechanical compression.

7.1.2 Composite Foams Formed by Carbon Material

Cellulose has become an excellent candidate for EMW-absorbing materials due to its low cost, biodegradability, renewability, and excellent elastic modulus to build a 3D network structure. However, due to the low dielectric constant of biomass carbon, the ability to absorb EMW is limited, and highly conductive carbon nanotubes (CNTs) are often added to modify the dielectric properties of biomass carbon to enhance its microwave absorption performance. Shi's group fabricated the ultralight nitrogen-doped rGO/multi-walled carbon nanotubes (MWCNTs) composite aerogels via a facile hydrothermal and lyophilization two-step strategy [5]. Specifically, the composite aerogel with long MWCNTs and doped nitrogen content of 7.07 wt% (bulk density ~7.2 $mg\,cm^{-3}$) presented the optimal RL of −53.3 dB at 6.5 GHz (C-band) and a broad absorption bandwidth of 5.2 GHz (11.1–16.3 GHz) under a small matching thickness of merely 2.0 mm and a low filler content of 15 wt%. Interestingly, the RL values were all less than −20 dB with matching thicknesses changing from 1.5 to 4.5 mm. Defects and chemical groups like —OH and —COOH on the surfaces or edges of nitrogen-doped rGO and MWCNTs act as the polarized centers, which can introduce dipole polarization that will further attenuate the incident waves. Owing to the difference in electronegativity between carbon and nitrogen atoms, the C—N electric dipoles could be formed that would be rotated under the alternating electromagnetic (EM) field in order to produce dipole polarization loss and further attenuate incident EMW. Besides, the doped nitrogen atoms as the donor dopants could offer additional electrons for higher conductivity, which improves the conduction loss. Lastly, abundant defects and interfaces existing in the porous structure can form multiple polarizations that are beneficial to the dissipation of EM energy. This work was helpful for designing and fabricating graphene-based 3D structure composites as lightweight and high-efficiency EMW absorbers. Wu's group successfully prepared 3D carbon foams derived from CNTs/cellulose foam through a freeze-drying and high-temperature annealing process for the first time (Figure 7.2a) [6]. Microwave absorption performance can be controlled by adjusting the carbonization temperature and constructing nanointerfaces. When the carbonized temperature was 700 °C, the cellulose-based foam obtained the an RL_{min} of −55.4 dB at 2.11 mm and an EAB of 2.35 GHz. The high conduction loss is owing to the 3D conductive network formed by one-dimensional CNTs and cellulose, which effectively provide paths for electronic directional transmission. Secondly, dipole polarization is caused by defects and residual oxygen-containing functional groups. Carbon nanotube–cellulose interfaces are considered capacitor structures, and charged carriers will be assembled at these nanointerfaces. The jump of electrons on the nanocapacitors leads to an increase in electrical conductivity, further improving the interface polarization. In terms of practical application, the carbon foam exhibits good thermal insulation and hydrophobicity, guaranteeing the stable and durable EMW absorption applications in the field of resist deformation, water or humid environments, and high-temperature attacks. Zeng's group prepared a new hybrid carbon aerogel sphere composed of cellulose nanofiber-derived carbons (CNFCs) and graphene nanosheets by using drip drying and carbonization treatment (Figure 7.2b) [7]. The CNFC prevents the agglomeration of the graphene layers and interacts with graphene to construct the aerogel spheres with tunable dielectric properties. The rGO

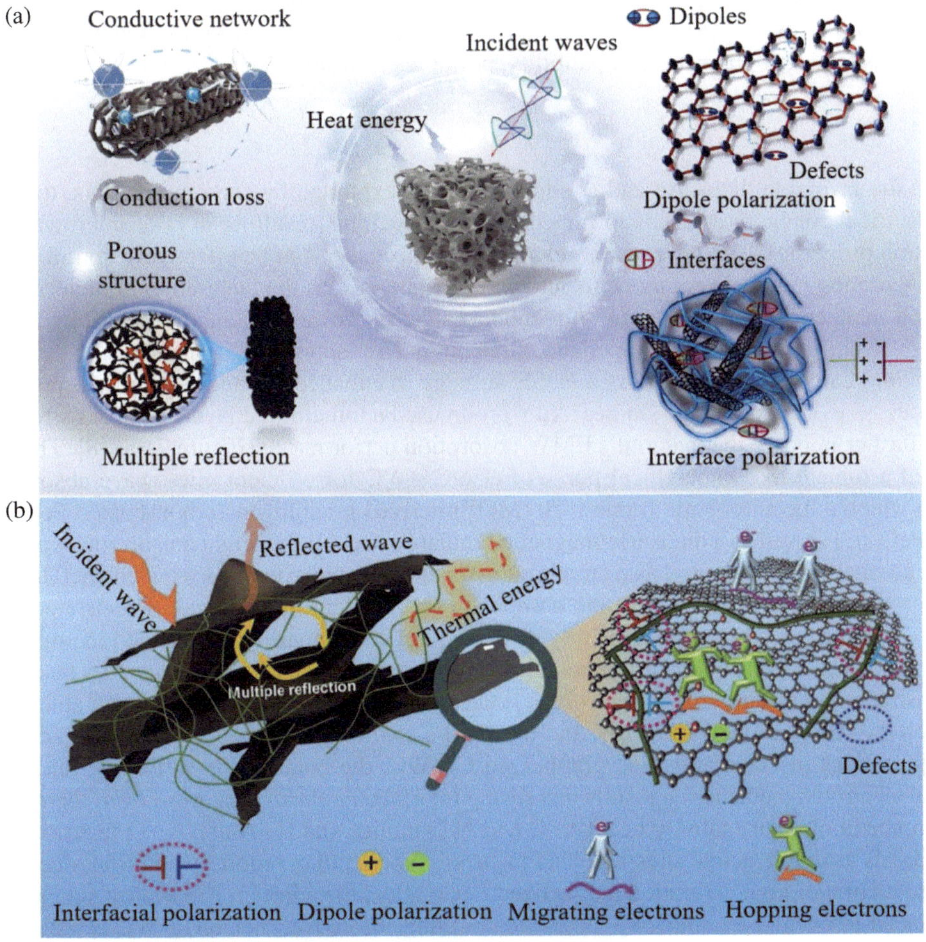

Figure 7.2 (a) Microwave absorption mechanism of carbonized cellulose-based foam. Source: Zhao [6]/with permission of Elsevier. (b) Diagrammatical illustration of EMW-absorbing mechanism of rGO/CNFC aerogel spheres. Source: Zhang [7]/with permission of Springer Nature.

reduces the volume shrinkage of the CNFC aerogel spheres in the carbonization process, contributing to the preparation of ultralight high-porosity aerogel spheres. In addition to the high conduction loss capability derived from the interconnected hybrid carbon skeletons, abundant CNFC/graphene interfaces contribute to high interfacial polarization loss of incident EMWs. Moreover, the defects and polar functional groups of the aerogel spheres are instrumental in enhancing the polarization loss capability. Together with the effective multiple reflections and scatterings of incident EMWs caused by the porous structure, which can enhance the interactions between the skeletons and EMWs, the hybrid carbon-based aerogel spheres show excellent EMW absorption performance. With a filler ratio of merely 3 wt%, the RL_{min} could reach −70.44 dB, and the EAB was 6.16 GHz for the aerogel spheres, outperforming other biomass-derived EMW-absorbing materials ever reported. This work not only develops a simple, facile, ecofriendly, and scalable preparation approach for lightweight biomass-derived aerogel spheres but also, through a rational design of components and microstructure, provides the sustainable high-performance EMW absorption materials for applications in electromagnetic compatibility, portable electronics, and aerospace.

7.1.3 Composite Foams of Carbon Material and Magnetic Metal

Despite the favorable dielectric and conduction losses exhibited by carbon materials, their high electrical conductivity leads to impedance mismatching, resulting in poor microwave absorption properties. The microwave absorption performance can be effectively optimized by incorporating magnetic components into carbon materials, as the integration of dielectric units and magnetic components not only enhances the microwave attenuation capacity but also significantly improves the impedance matching. The incorporation of magnetic components into carbon materials offers a viable strategy to enhance the microwave absorption properties, as evidenced by the studies. Xu's group used a tomato-like layered porous structure to further improve the material's EMW absorption performance (Figure 7.3a) [8]. They prepared a tomato-like hierarchical porous FeCo/C@WC for efficient microwave absorption, in which a metal–organic framework (MOF)-derived graphitized carbon-coated FeCo alloy (FeCo/C) served as a dielectric/magnetic regulator, and a lightweight anisotropic wood carbon aerogel (WC) was used as a structural assistant. The macroporous wooden shell and mesoporous carbon nanocage gave the composite an excellent hierarchical dielectric coupling network. Meanwhile, the uniformly dispersed FeCo nanoparticles enhanced the magnetic loss to the carbonaceous substrate, allowing for better impedance matching. As a result, the constructed multidimensional FeCo/C@WC exhibited superior EMW-absorbing ability, in which the RL_{min} reached −47.6 dB, and the EAB was up to 8.9 GHz. Due to the different metal properties of FeCo alloy and graphitized C@WC, the polarization of FeCo/C under external electromagnetic action results in a strong interfacial polarization. Moreover, there is a difference in electronegativity between C and N/O atoms, and C—N and C—O functional free radicals as dipole active sites will lead to strong dipole polarization. In addition, due to the introduction of high-saturation magnetized FeCo alloy, the adjacent nano-alloy magnets are induced to each other, showing magnetic coupling. Moreover, FeCo/C@WC reveals a layered porous network from micropores to macropores. At the same time, MOF-derived graphitized carbon and FeCo alloy are introduced to make the tomato-like FeCo/C@WC prepared with good electrical conductivity. It could be firmly believed that the structural engineering of wood-based materials would become a promising area of research in advanced multidimensional hierarchical materials. Liu's group introduced CuNi nanoparticles of different compositional ratios into the pores of carbon foam to obtain CuNi alloy/CF through a simple carbonization process [10]. In the synthetic procedure, Ni^{2+} and Cu^{2+} were first adsorbed onto the melamine foam (MF) to form Ni^{2+}–Cu^{2+}/MF composites, which were then subjected to the pyrolysis treatment, resulting in the production of mesoporous CuNi alloy/CF. The RL_{min} of CuNi11(the molar ratio of Cu^{2+} to Ni^{2+} is 1:1)/CF was −50.20 dB at a thin thickness of 1.6 mm, and the effective bandwidth ranged from 3 to 18 GHz at the thickness of 1–5 mm. When the EMW entered the CuNi11, the high conductivity of the CF contributed to conductive loss and produced a micro current. Benefiting from the 3D network structure, the EMW can reflect many times within the carbon framework. Due to the electronegativity difference between carbon and metal alloy, the interfacial polarization was induced by the electromagnetic field on their interfaces. Moreover, the magnetic loss mechanism of CuNi11 is eddy current loss. In a word, the as-prepared CuNi/CF can be considered as a potential

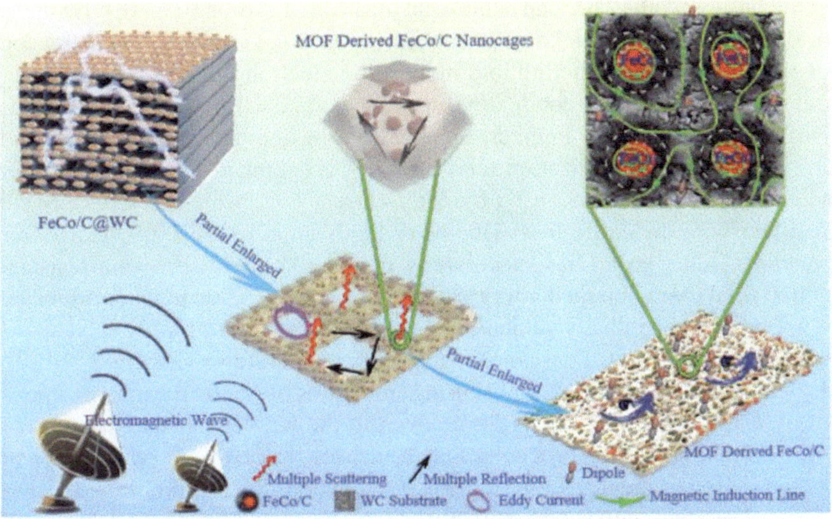

(a)

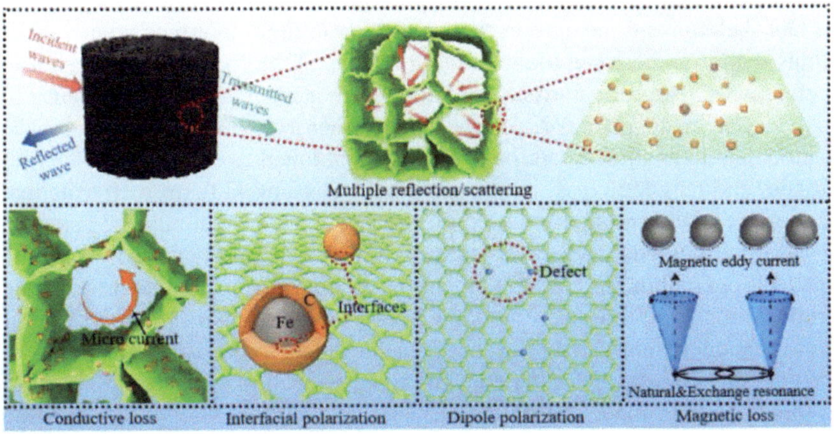

(b)

Figure 7.3 (a) Microwave absorption mechanism of FeCo/C@WC. Source: Xiong [8]/with permission of Royal Society of Chemistry. (b) Schematic illustration of the different microwave absorption mechanisms in GN/Fe@C. Source: Li [9]/with permission of American Chemical Society.

candidate for EMW absorption. Similarly, Yue's group designed a three-dimensional carbonized melamine foam (CMF) with double coatings (CMF@FeNi-SiO_2-CNTs) [11]. The experimental results showed that the dielectric properties of CMF@SiO_2-CNTs can be tuned by changing the length and growth density of CNTs in the hybrids. The introduction of a magnetic FeNi layer can help further improve dielectric permittivity and widen the EAB of the hybrids by increasing magnetic loss. With a thickness of 2.5 mm, the CMF@FeNi-SiO_2-10CNTs single-layer absorber has an ultrawide EAB of 6.2 GHz (9.8–16 GHz). Thus, based on the hybrids with different electromagnetic properties, a triple-layer absorber model is proposed in which there is an impedance layer (CMF@SiO_2-5CNTs), an absorption layer (CMF@FeNi-SiO_2-10CNTs), and a reflection layer (CMF@SiO_2-30CNTs). The calculation results show that the proposed absorber can achieve an ultrawide EAB of 10.6–18 GHz and a RL_{min} of −64 dB. By optimizing the thickness of each layer, the EAB can be further extended to 14.7 GHz (3.3–18 GHz). The conducting loss induced by electron migration plays a significant role in absorption in the 3D conducting network composed of CNTs and porous carbon foam. Furthermore, the large number of defects from doping other elements and functional groups causes dipole polarization relaxation in CNTs; meanwhile, the magnetic loss of the FeNi coating is primarily caused by eddy current loss, which absorbs EMWs via a synergistic effect with dielectric loss. This work holds great promise for microwave absorption applications. Liu's group has successfully prepared a novel corrosion-resistant GN/Fe@C foam (the final foam obtained after treating the graphene oxide powder is labeled GN) by self-assembly of iron phthalocyanine/Fe_3O_4 (FePc hybrid) on graphene skeletons under solvothermal conditions and then annealing at high temperatures (Figure 7.3b) [9]. The optimal RL_{min} of ultralow GN/Fe@C (11.36 mg cm^{-3}) was −72.46 dB, and the corresponding EAB was 4.32 GHz (7.68–12.00 GHz), covering the whole X-band. In terms of dielectric loss, the different relaxation processes may be caused by the dipole polarization of abundant defects in GN foam under the altering electromagnetic field. In addition, there is conductive loss caused by interconnected conductive graphene networks. At the same time, more interfaces can be formed between the Fe core and the carbon shell in Fe@C particles and between the GN skeleton and the carbon shell, resulting in more interface polarization. Multiple reflections/scattering between the internal interfaces of the foam skeleton can also dissipate EMW effectively. The eddy current, natural resonance, and exchange resonance of the main magnetic loss mechanism, and dielectric loss together make GN/Fe@C foam have excellent attenuation characteristics. Therefore, the aforementioned two main facts, including the dielectric loss and magnetic loss, jointly endow the GN/Fe@C foam with excellent attenuation characteristics. Liu's group used a simple mechanical mixing to replace common complex particle loading methods, and the complexation reaction between cobalt ions and chitosan achieved a close combination [12]. With subsequent ice-templated freezing casting and one-step carbonization, biomimetic porous cobalt/carbon aerogels with strong magnetoelectric coupling were facilely fabricated. The aerogel achieved an impressive EMW absorption performance of −75.8 dB at a low filling content (10%), as well as an EAB of 8.53 GHz. In addition, the in situ growth made the cobalt particles and carbon aerogel closely bind together, which improved the interfacial intensity, thus enhancing the interface polarization and attenuating the EMW. Besides, the C—O functional groups and C—N derived from chitosan also contributed to the dipole polarization to consume EMW. In terms of magnetic loss, a wide resonance peak could be observed in the high-frequency range, implying the presence of exchange resonance loss.

7.1.4 Composite Foams of Carbon Material and Metal Oxides

Magnetic metals improve the impedance matching of carbon materials by providing good magnetic losses, but the chemical stability is poor. Therefore, magnetic ferrites with higher environmental stability and other low dielectric metal oxides provide a better choice. The addition of metal oxides to carbon materials can also effectively enhance the EMW-absorbing properties of the materials. Wu's group has, for the first time, synthesized magnetic porous carbon foams derived from fish skin using a simple hydrothermal method [13]. Then, a series of Fe_3O_4@carbon (FC) composites were obtained through a simple reflux and annealing process. The FC-3 showed enhanced performance compared to other composites, with the RL_{min} reaching −47.3 dB at a thickness of 1.9 mm, and the EAB achieved 5.68 GHz at a matching thickness of 2.2 mm. In this case, the dipolar polarization is mainly attributed to the porous carbon, and the interface polarization originates from the interface between any two parts of Fe_3O_4, porous carbon, air, and paraffin. Eddy current is the dominant magnetic loss mechanism of the FC samples. It also paves the way for animal organs to be used as raw materials for the manufacture of high-performance EMW absorbents. Chen's group rationally designed a uniquely porous carbonated rGO/Fe_3O_4 (C-rGO/Fe_3O_4) carbon foam with hollow microspheres and a concave-convex microstructure through Pickering emulsion gel technology combined with an annealing process (Figure 7.4a) [14]. The C-rGO/Fe_3O_4 carbon foam possesses excellent comprehensive EMW absorption performance (RL_{min} of −57.50 dB and EAB of 6.72 GHz) and exhibits convenient RL value regulation (when varying compressive strains). The EMW absorption of the C-rGO/Fe_3O_4 carbon foam mainly depends on dielectric loss and magnetic loss. First, the conductive rGO layer was welded by CNF and paraffin wax to form a conductive network structure. Then, the magnetic Fe_3O_4 nanoparticles were introduced into the carbon foam structure to provide a magnetic loss, which formed a good impedance match with the dielectric loss. Due to rGO's disordered characteristics and structural defects, dipole polarization arose in the alternating electromagnetic field, and the interface polarization between rGO and the non-conductive Fe_3O_4 nanoparticles contributed to the dielectric loss. Thus, the incident EMWs experienced multiple scattering and reflections in the concave-convex microstructure and hollow microspheres of the carbon foam, resulting in higher energy dissipation. This provides a promising platform for practical applications in adjustable EMW absorption. Chen's group successfully synthesized a 3D hierarchical $NiCo_2O_4$ nanoflowers/rGO composite foam by a simple method under gentle conditions [16]. When the addition amount of $NiCo_2O_4$/rGO foam (the mass ratio between $NiCo_2O_4$ and RGO is 1:1) in paraffin wax was 5 wt%, the RL_{min} can reach −52.2 dB with a thin thickness of merely 2.6 mm, and the EAB is 7.04 GHz that covers the whole Ku band (10.96–18 GHz). The conductive network structure of the 3D $NiCo_2O_4$/rGO promotes the electron transport, which is favorable for EMW absorption. Meanwhile, the accumulation of charges at the multi-interfaces can generate interfacial polarization, favoring the improved EMW dissipation behavior. On the other hand, the existence of many defects and functional groups on $NiCo_2O_4$ and RGO gives rise to an uneven distribution of charges, which can induce dipolar polarization. Finally, the eddy current effect and magnetic resonance of $NiCo_2O_4$ can also lead to the attenuation of EMW. Therefore, the synergistic effects of electromagnetic loss mechanisms promote the absorption of EMW. Besides, spinel cobalt-zinc ferrite plays an important role as an EMW absorption material because of its remarkable properties, such as high-saturation magnetization, strong anisotropy, good mechanical hardness, and chemical stability. Li et al. synthesized the nitrogen-doped rGO/cobalt-zinc ferrite (NRGO/$Co_{0.5}Zn_{0.5}Fe_2O_4$) composite aerogels through a solvothermal and subsequent hydrothermal self-assembly two-step method (Figure 7.4b) [15]. The as-synthesized NRGO/$Co_{0.5}Zn_{0.5}Fe_2O_4$ composite aerogels have a very low density (12.1–14.6 mg cm^{-3}) and good compression recovery. Impressively, excellent EMW absorption performance could be achieved through facilely regulating the additive volume of ethylenediamine and filler contents. The composite aerogel with a doped nitrogen content of 2.5 wt% displayed the RL_{min} of −66.8 dB at a thickness of 2.6 mm and the EAB of 5.0 GHz under an ultrathin thickness of merely 1.6 mm. Many defects and residual functional groups on the surfaces or edges of NRGO act as polarized centers that can result in the imbalance of the dipole moment, which will induce the formation of dipole polarization belonging to dielectric loss that further

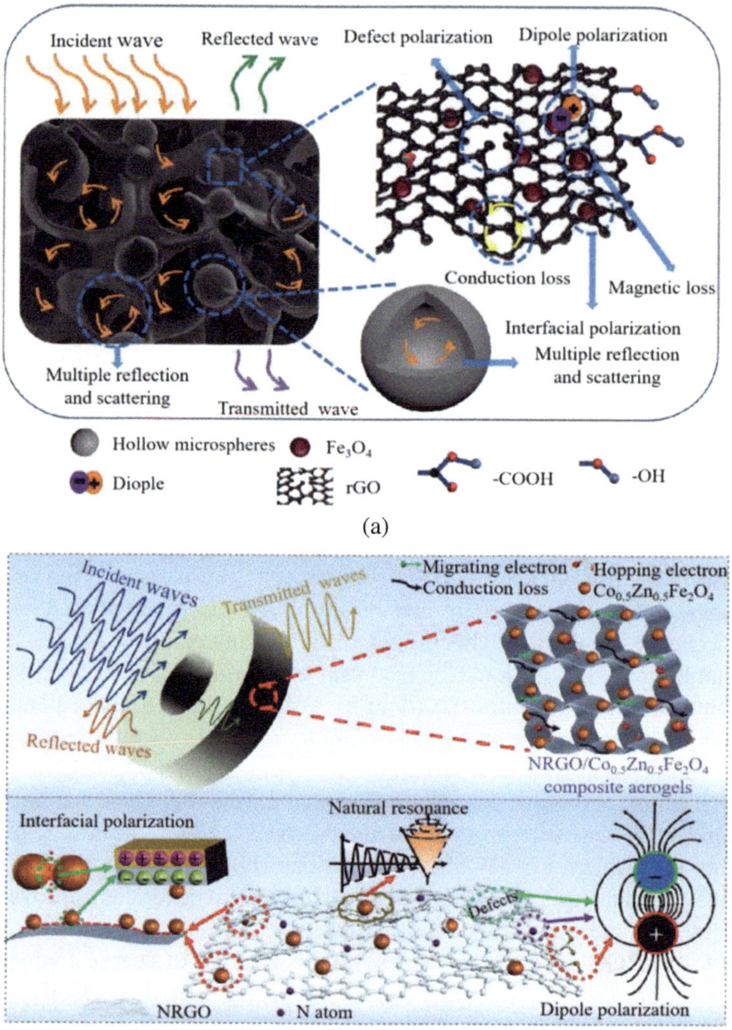

Figure 7.4 (a) Schematic representation of the EMW absorption mechanism for the C-C-rGO/Fe_3O_4 carbon foam. Source: He [14]/with permission of Royal Society of Chemistry. (b) The proposed EMW absorption mechanisms of NRGO/$Co_{0.5}Zn_{0.5}Fe_2O_4$ composite aerogels. Source: Shu [15]/with permission of Royal Society of Chemistry.

attenuates the incident waves under the alternating electromagnetic field. Besides, owing to the difference in the electronegativity of carbon and nitrogen atoms, numerous C—N electric dipoles form and are repeatedly rotated in order to induce dipole polarization loss that can dissipate the incident EMW. In addition, the doped nitrogen atoms as the donor dopants can offer additional electrons for higher electrical conductivity that will improve the conduction loss. The uniform distribution of $Co_{0.5}Zn_{0.5}Fe_2O_4$ microspheres loaded onto the surfaces of NRGO can form abundant heterogeneous interfaces. As a result, space charges can gather at a heterogeneous interface and generate interfacial polarization losses by arranging polar bonds or charges, thus attenuating the power of the incident wave. Finally, the ferromagnetic behavior of composite aerogels derived from magnetic $Co_{0.5}Zn_{0.5}Fe_2O_4$ microspheres could generate magnetic loss mainly in the form of natural resonance, which also contributes to the attenuation of waves under the alternating electromagnetic field.

Because of its low density, remarkable stability, and easy synthesis, multiphase and various forms of MnO_2 can be used as an effective candidate to adjust EM parameters. Zou's group prepared MnO_2-deposited graphene foams (MnO_2/GFs) using a simple hydrothermal method [17]. Benefiting from the well-designed interconnected 3D porous network structure and strong interfacial polarization between MnO_2 nanoparticles and the graphene skeleton, the MnO_2/GFs nanocomposite achieves the strongest electromagnetic absorption with an RL_{min} of −53 dB, and the EAB is up to 7.0 GHz. The strong magnetic attenuation capability mainly comes from natural ferromagnetic resonance. Under the alternating electromagnetic field, the accumulation of charges at the interfaces between the graphene skeleton and manganese dioxide nanoparticles could result in interfacial polarization. In addition, there are abundant defects and groups on the graphene skeleton surface, which can generate polarization relaxation and electron dipole polarization, respectively. Therefore, the synthesized MnO_2/GFs composites have the potential for practical application as wide absorbing bandwidth, highly efficient and lightweight EMW absorbers. Wu's group prepared MnO_2/carbon foam (MCF) composites with tunable morphology by a facile hydrothermal and annealing process [18]. The RL_{min} of MCF-4 composite reaches −65.18 dB at 3.19 mm, and the EAB covers the entire X-band. The 3D CF matrix acts as a skeleton to provide electron migration and hopping, resulting in conductive losses. MnO_2 nanopowder is beneficial for producing multiple scattering and extending the propagation path of EMWs. The polarization loss caused by defects is the main loss mechanism that affects the excellent EMW absorption performance, in which the lattice defects play a leading role. Li's group developed N-doped magnetic carbon foam (N-MCF) by heating the prepared carbon foam at high temperature [19]. N-doping content is determined by the mass ratio of adsorbed Rhodamine B versus MCF. By adjusting the N-doping content, the impedance matching of N-MCFs can be optimized. N-MCF-2 with a loading rate of 15% possesses the lowest RL_{min} value of −60.46 dB with a thickness of only 1.70 mm, and the corresponding effective bandwidth is 14.29 and 13.79 GHz for −10 and −20 dB, respectively. The abundant defects caused by N-doping not only act as the dipole centers for electron hopping and migrating following the distorted skeletons but also bring out both defect polarization and dipole polarization effects, enhancing the dielectric loss. In details, at ~300 °C, phenolic resin is partially carbonized, and $Fe(acac)_3$ is pyrolyzed to form Fe_3O_4. As the pyrolysis temperature further increases to 1000 °C, the temperature increase inside the resin foam should be slightly retarded due to the porous structures. And at ~600 °C, Fe_3O_4 particles are further pyrolyzed to produce α-Fe and Fe_2O_3. In addition, there are a large number of heterogeneous interfaces between paraffin wax, g-C_3N_4/carbon skeleton and in situ formed α-Fe/Fe_3C/Fe_2O_3 composites, which can be used to attenuate electromagnetic field induced current and convert EMW energy into heat energy. Meanwhile, the α-Fe/Fe_3C/Fe_2O_3 composites activate complex permeability from exchange resonance, eddy current resonance, and natural resonance. And due to the synergistic effect of magnetic loss and dielectric loss, excellent impedance matching is achieved, ensuring less reflection and more transmission of EMW energies at the surface of N-MCFs. This work provides a possible direction for the reasonable disposal of used dye-adsorbent that has adsorbed N-containing organic dyes.

7.1.5 Composite Foams of Carbon Material and Ceramic Materials

Notably, SiC, especially SiC nanowires (SiC_{nw}s), is considered an excellent candidate due to its high aspect ratio, abundant stacking faults, exceptional thermal and chemical stability at room temperature, and even at high temperatures, and adjustable electrical conductivity. It is revealed that SiC_{nw}s with abundant stacking faults, twinning interfaces, and bridged junctions play an important role in the enhanced electromagnetic absorption performance, in addition to the contribution of interconnected graphene networks. Yin's group prepared 3D flexible foams consisting of rGO and in situ grown SiC_{nw}s using freeze-drying and carbothermal reduction processes [20]. Furthermore, rGO/SiC_{nw}s foams were infiltrated into the dilute poly(dimethylsiloxane) for the investigation of dielectric and EMW absorption properties. The rGO/SiC_{nw}s foam in a poly(dimethylsiloxane) matrix achieves effective absorption in the entire X-band (8.2−12.4 GHz) with a thinner thickness (3 mm) compared to pure rGO foam. The high surface-to-volume ratio of 2D rGO sheets and 1D SiC_{nw}s gives rise to the accumulation of charges at the interfaces, leading to interfacial electron polarization. Second, the hopping of electrons occurs in the bridged rGO and SiC_{nw}s, which attributes to conductive loss. More importantly, the multiple polarization processes arising from the high-density stacking faults and the twinning interfaces in SiC_{nw}s play an important role in the enhanced EM dissipation of rGO/SiC_{nw}s foams. It is revealed that SiC_{nw}s with abundant stacking faults, twinning interfaces, and bridged junctions play an important role in the enhanced electromagnetic absorption performance in addition to the contribution of interconnected graphene networks. Chen's group, for the first time, synthesized a 3D hybrid foam of SiC_{nw}s with a nanostructured, layered carbon film by combining one-way freeze-drying and carbonization techniques (Figure 7.5a) [21]. EAB of the SiC_{nw}s/C foam was observed to cover the whole X-band frequency range of 8.2–12.4 GHz, with RL values reaching −31 dB for an optimum thickness of 3.3 mm. Therefore, the microwave absorption performance of the hybrid foam can be optimized by adjusting the degree of crystallization of the carbon film achieved by foam heat treatment. As the incident microwave radiation interacts with the material, tremendous resistance-inductance-capacitance-coupled circuits and time-varying electromagnetic field-induced currents occur on the cell walls and struts of the foam. Such long-range-induced currents quickly attenuate in the resistive network and convert into thermal energy, resulting in the rapid massive decay of the incident EMW. Secondly, when an incident EMW interacts with the SiC_{nw}s/C hybrid foam of the present study, extensive interfacial electron polarization occurs throughout the vast interfacial area between SiC_{nw}s and carbon phase, leading to interfacial polarization relaxation. The extensive interface area increases the EM transmission path, which in turn increases the possibility of attenuation of the incident wave. The defect polarization from residual groups and interfacial polarization from amorphous carbon and microcrystal carbon, as well as the multiple polarizations arising from the high stacking fault density and the twinning interfaces in SiC_{nw}s, also play an important role in the enhanced EM dissipation of SiC_{nw}s/C hybrid foams. The material is suitable for attenuating electromagnetic interference in modern electronics, modern military applications, weapon systems, and information countermeasures. Chen's group prepared a double-network nested foam composite with tunable EMW absorption by carbonization, impregnation, and chemical vapor infiltration methods (Figure 7.5b) [22]. The SiC_{nw}/CF composite prepared with a lower concentration of Ni catalyst possessed an excellent microwave absorption properties with the RL_{min} of −36.31 dB with a matched thickness of 2.60 mm, accompanied by the EAB of 4.52 dB. With the introduction of the SiC_{nw} network,

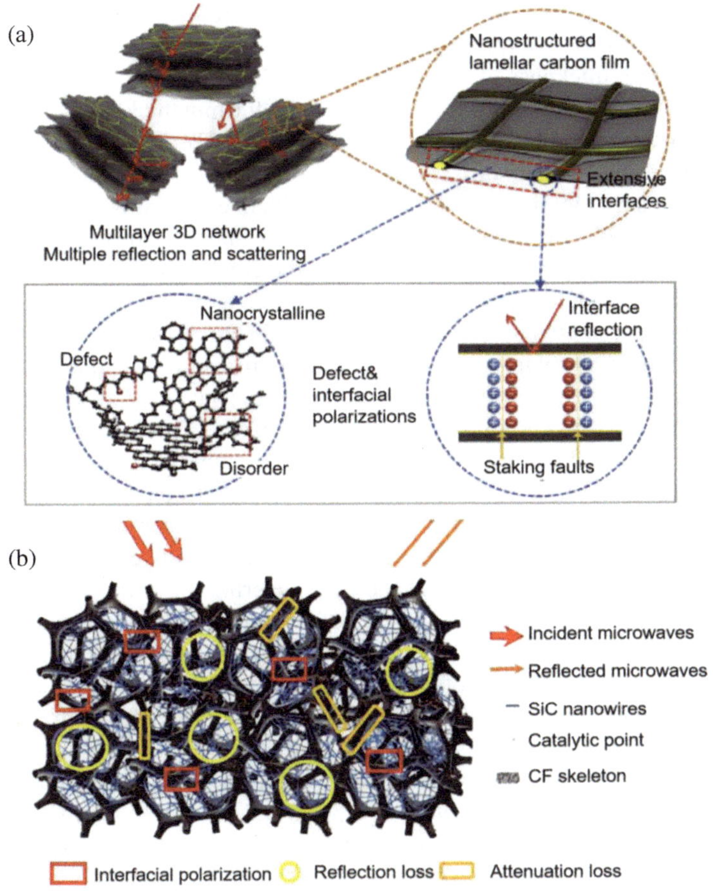

Figure 7.5 (a) A schematic for the EMW absorption mechanisms of SiC_{nw}s/C hybrid foams. Source: Xiao [21]/with permission of Elsevier. (b) Schematic of the microwave absorption mechanism of double networks. Source: Ye [22]/with permission of Royal Society of Chemistry.

more microwaves are absorbed at first due to the enhanced impedance matching. Multiple contact interfacial polarization, as well as the dipole polarization caused by the conductivity difference of the double networks, occurs, which is indispensable for ideal microwave absorption. In addition, the rod-shaped SiC_{nw}s of S1 (catalyst concentration is $0.01\,mol\,L^{-1}$) are long and thin, which divide the original pores into smaller ones. The greater the number of the SiC_{nw}s as well as their length, the smaller the pore size, which results in larger numbers of skeleton reflectors. When the incident microwaves enter the composite, part or all of the waves may be reflected, absorbed, or transmitted by the walls of the new pores. This work proved that the SiC_{nw} network has a significant effect on the electromagnetism parameters and thus on the electromagnetic absorption properties. Wang's group developed a series of lightweight SiC@C nanowire foams (SCNFs) constructed of a large number of ultra-long SiC@C nws by a simple glucose solution permeation-carbonization method [23]. The density of the SCNF could be adjusted by changing the amount of the glucose-coated

SiC_{nw} aerogel sample placed in the partial-hot-pressing mold. SCNF108, with a density of 180 mg cm^{-3}, exhibited superior EMW absorption performance, that is, an RL_{min} of −52.5 dB at a frequency of 11.5 GHz and a thickness of 3 mm, and EAB (7.9–18 GHz) covering the whole X- and Ku-bands. SCNF036-9.6 (SCNF036-9.6 sample with 9.6 mm thickness) exhibits a mechanically controlled absorption band ranging from 2.9 to 18 GHz at a compression strain range of 0–66.7%, covering more than 93% of the entire radar bandwidth. At the macroscale, the highly porous structure and the assembled conductive continuous network provide a longer transmission path for the migration of free carriers, and the repeated reflection and scattering of EMWs at the interface greatly improve the conductance loss ability. At the nanoscale, there are a lot of defects in the SiC@C nanowire assembly system, and there are abundant phase interfaces between carbon shell and SiC core, resulting in dipole polarization relaxation and interface polarization. Therefore, multiscale dissipation paths in the SCNF hierarchy work together to achieve broadband and efficient EMW absorption. Wang's group developed a hierarchical carbon fiber reinforced SiC/C (CF-SC) aerogel via a simple method, in which carbon fibers are incorporated into the hydrogel using a free radical polymerization reaction to form a 3D framework, and SiC is introduced through a carbothermal reduction reaction [24]. Remarkably, the lightweight CF-SC aerogel achieves exceptional absorption performance compared to all reported carbon- and ceramic-based materials, with an RL_{min} of −52.6 dB and an EAB of 8.6 GHz covering the entire Ku-band at the thickness of 3.2 mm. A wide heterogeneous interface is formed between the porous skeleton and the fibers and between C and SiC inside the material, which enhances the interfacial polarization and promotes the dissipation of EMW. Secondly, the conductive network formed by the porous skeleton and fibers provides a longer transmission path for migrating electrons, which enhances the conductivity loss ability. The density of SC5 intermediate defects increases with the increase of SiC content, and this defect causes charge accumulation to produce defect dipoles, resulting in polarization of dipoles and consumption of EMW. This work paves a new avenue to develop highly efficient EM wave absorption materials with hierarchical structures.

7.1.6 Composite Foams of Carbon Material and MXene

The high electric conductivity and abundant hydrophilic surface groups of MXene nanosheets endow them with strong EMW dissipation ability, and thus they are very suitable for fabricating high-performance EMW-absorbing materials. In terms of EMW absorption, not only does the nanosheet feature of the MXene with a large specific surface area contribute to multiple scattering for enhanced dissipation, but the abundant functional groups also generate strong polarization loss. For the first time, Yin's group selected $Ti_3C_2T_x$ MXenes as the dielectric mediator to prepare an rGO/$Ti_3C_2T_x$ hybrid foam with a hollow-core–shell architecture and controllable complex permittivity through self-assembly and sacrificial template processes [25]. rGO was grafted on the outer surface of $Ti_3C_2T_x$ spherical nucleus to form a unique heterostructure. The rGO/$Ti_3C_2T_x$ foam possesses excellent EMW absorption performance, superior to all reported foam-based counterparts. The EAB covers the whole X-band at 3.2 mm, while the density is merely 0.0033 g cm^{-3}, and its specific EMW absorption performance (SMAP = RL (dB)/Thickness (cm)/Density (g cm^{-3})) value exceeds 14,299.2 dB cm^{-2} g^{-1}. The macroporous structure, coupled with the core–shell feature, is beneficial for optimizing impedance matching, and more EM scattering occurs on the cell walls and inside the hollow spheres. Furthermore, the abundant heterointerface formed by flexible rGO flakes wrapping $Ti_3C_2T_x$ spheres, and defects such as boundaries, stacking faults, and surface functional groups, enhance the polarization effect. Meanwhile, the charge carriers migrate within the conductive $Ti_3C_2T_x$ spheres and hop to rGO flakes via the heterointerface and defects under the alternated EMW field, forming plentiful field-induced microcurrents, giving rise to the conduction loss. This work will guide future exploration on designing high-performance EMW absorption materials based on MXene materials. Li's group took a different approach to preparing carbon materials that were combined with MXene [26]. They assisted in the preparation of 3D MXene/rGO porous foams by unidirectional freeze casting. Combined with the excellent electrical properties of MXene and RGO, the resultant lightweight (~6 mg cm^{-3}) MXene/rGO composite foam (MGF) with a free-standing configuration exhibits anisotropic microwave absorption performance. At a sample thickness of 5 mm, MGF-1/2 shows an RL_{min} of −26.79 dB at 4.39 GHz along the perpendicular direction that relates to the unidirectional freeze-casting direction, while along the parallel direction, materials with 3 mm thickness reach the EAB of 8.34 GHz. The aligned microstructure of the foam avoids the agglomeration of MXene nanosheets, which can provide a larger surface region for the effective attenuation of EMW through interface polarization. Since there is the existence of abundant surface termination groups and high in-plane electric conductivity, $Ti_3C_2T_x$ and rGO nanosheets are found to present greater polarization relaxation loss ability and in-plane conductive losses compared to those within the out-of-plane direction. Due to the existence of the unidirectional aligned $Ti_3C_2T_x$ and rGO nanosheets in the parallel direction, the parallel direction is the tunnel EMW channel, and the perpendicular direction is the $Ti_3C_2T_x$ EMW barrier, thereby leading to greater attenuation ability within the perpendicular direction because the EMW has a transverse wave characteristic. Therefore, the lightweight MXene/rGO nanocomposite foam that has the controllable 2D material alignment can be further researched within the usage of EMW absorption. Wu's group prepared MXene/cellulose nanofiber ($Ti_3C_2T_x$/CNF) aerogels by selective etching and freeze-drying, revealing the effect of $Ti_3C_2T_x$ content on the evolution of EMW absorption properties (Figure 7.6a) [27]. With both improved impedance matching and enhanced interfacial polarization and conduction loss, an optimal RL_{min} of −44.9 dB and a wide EAB of 5.7 GHz at a thickness of 2.0 mm can be achieved with the addition of 50 wt% $Ti_3C_2T_x$. The unique porous structure and large surface areas contribute to multiple scattering of incident EMW for enhanced attenuation. At the same time, $Ti_3C_2T_x$ nanosheets have excellent inherent conductivity and their interconnections form a continuous network, providing a large number of conductive paths for enhancing electron migration and conductive loss. In addition, polarization relaxation also has a synergistic effect on EMW dissipation, including interfacial and dipole polarization. They also studied the absorption capacity of $Ti_3C_2T_x$/CNF aerogels after multiple deformations by applying repeated stress to them. Compared with the performance at the natural state, satisfactory absorption can be maintained due to the stress dilution by the resilient porous structure and interlayer slipping by the existence of $Ti_3C_2T_x$ nanosheets. Consequently, this work provides not only an effective strategy for composite aerogels with efficient EMW absorption but also the design of flexible devices with performance stability for different applications. Similarly, Li's group proposed to prepare 3D Mxene@gelatin (M@G) nanocomposite aerogel by unidirectional freeze casting using gelatin molecules as "chemical glue" [29]. $Ti_3C_2T_x$ MXene nanosheets are well aligned in M@G nanocomposite aerogel, resulting in greatly enhanced yet anisotropic mechanical properties. The favorable high electrical conductivity of $Ti_3C_2T_x$ nanosheets and the unidirectional

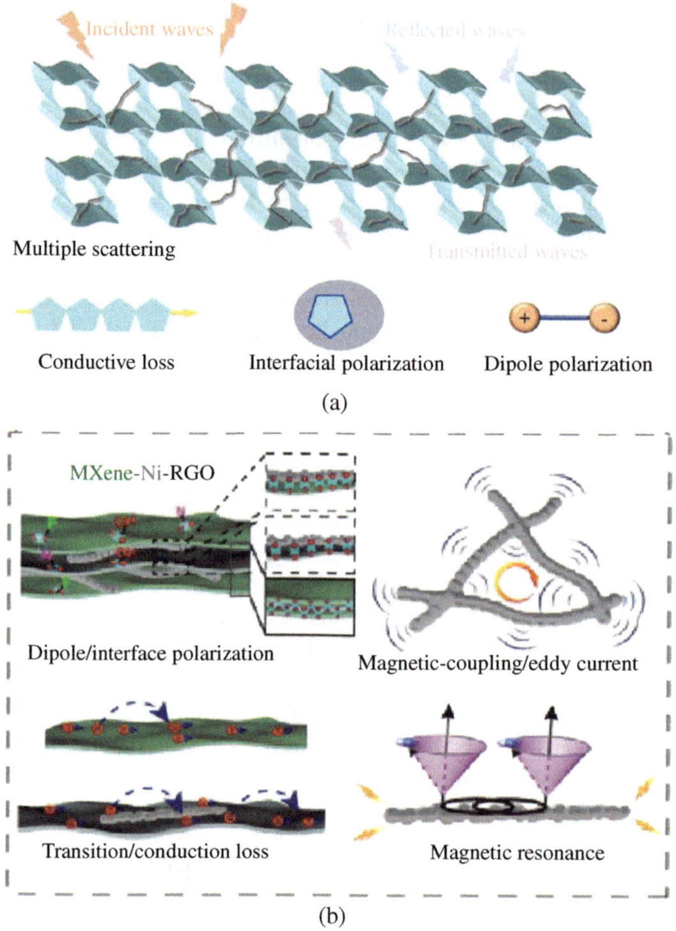

Figure 7.6 (a) Schematic illustration showing the mechanisms involved in the absorption of EMW for the $Ti_3C_2T_x$/CNF aerogels. Source: Qian [27]/with permission of American Chemical Society. (b) Schematic illustrations of the absorption mechanism. Source: Liang [28]/with permission of American Chemical Society.

aligned microstructure of the M@G aerogel are beneficial for the EMW absorption. The anisotropic EMW-absorbing material capability is evidently revealed in RL_{min} of M@G-15 (the mass ratio of $Ti_3C_2T_x$ nanosheets is 15%) in the vertical and parallel directions. When irradiated in the parallel direction, M@G-15 shows no EAB in the whole 2–18 GHz frequency range. Only a −5.5 dB RL_{min} at 12.04 GHz is observed with a sample thickness of 3 mm. This poor absorption performance is attributed to the low permittivity and weak dielectric dissipation caused by low $Ti_3C_2T_x$ content and the unidirectional alignment parallel to the EMW propagation direction. As gelatin is nonconductive, $Ti_3C_2T_x$ nanosheets dominate the dielectric polarization. Since an EMW is a transverse wave, when it incident in the vertical direction, strong electronic polarization in the parallel direction is expected under an alternating electric field, which results in high permittivity. On the contrary, when an EMW incident in the parallel direction, the random distribution of $Ti_3C_2T_x$ nanosheets in this direction limits the orientation of the electric dipole, and thus the aerogels show relatively weak dielectric polarization and lower permittivity values. Meanwhile, more semicircles can be distinguished in the ε'–ε'' plots for the vertical direction compared with the parallel direction, indicating that the M@G aerogels show more polarization relaxation processes and stronger polarization loss ability in the vertical direction. Therefore, the lightweight M@G nanocomposite aerogel with controllable MXene alignment is very promising in EMW absorption applications. Feng's group constructed a 3D dielectric $Ti_3C_2T_x$ MXene/rGO aerogel anchored with magnetic Ni nanochains via a directional-freezing method followed by the hydrazine vapor-reduction process (Figure 7.6b) [28]. Interestingly, the prepared ultralight Ni/MXene/rGO (NiMR-H) aerogel (6.45 mg cm^{-3}) delivers the best EMW performance in reported MXene-based absorbing materials up to now, with an RL_{min} of −75.2 dB (99.999,996% wave absorption) and the broadest EAB of 7.3 GHz. On the macroscopic level, the excellent impedance matching resulting from the mild hydrazine vapor reduction and the regular porous cell structure induces more EMWs to enter the NiMR-H aerogel instead of being reflected from the surface. Then, the incident EMWs are captured and attenuated by the 3D electric/magnetic coupling network in the cell space, which is further intensified by the internal multiple scattering. On the microscopic level, in the multiple cell walls composed of MXene, Ni nanochains, and rGO, the EMWs can be depleted gradually under the synergistic dielectric losses (multiple heterogeneous interface polarizations, dipolar polarization, and conduction loss) and magnetic losses (magnetic resonance, magnetic coupling effect, eddy current loss, and so on). Furthermore, the excellent structural robustness and mechanical properties, as well as the high hydrophobicity and heat insulation performance (close to that of air), guarantee the stable and durable application of the NiMR-H aerogel for EMW absorption to resist deformation, water or humid environments, and high-temperature attacks.

7.2 Ferrite-Based Foam for EMW Absorption

Ferrites, such as MFe_2O_4 (M = Fe, Co, Ni, Cu, Zn, etc.), have been widely used for EMW absorption owing to their advantageous magnetic loss, abundant resources, and environmental benignity. However, the inherent drawbacks such as high density, low permittivity, and easy magnetic agglomeration block ferrites from meeting the requirements of advanced absorbents for lightweight and high-efficiency absorption. The traditional strategy is combining ferrites with high conductivity materials, especially carbon materials, to reduce the density, improve the absorption performance, and resist the magnetic agglomeration of pure ferrites. Unfortunately, these schemes have not fundamentally addressed the issue for the impedance mismatch and low permittivity of ferrites. Integrated foam-type ferrites have the advantages of lightweight characteristics, and furthermore, the high porosity not only improves the impedance matching of materials but also prolongs the transmission path of EMW within materials, thus enhancing scattering loss. Accordingly, a series of strategies have been developed to prepare ferrite foam with the expectation of optimizing the intrinsic dielectric properties of ferrites.

The morphology strategy has been applied for the regulation of the electromagnetic parameters of ferrite. Regarding the inherent drawbacks of pristine Fe_3O_4, such as high density, low permittivity, and easy magnetic aggregation, a design mentality of constructing a 3D structure shaped by curled two-dimensional (2D) porous surfaces was proposed in our previous study (Figure 7.7a) [30]. A series of Fe_3O_4 foams with a unique 3D structure shaped by the folding of 2D curved surfaces were prepared using a facile solgel auto-combustion method. The 2D porous surfaces are formed by stacked planar Fe_3O_4 without obvious granular structures, demonstrating that Fe_3O_4 grains have been integrated to form a planar structure, which is convenient for the electron transfer between Fe^{3+} and Fe^{2+}, and thus the complex permittivity has been significantly increased. Moreover, the 3D structure overcomes the easy agglomeration issue of 2D materials and meanwhile maintains their conductivity. The density of Fe_3O_4 foam is even low at $0.03368\,g\,cm^{-2}$, and the specific surface area is as high as $61.64\,m^2\,g^{-1}$. Impressively, the EAB of S0.5 is as wide as 6.24 GHz at a small thickness of 1.8 mm, which outperforms all reported pure Fe_3O_4 absorbents. The defect-induced polarization, caused by oxygen vacancies and lattice defects, together with interfacial polarization, is the main loss mechanism. $CoFe_2O_4$ foam with a similar microstructure was obtained using the same strategy; electromagnetic parameters were adjusted through morphology regulation [33]. Owing to the appropriate impedance matching and conductance loss provided by moderate complex permittivity, the EAB of S0.5 was as high as 7.3 GHz, exceeding those of most $CoFe_2O_4$-based absorbents. Moreover, the EAB of S1.5 reached 5.0 GHz (8.9–13.9 GHz), covering

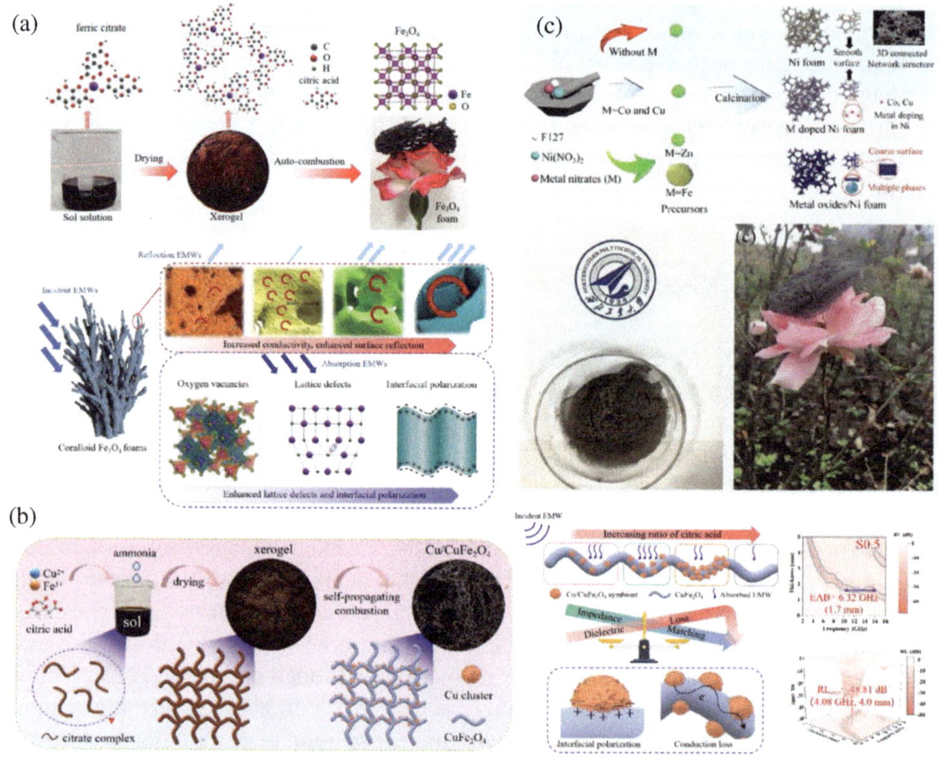

Figure 7.7 (a) The preparation process and EMW loss mechanism diagram of Fe_3O_4 foams. Source: Chang [30]/with permission of Elsevier. (b) Schematic illustration of the fabrication and EMW dissipation mechanism of $Cu/CuFe_2O_4$ foams. Source: Chang [31]/with permission of Elsevier. (c) Schematic diagram of the synthesis of Ni foam-based materials and digital image of Ni foam. Source: Qin [32]/with permission of John Wiley & Sons.

most of the X-band, owing to the intense polarization provided by lattice defects and the heterogeneous interface. The 3D foam structure circumvented the high-density and magnetic agglomeration issues of $CoFe_2O_4$ nanoparticles, and the good conductivity of 2D curved surfaces could effectively elevate the complex permittivity to ameliorate the dielectric loss of pure $CoFe_2O_4$.

In situ component regulation is another feasible strategy to improve the intrinsic complex permittivity and absorption performance of pure ferrite. We propose an in situ symbiotic strategy for metallic Cu with ferritic $CuFe_2O_4$ to increase the intrinsic complex permittivity of $CuFe_2O_4$ (Figure 7.7b) [31]. The Cu/$CuFe_2O_4$ composites were synthesized using a facile and low-energy solgel self-propagating combustion, and the metallic Cu content was adjusted by changing the ratio of the reductant (citric acid) to the oxidant (ferric nitrate). The symbiosis and coexistence of metallic Cu with ferritic $CuFe_2O_4$ increase the intrinsic complex permittivity of $CuFe_2O_4$, which can be regulated by changing the metallic Cu content. Moreover, the unique ant-nest-like microstructure overcomes the issue of magnetic agglomeration. Because of the favorable impedance matching and strong dielectric loss (interfacial polarization and conduction loss) provided by the moderate metallic Cu content, S0.5 concurrently displays broadband absorption with an EAB of 6.32 GHz at an ultrathin thickness of 1.7 mm and strong absorption relying on RL_{min} of −48.81 dB at 4.08 GHz and 4.0 mm.

Surprisingly, the exchange of subject and object materials, namely metal as the substrate for the growth of ferrite, also realized ultrawide absorption for EMW. Focusing on the skin effect and high-density issues of metal-based EMW-absorbing materials, our group fabricated NiO/$NiFe_2O_4$/Ni component foams, which in situ grew $NiFe_2O_4$ on Ni foam by a facile and ecofriendly leaven dough route to overcome the inherent drawbacks of metal-based EMW absorbents and also provided an effective alternative to high-performance metal-based EMW absorbents (Figure 7.7c) [32]. Impedance matching originating from the foam structure and energy dissipation from interfacial polarization and defect-induced polarization provided by NiO/$NiFe_2O_4$ mainly contribute to its ultra-broadband EMW absorption performance. In detail, the foam structure allowed the entrance and propagation of incident EMW inner absorbers. Innovatively, metal oxide/Ni foam chain-like skeletons provided an EM wave reflection on the surface of metal Ni and two times of EM wave loss from interfacial polarization and defect-induced polarization provided by NiO/$NiFe_2O_4$ composites due to the in and out movements of the metal oxide layer. Benefiting from this materials design, a wide absorption bandwidth (14.24 GHz) and a thin matching thickness (0.6 mm) were achieved by a lightweight (0.06 g cm^{-3}) $NiFe_2O_4$/NiO/Ni foam composite (Fe 5–20 wt% filling). This work sheds light on the novel view of the synergistic effect of structure and components on EMW absorption behaviors and demonstrates a new pathway for the preparation of lightweight and high-performance metal-based EMW absorbers.

Oxygen vacancy engineering was adopted by Ding's group to promote the poor intrinsic dielectric properties and limited absorbing bandwidth of ferrite at high-frequency bands (8–18 GHz). Ferrite composite foams (NFC) with a high surficial oxygen vacancy concentration (48.1%) were prepared by the NH_4NO_3-assisted solution-combustion strategy and solid-phase sintering [34]. The concentration of surficial oxygen vacancy can be effectively regulated by the addition of NH_4NO_3, which facilitated to achieve balanced impedance matching. Therefore, the EAB of the NFC covered the whole X-band and Ku-band, and conduction loss and defect polarization relaxation realized by surficial oxygen vacancies were confirmed to be the key to such wide EAB for the NFC at 8–18 GHz. For one thing, the oxygen vacancy can provide shallow donors, and these shallow donors can lead to an increase in the density of states near the Fermi level, which can increase the number of available charge carriers. As a result, the addition of oxygen vacancy can greatly enhance the electrical conductivity, thus improving the dielectric loss. Additionally, the oxygen vacancy can act as unsaturated active centers, inducing excess electrons in the surroundings. This defect-induced electric dipole formation is expected to produce defect-induced polarization loss under external electromagnetic fields, thereby increasing the polarization loss. In addition, oxygen vacancies can effectively regulate the electromagnetic parameters of the ferrites. The altered dielectric constant facilitates the ferrites to achieve a balance of electromagnetic parameters, which is facilitated to achieve satisfactory impedance matching. This work explores the important role of surficial oxygen vacancy engineering in regulating the dielectric parameters of the ferrites and provides a feasible strategy for surficial oxygen vacancy engineering for wide EAB electromagnetic performance at high frequencies.

7.3 SiC-Based Foam for EMW Absorption

Despite their high performance in EMW absorption capacity at room temperature, neither carbon-based materials nor magnetic ferrites are suitable for high-temperature stealth. The main reason is that carbon-based materials mainly consume EMW to convert EM energy into heat energy. However, it is difficult for them to be used in high-temperature environments, and restricted by the Curie temperature, magnetic ferrites are hardly satisfying the practical applications in high-temperature fields. Compared with carbon-based materials and ferrites, SiC-based materials, as dielectric absorbents with tunable dielectric properties and the advantages of chemical and high-temperature stabilities, can meet the application requirements of high-temperature stealth.

In recent years, the newly developed SiCnw aerogel foam has shown high-efficiency heat insulation properties and excellent absorbing properties, which not only overcome the brittleness problem of traditional aerogels but also show excellent flexibility by using SiCnw to cross-link with each other in space to form a network structure. Therefore, an ultralight and high-strength SiCnw@SiC foam with highly efficient EMW absorption and heat insulation properties was successfully synthesized by Wang's group using the template sacrifice method and chemical vapor deposition process (Figure 7.8a) [35]. The microstructure is a novel double-network structure, which is formed by the coupling of the morphology-controlled SiCnw and the SiC skeleton. The introduction of SiCnw can not only provide more interface polarization and dielectric loss to the SiC foam, which greatly enhances the microwave absorption capacity of the composite foam, but also can enable it to act as an excellent radiation absorbent, which can effectively reduce the thermal conductivity of the foam, especially at high temperatures. Therefore, a RL_{min} of −52.49 dB was achieved at 2.82 mm with an EAB of 5.6 GHz. As the length/diameter ratio of SiCnw decreases, the composite foam exhibits excellent high-temperature thermal insulation and mechanical properties, and the thermal conductivity is only 0.304 W mK^{-1} at 1200 °C and the compressive strength reaches 1.53 MPa. This multifunctional SiCnw@SiC foam has potential applications in microwave absorption and high-temperature heat insulation in harsh environments. A "wire-on-sphere" hierarchical SiC_{nw}s/SiC whiskers (SiCw) foam was developed by Ye's group using a 3D printing method for efficient high-temperature EMW absorption (Figure 7.8b) [36]. SiCnw was assembled onto porous SiCw spheres via spray drying, and then SiCw spheres were

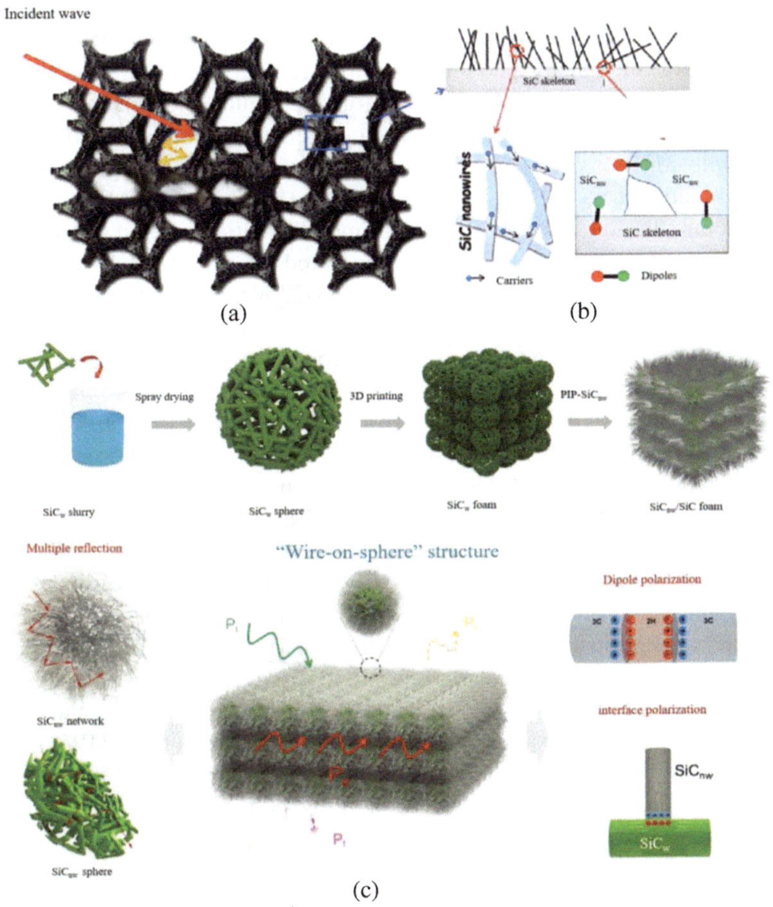

Figure 7.8 (a) Placeholder Text Schematic diagram of the microwave-absorbing mechanism for the SiCnw@SiC foam. Source: Su [35]/with permission of American Chemical Society. (c) Preparation process diagram and EMW absorption mechanisms of the "wire-on-sphere" hierarchical structure of SiCnw/SiC foam. Source: Lv [36]/with permission of Elsevier.

3D printed into SiCw foam. In this study, the catalyst-free precursor infiltration pyrolysis process was first developed to ensure that SiCnw can intentionally grow in situ on the surface of SiCw spheres, thereby achieving "wire-on-sphere" hierarchical structure foam. At room temperature, the EAB and RL_{min} of SiCnw/SiCw foam can reach 4 GHz and −57 dB, respectively. At 600 °C, the EAB and RL_{min} were 3 GHz and −15 dB, respectively. Furthermore, even when oxidized at 1000–1500 °C, SiCnw/SiCw foam can still retain EAB ranging from 2.7 to 3.9 GHz and RL_{min} ranging from −16 to −64 dB because the formation of a SiO_2 layer with an appropriate thickness can boost interfacial polarization and regulate impedance matching. The SiCnw/SiCw foam also shows the flexural strength as high as 17.05 MPa. All results demonstrate that the SiCnw/SiCw foam is a promising EWA material for applications of harsh environments. The material-independent "wire-on-sphere" hierarchical structure and its novel preparation process should find the widespread use in the design and fabrication of high-efficiency EWA micro-nano materials.

Although SiC is an excellent dielectric loss EMW absorption material with stable physicochemical properties, the lack of multiple loss mechanisms in single-component SiC and the inherent dielectric constant of SiC make its conductivity low, causing it difficult to achieve a wide bandwidth absorption in single-component SiC. Moreover, multifunctional nanofiber ceramic aerogels are highly desirable for applications in extreme environments; however, the integration of multiple functions in their preparation is extremely challenging. Fan's group fabricated a multifunctional SiC@SiO_2 nanofiber aerogel (SiC@SiO_2 NFA) with a 3D porous cross-linked structure through a simple chemical vapor deposition method and subsequent heat-treatment process (Figure 7.9a) [37]. The as-prepared SiC@SiO_2 NFA exhibited an ultralow density (~11 $mg\,cm^{-3}$), thermal superinsulation (0.027 $W\,m^{-1}\,K^{-1}$), great recoverable compressibility (repeated full recovery from 60% strain), good thermal and chemical stabilities, and significant strain-dependent piezoresistive sensing behavior. Furthermore, a superior EMW absorption performance with RL_{min} of −50.36 dB and EAB of 8.6 GHz was obtained. The incident EMW was attenuated by multiple reflections among the pores, resulting in the conversion of EM energy into heat for dissipation. The conduction loss was caused by the converted energy of the EMW into an electric current when it propagated in the SiC@SiO_2 nanofibers. When the generated current was transported along the nanofibers, Joule heat was generated due to the resistance of the SiC@SiO_2 nanofibers, which consumed the EMW energy. Oxygen vacancies are introduced during the oxidation process to form defect sites in the SiC@SiO_2 nanofibers. The charge carriers can be trapped at these defect sites, leading to an imbalanced charge distribution. The resulted polarization and EMW energy loss then occur. Furthermore, the interfacial polarization and dipole relaxation induced by the SiO_2 nanolayers can improve the dielectric loss. The SiC cores of the SiC@SiO_2 nanofibers undergo dipole polarization and relaxation processes under changes in the EMW field, consuming the EMW energy. In addition, transverse electric fields are formed inside the SiC@SiO_2 nanofibers under the action of the dipoles, and the electrons are affected by the

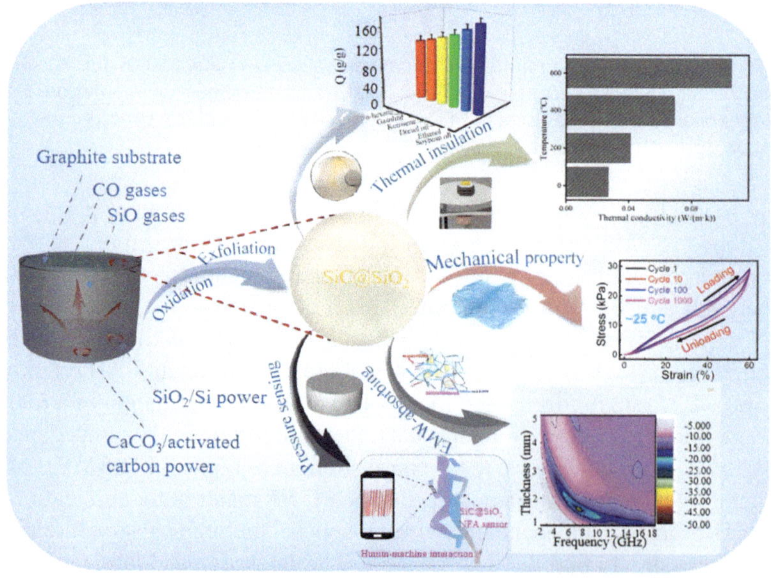

Figure 7.9 Preparation process, EMW absorption performance, and multifunctional applications of the SiC@SiO_2 NFA. Source: Song [37]/Springer Nature/CC BY 4.0.

transverse electric field during the movement process, which increases the transmission path and further consumes the EM energy. Thus, the synergistic effect of multiple reflections, conduction losses, defect-induced polarization, interfacial polarization, and dipolar polarization together enables the excellent EMW-absorbing property of the $SiC@SiO_2$ NFA. This work provides a new reference for the development of materials with rational microstructure construction while achieving thin thickness, strong absorption, and broadband absorption.

7.4 Conductive Polymer Composites Foam for EMW Absorption

Compared with ponderous ferrites, metallic magnets, ceramics, and their hybrids, 3D porous foam materials based on polymers have recently attracted increasing attention because of their low density and impressive mechanical and thermal stability. Impedance matching in the material and numerous reflections inside the material both benefit from the porous wall interface [38]. However, pure polymers may not have appreciable EMW absorption due to low electrical conductivity and magnetic properties [39]. Therefore, conducting materials with high aspect ratios and mechanical strengths, such as carbon black, rGO, and CNTs, are used to construct electrically CPC.

For polymer-based composite foams, improving the surface impedance matching can effectively promote the incident EMW entering the materials to the maximum extent. Furthermore, the optimization of impedance distribution and microstructure of the materials facilitates the efficient attenuation of EMW entering the interior of the material. Therefore, a suitable impedance matching distribution and microstructure in polymer-based porous nanocomposites contribute to enhancing the dissipation of incident EMW. In comparison with the isotropic random pore structure, the construction of an ordered lamellar pore structure has also proved to be a method to facilitate the efficient attenuation of incident electromagnetic energy in past years. Using a bidirectional freezing process, ultralight polyimide-based graphene foam with an ordered lamellar structure was precisely designed and controllably constructed [40]. More lamellar interfaces formed inside the foam per unit volume effectively facilitate the layer-by-layer dissipation for the vertical incident EMW, thereby endowing the foam with efficient broadband electromagnetic absorption performance. More importantly, electromagnetic absorption performance can be controllably adjusted by optimizing impedance distribution and microstructure of skeletons. As a result, the optimized foam with an ultralow density of 9.10 mg cm^{-3} presents an RL_{min} of −61.29 dB at 9.25 GHz and an EAB of 5.51 GHz (7.06–12.57 GHz, covering the whole X-band) when the thickness is 4.75 mm. A strategy of constructing a microcellular structure in the CPCs was proposed by Che's group to solve the secondary EM radiation problem of CPCs (Figure 7.10a) [41]. A simple microcellular injection molding method was used to fabricate highly compressible thermoplastic urethane (TPU)/CNT composite foams, which also had increased conductivity with an increase in CNT content. Compared to unfoamed composites, foamed composites exhibited higher conductivity and EM attenuation properties because of the presence of a microcellular structure. Moreover, the TPU/CNT foam with 4 wt% CNTs demonstrated strong EM dissipation and an optimal RL value of −30.4 dB. Furthermore, stimulated by thermal heating and cyclic compression, EM attenuation was observed to increase because of the higher conductivity. Note that the TPU/CNT foam with 4 wt% CNTs having a small thickness of 1.3 mm when treated at 333 K had the highest EM dissipation and the lowest RL value of −51.8 dB. Enhanced polarization, conductive losses, and multi-scattering were responsible for the increased EM absorption. This behavior is attributed to the movement

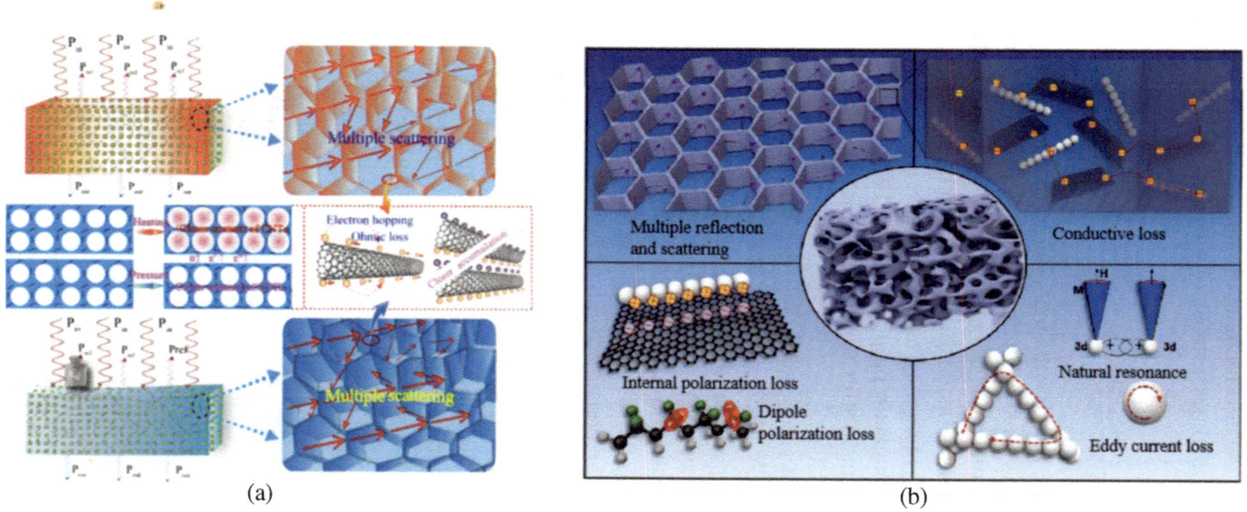

Figure 7.10 (a) Mechanism of improved microwave absorption of TPU/CNT composite foams treated by thermal heating and compression: The strengthened absorption mechanism of foam stimulated by thermal heating and the improved absorption mechanism of foam stimulated by compression. Source: Zhao [41]/with permission of American Chemical Society. (b) The preparation process diagram and microwave dissipation mechanism of GN/Ni PVDF foam. Source: Cui [42]/with permission of Elsevier.

of CNTs within the TPU elastomer walls via thermal or compression stimulation. For the delicate control of 3D microcellular structure by decoration dielectric/magnetic components, their group also fabricated the hierarchical structure of 3D poly(vinylidene fluoride) (PVDF) foam decorated with an ultralow content of graphene nanoplates and Ni nanochains toward wideband EMW absorption (Figure 7.10b) [42]. The correlation between microcellular structure and electrical conductivity/microwave absorption was systematically investigated. After introducing the conductive graphene nanoplates and magnetic Ni nanochains, a strong enhancement of dielectric loss, stemming from the presence of conduction loss and polarization loss of 3D network, and magnetic loss, originating from strong eddy current loss and the natural resonance, was observed. Meanwhile, a high void fraction (~92%) greatly optimized the impedance matching of incident waves and favored the multi-scattering. As a result, the superior microwave absorption was obtained, in which the EAB covered the whole K band (18–26.5 GHz) under the thickness of 3 mm. This work provides an ameliorated strategy for porous foams applied in microwave absorption.

With the application scenarios becoming more complex and variegated, designing the advanced high-performance EMW absorption materials for functional integrations (e.g. excellent electromagnetic protection, heat conduction, sound absorption, and mechanical robustness) is the direction of further development. Our group proposed a foam-type microwave absorption device consisting of melamine and phenolic aldehyde amine as the 3D foam skeleton and cementing compound, respectively, as well as the inorganic fillers of graphite as dielectric loss components and heat-conducting bridge via a universal dipping and drying process at room temperature (Figure 7.11a) [43]. Melamine and phenolic aldehyde amine worked as the skeleton and cementing compound, respectively, in which graphite was embedded in the cementing compound interconnected into the mesoscopic 3D electric conductive and heat conductive network. Interestingly, the prepared flexible graphite/melamine foam (CMF) delivers a great EWA performance, with an EAB of 9.8 GHz, an ultrathin thickness of 2.60 mm, and a strong absorption RL of −41.7 dB. Moreover, the CMF possesses porosity and flexibility, endowing it with sound absorption ability. The CMF is unique in its integration of EWA, heat conduction, sound absorption, and mechanical robustness, as well as its cost-effective and scalable manufacturing. These attributes make CMF promising as a multifunctional device widely used in communication base stations, servers, and chip protection. An ultra-flexible silver/iron nanowire decorated melamine composite foam (AgFe-MF) for high-efficiency EMW absorption and thermal management was prepared via a simple dip-coating process by Chen's group [44]. Owing to optimal impedance matching, synergistic dielectric and magnetic losses, as well as a three-dimensional porous structure, the AgFe-MF with an ultralow filler content (0.22 vol%) exhibited an outstanding minimum RL of −69.61 dB, and the best EAB could reach up to 6.37 GHz. Importantly, the EAB of the long-time working AgFe-MF was enhanced to 7.01 GHz after 1000 compress-release cycles under 40% strain. Additionally, it also featured considerable Joule heating capacity and achieved a saturation temperature of over 85.7 °C under a 2.6 V. The impressive thermal isolation and long-term stability ensured the safety of its use as a portable heater. Therefore, this work will provide a vital insight for fabricating smart wearable electronic devices with integrated anti-electromagnetic radiation and personalized thermal management performances toward potential thermal and health threats.

High-performance electromagnetic wave-absorbing (EMA) materials used in high-temperature environments are of great importance in both civil and military fields. For example, the surface temperature of a high-speed reconnaissance plane could reach 300 °C when

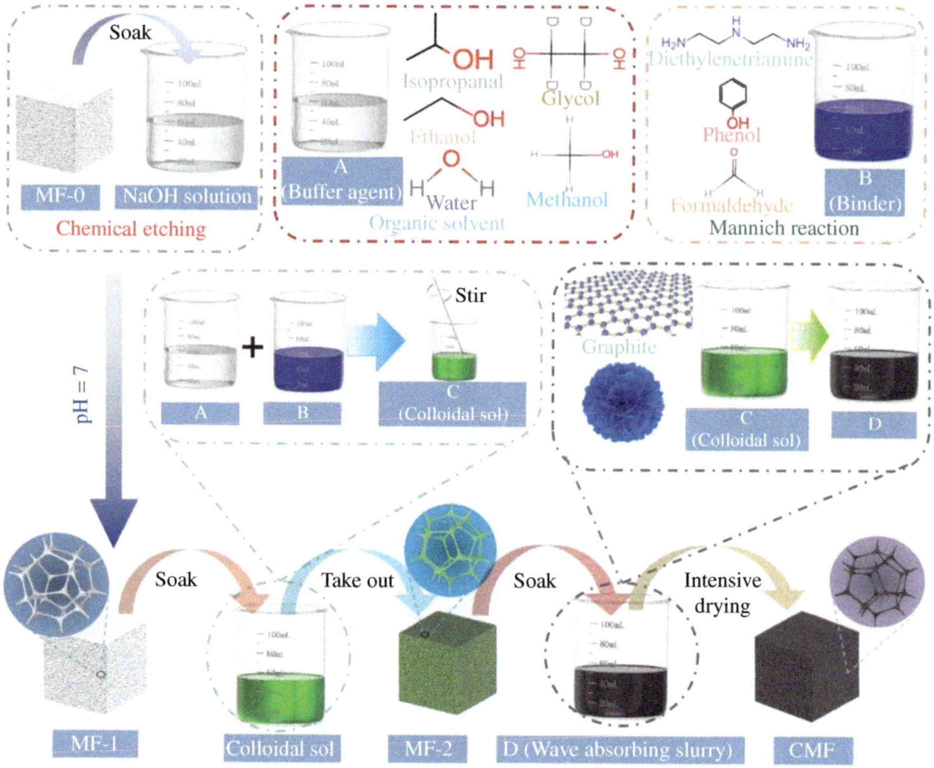

Figure 7.11 The preparation process of CMF via an organic polymer assisted in direct adsorption strategy. Source: Li [43]/John Wiley & Sons.

it is cruising with a speed of Mach 3; the temperature of stealth satellites can reach nearly 200 °C under sunshine radiation; stealth artillery launcher systems should also bear 200 °C during firing [45]. In these cases, these room-temperature-usable EMA materials may get oxidized and degraded, finally weakening their EMA performance [46, 47]. Therefore, EMA materials with high-temperature resistance are in urgent need. Huang's group developed the ultralight graphene/polyaramid (PA) composite foam for wideband EMW absorption in both gigahertz and terahertz bands, with a higher service temperature of 300 °C [48]. It is found that strong interfacial π–π interactions are spontaneously constructed between graphene and polyaramids during the foam preparation process. This endows the foam with two advantages for its EMA performance. First, the π–π interactions trigger the interfacial polarization for enhanced microwave dissipation, as confirmed by the experimental and simulation results. The composite foam with an ultralow density of 0.0038 g cm^{-3} showed an RL_{min} of −36.5 dB and an EAB of 8.4 GHz between 2 and 18 GHz bands. Meanwhile, excellent terahertz (THz) absorption is also achieved, with EAB covering the entire 0.2–1.6 THz range. Second, the interfacial π–π interactions promote PA to present a unique in-plane orientation configuration along the graphene surface, thus making PA an effective antioxidation barrier layer for graphene at high temperatures. The EMA performance of the foam could be completely preserved after 300 °C treatment in an air atmosphere. Furthermore, the composite foam exhibits multifunctions, including good compressive, thermal insulating, and flame-retardant properties. This study could provide useful guidance for the design of next-generation EMA materials used in harsh environments. For more precisely controlling the size and shape of pores and also the orientation of the fillers, Pauline et al. prepared CNT and $CuFe_2O_4$ nanoparticle-based porous polydimethylsiloxane (PDMS) composite foams using the baking soda template method [49]. The PDMS foam acts as the 3D porous network, CNT gives the required conductive path, and magnetic $CuFe_2O_4$ provides the necessary magnetic loss. As a result, the EMI SE is up to 32 dB at a thickness of 3 mm. In addition, the thermal insulation properties are also reported, and the resultant material shows a moderately low thermal conductivity of $0.046\,W\,m^{-1}\,K^{-1}$. These characteristics of the composite foam highlight the strong absorption-dominated EMI shielding and thermal insulation performance, allowing it to successfully meet a wide range of applications .

References

1 Huang, X. (2021). Design of cellular structure of graphene aerogels for electromagnetic wave absorption. *Chem. Eng. J.* 426 (15): 131894.

2 Chen, X. (2022). Morphology control of eco-friendly chitosan derived carbon aerogels for efficient microwave absorption at thin thickness and thermal stealth. *Green Chem.* 24: 5280–5290.

3 Wang, Z. (2018). Ultralight, highly compressible and fire-retardant graphene aerogel with self-adjustable electromagnetic wave absorption. *Carbon* 139: 1126–1135.

4 Liu, P. (2019). Synthesis of lightweight N-doped graphene foams with open reticular structure for high-efficiency electromagnetic wave absorption. *Chem. Eng. J.* 368 (15): 285–298.

5 Shu, R. (2021). Synergistically assembled nitrogen-doped reduced graphene oxide/multi-walled carbon nanotubes composite aerogels with superior electromagnetic wave absorption performance. *Compos. Sci. Technol.* 210 (7): 108818.

6 Zhao, X. (2022). Environmentally friendly a multifunctional cellulose-based carbon foam for superior electromagnetic wave absorption performance. *Compos. Commun.* 35: 101320.

7 Zhang, R. (2023). Ultralight aerogel sphere composed of nanocellulose-derived carbon nanofiber and graphene for excellent electromagnetic wave absorption. *Nano Res.* 16 (5): 7931–7940.

8 Xiong, Y. (2020). Implanting FeCo/C nanocages with tunable electromagnetic parameters in anisotropic wood carbon aerogels for efficient microwave absorption. *J. Mater. Chem. A* 8: 18863–18871.

9 Li, S. (2022). Corrosion-resistant graphene-based magnetic composite foams for efficient electromagnetic absorption. *ACS Appl. Mater. Interfaces* 14: 8297–8310.

10 Lyu, L. (2021). CuNi alloy/carbon foam nanohybrids as high-performance electromagnetic wave absorbers. *Carbon* 172: 488–496.

11 Zou, Y. (2023). Triple-layer structure of carbon foam coated with carbon nanotubes and FeNi alloy for high-performance electromagnetic wave absorption. *J. Alloys Compd.* 946 (15): 169404.

12 Wang, S. (2024). Facile fabrication of biomass chitosan-derived magnetic carbon aerogels as multifunctional and high-efficiency electromagnetic wave absorption materials. *Carbon* 216 (5): 118528.

13 Zhou, X. (2019). Synthesis of Fe_3O_4/carbon foams composites with broadened bandwidth and excellent electromagnetic wave absorption performance. *Compos. Part A-Appl. S.* 127: 105627.

14 He, Y. (2021). Multifunctional carbon foam with hollow microspheres and a concave-convex microstructure for adjustable electromagnetic wave absorption and wearable applications. *J. Mater. Chem. A* 9: 25982–25998.

15 Shu, R. (2021). Synthesis of nitrogen-doped reduced graphene oxide/cobalt-zinc ferrite composite aerogels with superior compression recovery and electromagnetic wave absorption performance. *Nanoscale* 9 (13): 11803–11810.

16 Sun, N. (2021). Facile synthesis of lightweight 3D hierarchical $NiCo_2O_4$ nanoflowers/reduced graphene oxide composite foams with excellent electromagnetic wave absorption performance. *J. Mater. Sci. Technol.* 91 (20): 187–199.

17 Zhang, Z. (2019). MnO_2 nanostructures deposited on graphene foams for broadband and lightweight electromagnetic absorption. *J. Alloys Compd.* 810 (25): 151744.

18 Deng, W. (2023). Morphology modulated defects engineering from MnO_2 supported on carbon foam toward excellent electromagnetic wave absorption. *Carbon* 206 (25): 192–200.

19 Lou, Z. (2021). Used dye adsorbent derived N-doped magnetic carbon foam with enhanced electromagnetic wave absorption performance. *J. Alloys Compd.* 854 (15): 157286.

20 Han, M. (2017). Flexible and thermostable graphene/SiC nanowire foam composites with tunable electromagnetic wave absorption properties. *ACS Appl. Mater. Interfaces* 9 (13): 11803–11810.

21 Xiao, S. (2017). Ultralight lamellar amorphous carbon foam nanostructured by SiC nanowires for tunable electromagnetic wave absorption. *Carbon* 122: 718–725.

22 Ye, X. (2019). Double network nested foam composites with tunable electromagnetic wave absorption performances. *Inorg. Chem. Front.* 6: 1579–1586.

23 Cai, Z. (2020). Hydrophobic SiC@C nanowire foam with broad-band and mechanically controlled electromagnetic wave absorption. *ACS Appl. Mater. Interfaces* 12 (7): 8555–8562.

24 Dai, D. (2023). Hierarchical carbon fiber reinforced SiC/C aerogels with efficient electromagnetic wave absorption properties. *Compos. Part B Eng.* 248 (1): 110376.

25 Li, X. (2018). Self-assembly core-shell graphene-bridged hollow MXenes spheres 3D foam with ultrahigh specific EM absorption performance. *Adv. Funct. Mater.* 28 (41): 1803938.

26 Ning, Y. (2022). Anisotropic electromagnetic absorption of the aligned $Ti_3C_2T_x$ MXene/RGO nanocomposite foam. *Compos. Sci. Technol.* 227 (18): 109609.

27 Qian, S. (2022). Flexible MXene/cellulose nanofiber aerogels for efficient electromagnetic absorption. *ACS Appl. Nano Mater.* 5 (7): 9771–9779.

28 Liang, L. (2021). Multifunctional magnetic $Ti_3C_2T_x$ MXene/graphene aerogel with superior electromagnetic wave absorption performance. *ACS Nano* 15: 6622–6632.

29 Yang, M. (2022). Anisotropic electromagnetic absorption of aligned $Ti_3C_2T_x$ MXene/gelatin nanocomposite aerogels. *ACS Appl. Mater. Interfaces* 12: 33128–33138.

30 Chang, Q. (2022). Microstructure induced dielectric loss in lightweight Fe_3O_4 foam for electromagnetic wave absorption. *iScience* 25 (3): 103925.

31 Chang, Q. (2023). Symbiotic strategy of Cu on $CuFe_2O_4$ realizing high-efficiency electromagnetic wave absorption. *J. Colloid Interface Sci.* 645: 841–849.

32 Qin, M. (2021). Lightweight Ni foam-based ultra-broadband electromagnetic wave absorber. *Adv. Funct. Mater.* 31 (30): 2103436.

33 Shi, B. (2023). Dielectric loss enhancement induced by the microstructure of $CoFe_2O_4$ foam to realize broadband electromagnetic wave absorption. *Int. J. Miner. Metall. Mater.* 30 (7): 1388.

34 Shao, C. (2023). Ultrahigh concentration surface oxygen vacancies in ferrite foams for broadened electromagnetic absorption bandwidth. *Mater. Res. Bull.* 167: 112411.

35 Su, K. (2021). Ultralight and high-strength SiCnw@SiC foam with highly efficient microwave absorption and heat insulation properties. *ACS Appl. Mater. Interfaces* 13 (18): 22017–22030.

36 Lv, X. (2022). 3D printing "wire-on-sphere" hierarchical SiC nanowires/SiC whiskers foam for efficient high-temperature electromagnetic wave absorption. *J. Mater. Sci. Technol.* 109 (20): 94–104.

37 Song, L. (2022). Multifunctional SiC@SiO_2 nanofiber aerogel with ultrabroadband electromagnetic wave absorption. *Nano Micro Lett.* 14: 152.

38 He, Q.M. (2022). Electricmagnetic-dielectric synergism and Salisbury screen effect in laminated polymer composites with multiwall carbon nanotube, nickel, and antimony trioxide for enhancing electromagnetic interference shielding. *Compos. Part A* 156: 106901.

39 Man, D. (2021). 3D structured graphene fluoride-based epoxy composites with high thermal conductivity and electrical insulation. *Compos. A: Appl. Sci. Manuf.* 149: 106585.

40 Zhang, Y. (2022). Ultralight and ordered lamellar polyimide-based graphene foams with efficient broadband electromagnetic absorption. *J. Mater. Sci. Technol.* 102: 97–104.

41 Zhao, B. (2020). Highly compressible polymer composite foams with thermal heating-boosted electromagnetic wave absorption abilities. *ACS Appl. Mater. Interfaces* 12 (45): 50793–50802.

42 Cui, W. (2023). 3D porous PVDF foam anchored with ultra-low content of graphene and Ni nanochains towards wideband electromagnetic waves absorption. *Carbon* 210 (15): 118070.

43 Li, Z. (2023). A regulable polyporous graphite/melamine foam for heat conduction, sound absorption and electromagnetic wave absorption. *Small* 20: 2305120.

44 Hang, T. (2024). Ultra-flexible silver/iron nanowire decorated melamine composite foams for high-efficiency electromagnetic wave absorption and thermal management. *J. Colloid Interface Sci.* 654 (15): 945–954.

45 Lafdi, K. (2008). Graphite foams infiltrated with phase change materials as alternative materials for space and terrestrial thermal energy storage applications. *Carbon* 46: 159–168.

46 Jian, X. (2019). High-temperature oxidation-resistant $ZrN_{0.4}B_{0.6}$/SiC nanohybrid for enhanced microwave absorption. *ACS Appl. Mater. Interfaces* 11: 15869–15880.

47 Luo, C. (2018). Graphene shield by SiBCN ceramic: a promising high-temperature electromagnetic waveabsorbing material with oxidation resistance. *ACS Appl. Mater. Interfaces* 10: 39307–39318.

48 Cheng, Z. (2022). Interfacial π–π interactions induced ultralight, 300 °C-stable, wideband graphene/polyaramid foam for electromagnetic wave absorption in both gigahertz and terahertz bands. *ACS Appl. Mater. Interfaces* 14 (2): 3218–3232.

49 Prabagar, C. (2023). Thermally insulating carbon nanotubes and copper ferrite based porous polydimethylsiloxane foams for absorption-dominant electromagnetic interference shielding performance. *Compos. Commun.* 42: 101691.

8

Integral Gel Electromagnetic Wave Absorption Materials

Jiaming Wen

MOE Key Laboratory of Material Physics and Chemistry Under Extraordinary Conditions, School of Physical Science and Technology, Northwestern Polytechnical University, Xi'an, China

CHAPTER MENU

With the development of electronic communication technology and the continuous expansion of application fields, there is an increasing demand for the control and management of electromagnetic waves (EMW), so materials with efficient EMW-absorbing properties have gained wide attention and applications. Integral gel EMW absorption materials are a new type of material with excellent EMW-absorbing properties, which include solid-medium carbon-based aerogels, MXene-based aerogels, polymer-based aerogels, metal-based aerogels, and ceramic-based aerogels. Advanced integral aerogel materials integrating material structure and functionality are very popular due to their lightweight, flexibility, high durability, and excellent mechanical properties. Many studies have reported that aerogels are widely studied structural materials with ultralow density, high porosity, and compressibility [1–3]. There are also hydrogels, ion gels, and organic gels, which are mainly liquid media. They have good flexibility and plasticity and can be pasted and covered on surfaces of different shapes and sizes to effectively achieve the absorption and dissipation of EMW [4–6].

The research and development of integral gel-based EMW absorption materials can not only meet the requirements of electronic communication equipment and radar systems for EMW interference and stealth performance but also play an important role in aerospace,

Electromagnetic Wave Absorbing Materials: Fundamentals and Applications, First Edition.
Edited by Hongjing Wu, Jun Luo, and Meiyin Yang.

military protection, electromagnetic shielding, radio wave control, and other fields. Therefore, the optimization of the properties, preparation process, and application of integral gel-based EMW absorption materials is of great theoretical and practical significance [7–9].

In this chapter, the research status and development trends of integral gel-based EMW absorption materials will be reviewed, focusing on the latest progress in their preparation processes and application prospects, aiming at providing references for researchers in related fields and promoting further development and application of integral gel-based EMW absorption materials. Whereas the previous chapters of this book have introduced carbon-based aerogels, Mxene-based aerogels, and polymer-based aerogels in detail, we will concentrate on metal-based aerogels, ceramic-based aerogels, and dielectric liquid medium gels.

8.1 Dielectric Liquid Medium Gel Electromagnetic Wave Absorption Materials

Microwave-absorption materials (MAMs) are widely studied, and in addition to their intrinsic properties, these materials often need to be further optimized through modification and microstructural design. While remarkable performances can be achieved through these methods, they may also lead to unpredictable, vulnerable, and inconsistently reproducible samples. Therefore, aside from the ongoing exploration of various absorption composites, there is potential to investigate new MAMs capable of achieving remarkable performance based on a singular or limited mechanism. In this context, dielectric liquid mediums and poly(ionic liquid)s (PILs) emerge as promising options, with an extensive range available for research. These options are devoid of defects and exhibit negligible interfacial polarization, and consequently, only dipole polarization and ionic conduction losses need to be taken into account [4, 5]. Furthermore, they offer additional advantages over traditional MAMs, including flexibility, transparency, biocompatibility, and the potential for mass production.

A range of gel-based EMW absorption materials can be obtained by binding liquids with different electromagnetic response properties in a functional polymer framework. Depending on the differences in liquid properties, gel materials can be classified as hydrogels (water), organic gels (organic solvents), and ionic gels (ionic liquids [ILs]). Additionally, the dielectric properties of gel-based absorbers are highly correlated with their inbuilt liquid polarity, ionic conductivity, and non-covalent interactions, without taking into account the effects of defects and vacancies [6, 10]. Then, based on the differences in electromagnetic responsiveness of different liquids, gel-based absorber materials are expected to achieve fine-tuning of EMW absorption performance in different frequency bands to meet specific needs. These gel-based EMW absorption materials, which are both transparent, flexible, and biocompatible, are expected to be widely used in precision instruments, wearable electronic devices, biomedicine, and other fields.

8.1.1 Progress in the Application and Research of Hydrogel EMW Absorption Materials

The dielectric properties of liquids have been a subject of interest for numerous years. Researchers have persistently investigated their dielectric relaxation behavior through experimental findings and theoretical calculations, providing valuable insights into their molecular structure. However, their potential for application in microwave absorption has been overlooked. Semiempirical studies have identified water, organic solvents, and ILs as promising liquid-absorbing mediums owing to their compatible dielectric properties (refer to Figure 8.1). Specifically, water, as a highly polar liquid, has been extensively recognized for its high static dielectric constant and strong polarization loss. Despite this, the potential of organic solvents and ILs in microwave (2–18 GHz) absorption remains relatively unexplored, presenting an area ripe for further investigation [4].

Moreover, liquids possess distinctive properties that set them apart from solids, such as flowability, transparency, compressibility, low density, and low shear strength. Nonetheless, it is essential to recognize that the amorphous nature and arbitrary fluidity of liquids can also limit their practical applications. To overcome this challenge, Fang and coworkers [6] utilized an optically transparent poly(methyl methacrylate) (PMMA) sandwich structure as the container to encase the active aqueous gel for microwave absorption, leveraging the flowability and adaptability of liquids to conform to the container's shape, as shown in Figure 8.2a. In a similar vein, Zhuang and coworkers [10] developed a swastika-shaped absorber by injecting water into a 3D-printed PLA-microchannel, achieving over 90% absorption in the frequency band of 9.3–49.0 GHz, as displayed in Figure 8.2b. While both studies leverage the dielectric properties of liquids to achieve exceptional absorption, this is done at the expense of the inherent characteristics of the liquid medium (refer to Figure 8.2a,b). In this context, polymer gels, with their cross-linked polymeric networks capable of retaining significant amounts of liquids while maintaining 3D hierarchical structures, emerge as a promising candidate for fully harnessing the advantages of liquids and exhibiting desirable microwave absorption performance. This innovative approach offers potential for enhanced microwave absorption technology, providing new opportunities for advanced functional material design and applications [12].

It is well known that polar water molecules with high polarization in hydrogels contribute to efficient microwave attenuation, but the combination between water molecules and polyacrylamide (PAM) will slow down the orientation polarization in PAM hydrogels, which

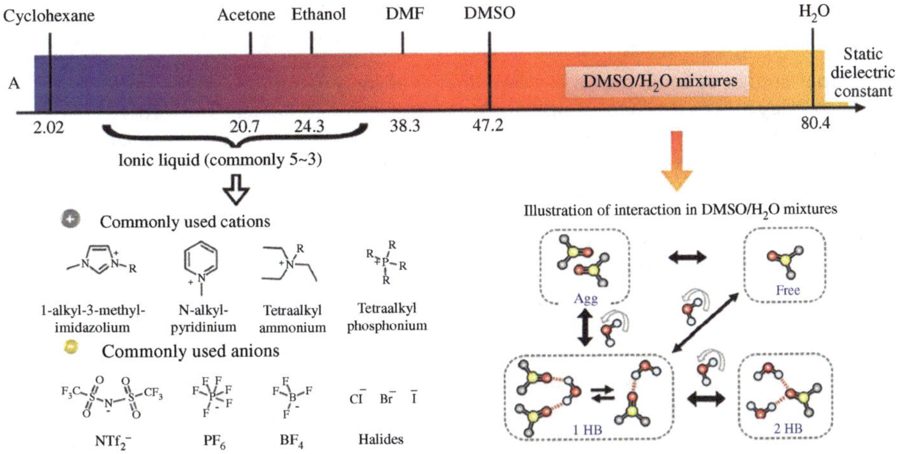

Figure 8.1 Dielectric properties of some representative liquids (water, organic solvents, and ILs) and schematic diagram of complex interactions in dimethyl sulfoxide (DMSO)/H_2O mixtures. Source: Oh et al. [4]/with permission of John Wiley & Sons.

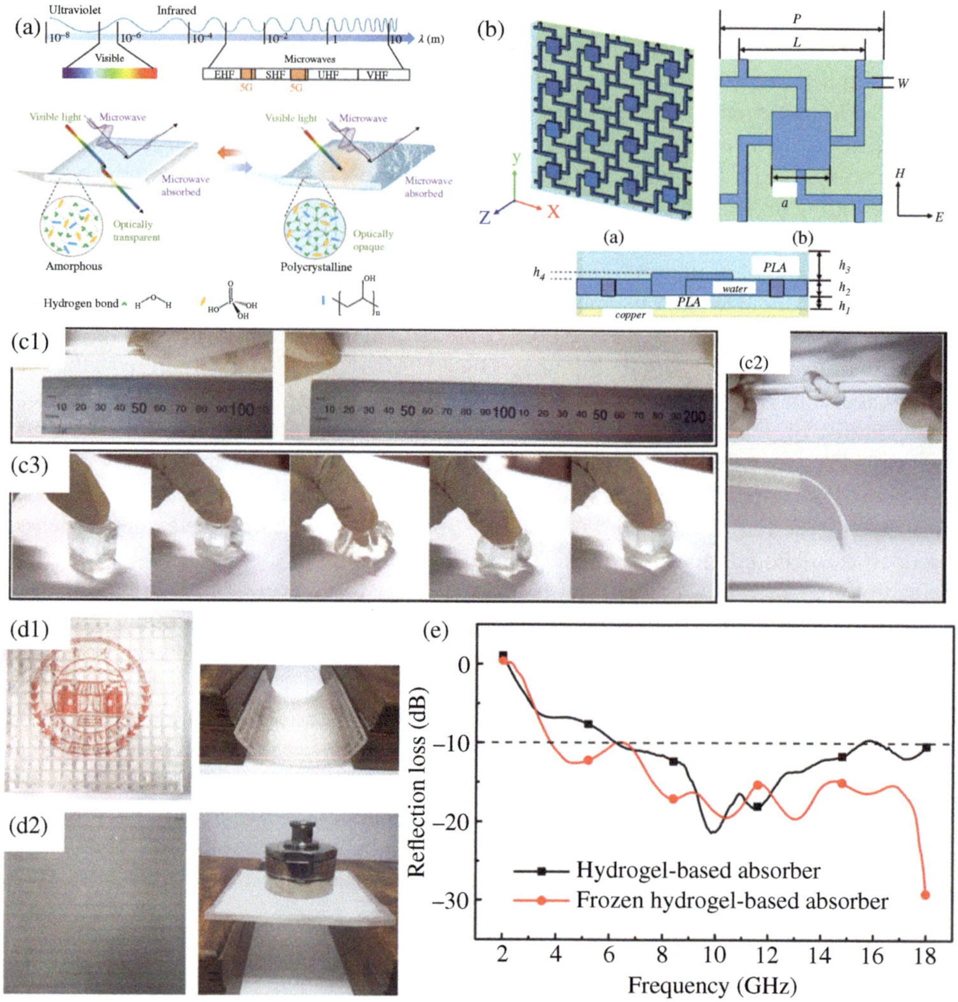

Figure 8.2 (a) Two examples of liquid media in a container as absorbers. Source: Song et al. [6]/with permission of John Wiley & Sons (b) Two examples of liquid media in a container as absorbers. Source: Reproduced with permission Zhang et al. [10]. Copyright 2020, IEEE. (c) The mechanical properties of PAM hydrogel: (c1) tensile performance, (c2) pressing resilience performance, and (c3) bending performance. The fabricated hydrogel-based absorber and measured reflection loss curves: (d1) the transparent and flexible absorber under room temperature conditions, (d2) the opaque and rigid absorber under frozen conditions, and (e) the measured reflection loss curves of the absorber under room temperature and frozen conditions. Source: Chen et al. [11]/with permission of Springer Nature.

provides a viable way to manipulate the dielectric constant. Zou and coworkers [11] has constructed broadband microwave absorbers based on PAM hydrogels to achieve smart switching properties for both optical and mechanical properties, as displayed in Figure 8.2c,d. As the temperature decreases, the flexible and transparent hydrogel will be transformed from an amorphous form to a polycrystalline state, which will then turn into a rigid and opaque state. By using hydrogels with such properties for the construction of EMW-absorbing structures, the resulting absorbers also exhibit optically and mechanically switchable properties and cover effective absorption in the range of 5.7–18 GHz, as presented in Figure 8.2e. Meanwhile, as research on hydrogel flexible smart sensors rapidly advances [13, 14], the multimodal sensing module comprises highly conductive, stretchable, self-adhesive, and biocompatible tannic acid (TA)-modified PAM-based hydrogel epidermal electrodes, as well as highly sensitive pressure sensors based on the structure of macroporous hydrogels (foam-PAM). In the future, the integration of intelligent monitoring of human function and drug-release functionality will be further elucidated through the use of microwave-fed actuation. This innovative approach holds great promise for advancing the field of smart sensor technology, offering new opportunities for enhanced healthcare monitoring and therapeutic applications.

8.1.2 Progress in the Application and Research of Ionic and Organic Gel EMW Absorption Materials

Semi-empirical studies have identified ILs as promising liquid-absorbing mediums due to their matching dielectric properties. Polar organic solvents, such as dimethyl sulfoxide (DMSO), glycerol, and ethylene glycol, exhibit similar polarization loss behavior in the GHz frequency range due to their dipolar molecules [15], and room-temperature molten salts, known as ILs, have tunable physicochemical properties, negligible vapor pressure, and high ionic conductivity [16, 17]. More importantly, their electromagnetic (EM) parameters can be predicted and designed based on polarity, ionic conductivity, temperature, and weight ratio.

Wang and coworkers have developed microwave-triggered IL-based hydrogel dressings with exceptional thermotherapy and transdermal drug delivery properties [18]. They specifically synthesized the hydrogels, known as VACPHs, by copolymerizing MW-responsive ILs (VBTMACl) and choline ILs (ChMACl), which also act as drug delivery agents. The IL hydrogel can increase the temperature by 26 °C within five minutes under microwave irradiation. Additionally, the IL hydrogel facilitates the transdermal delivery of drugs. Due to the strong polarity of the ammonium cation, VACPHs exhibit enhanced thermal conversion capability under MW radiation, leading to effective bacterial eradication, as illustrated in Figure 8.3a. This innovative approach holds great promise for advancing the field of medical dressings and therapeutic applications, offering new opportunities for improved patient care and treatment outcomes. The metallo-alginate hydrogel, as recently documented, represents a promising illustration of synergistic cancer therapy [19], as displayed in Figure 8.3b. Specifically, (1) the high concentration of Ca^{2+} within the hydrogel enhances heating efficiency and confines the heating zone; (2) an excess of extracellular Ca^{2+} can sensitize cancer cells to mild thermal damage by disrupting intracellular Ca^{2+} homeostasis; and (3) Mn^{2+} serves as an activator of the stimulation of interferon genes pathway, which can incite antitumor immunity to impede tumor growth. This innovative approach holds great promise for advancing the field of cancer therapy, offering new opportunities for enhanced treatment strategies and improved patient outcomes.

The advancement of microwave therapy has transformed the landscape of medical interventions, particularly in the realm of cancer treatment. However, the investigation of medical-grade MAMs is still in its early stages, and there is a pressing requirement to create tailored materials with reduced toxicity, drug-carrying capabilities, biocompatibility, and

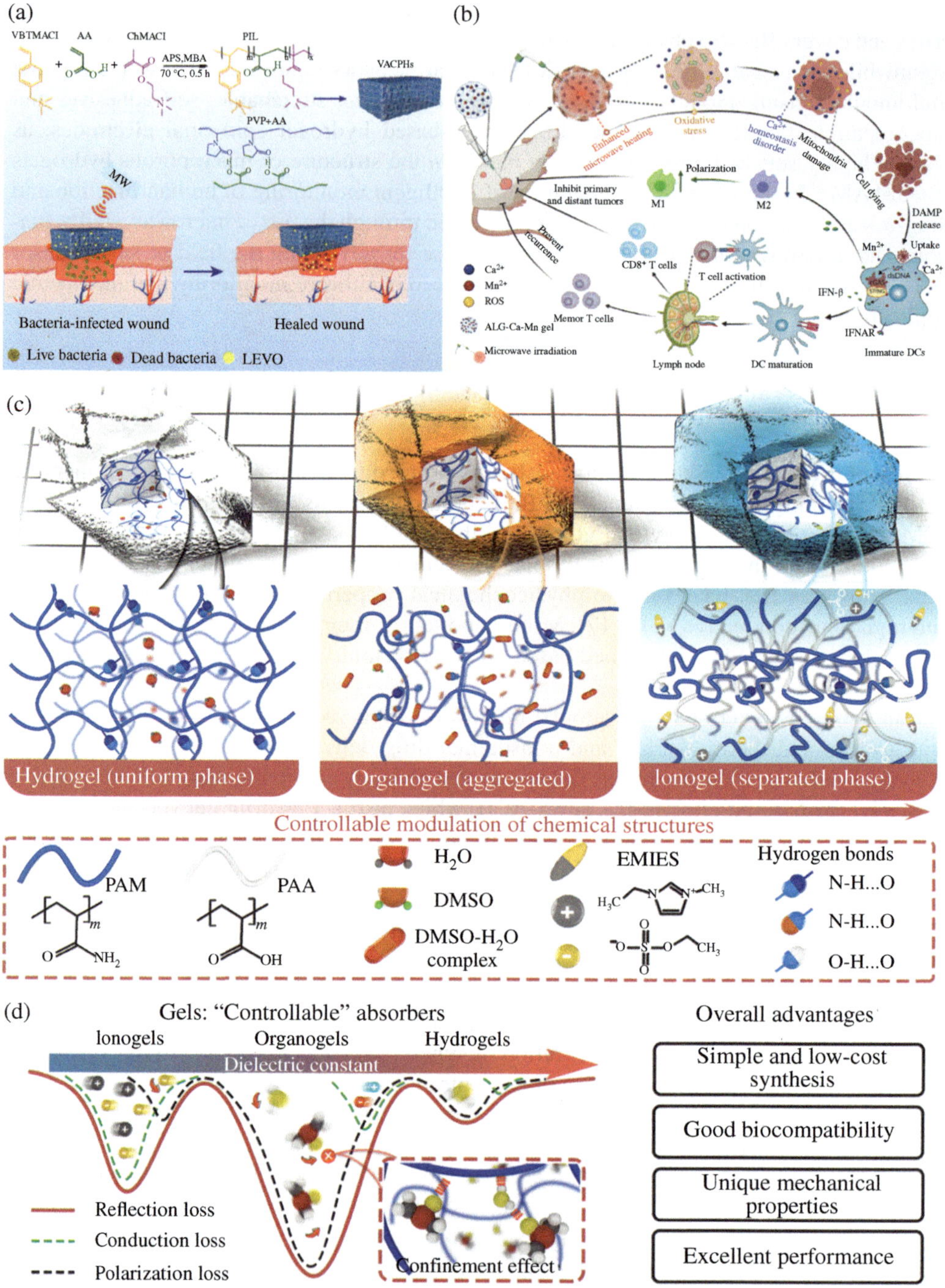

Figure 8.3 (a) The reactions involved in the preparation of the VACPH hydrogels and the schematic illustration of the all-in-one therapeutic system for the treatment of bacterial infections. Source: Gao et al. [18]/with permission of Elsevier. (b) A scheme illustrating the mechanism of metallo-alginate hydrogel in enhancing MWA treatment and boosting antitumor immunity. Source: Zhu et al. [19]/with permission of American Association for the Advancement of Science. (c,d) Schematic illustrations of the chemical structures of gels and the absorption characteristics of gel-based absorbers and their advantages. Source: Zhao et al. [20]/with permission of John Wiley & Sons.

improved response at low frequencies (2.45 GHz). Our team has recently utilized acrylamide (AM) [20], a representative soft polymeric monomer, to immobilize water, DMSO, and ILs, and has provided a comprehensive understanding of the absorption mechanisms of gel-based absorbers, as confirmed in Figure 8.3c. This study demonstrated that ionic gels and organic gels can exhibit full-wave absorption in the X-band, while hydrogels show promising shielding efficiency. As displayed in Figure 8.3d, these remarkable and adjustable performances are comparable to those of most traditional materials. Additionally, we have identified that solvent molecules play a pivotal role in determining the dielectric properties of gels, with hydro/organic gels being dominated by polarization loss and ionic gels being dominated by conduction loss. The gel stands out for its capability to cover a wide range of EM parameters with exceptional precision, making it a promising, controllable, and reliable resource for establishing an EM parameter database. Furthermore, the gel's ability to encompass a broad range of EM parameters with exceptional precision makes it a promising, controllable, and reliable resource for constructing an EM parameter database. This suggests that gel-based absorbers have the potential to revolutionize the field of EM materials and pave the way for the development of advanced and versatile applications in various industries.

8.1.3 Progress in the Research of Poly(Ionic Liquid) Gels

PILs, as a subset of polyelectrolytes, are formed through the polymerization of IL monomers, comprising a polymeric backbone with mobile counterions [21–23]. Ranging from monomers to oligomers to high-molecular-weight polymers, specific unique properties of ILs—such as minimal vapor pressure, thermal stability, nonflammability, design flexibility, and dielectric characteristics—are imparted to polymer chains [24, 25]. Additionally, PILs have the ability to form stable structures and morphologies that are unattainable via ILs. This makes PILs a promising candidate for utilization as MAMs, achievable through straightforward monomer polymerization to produce independent absorbers. Given that most PILs lack magnetic properties, their dielectric loss is predominant at microwave frequencies due to ionic conduction loss. In cases where the free counterions in the system are polar (e.g. $TFSI^-$, $[BF4]^-$, and $[PF6]^-$), dipole polarization becomes a significant factor; conversely (e.g. Na^+, H^+, and Cl^-) [26], it may be negligible. An additional appealing aspect of PILs is their solid-state nature, which circumvents issues such as evaporation, leakage, and instability commonly associated with gels.

Currently, there is only one report on the MW absorption performance of PILs. In 2008, Shen et al. published a report titled "Poly(ionic liquid)s as Optically Transparent Micro-EMW absorption Materials" [27]. This study investigated the complex permittivity of several PILs containing either imidazolium or ammonium cations, including P[VBBI][BF_4], P[VBBI][Sac], P[VBBI][$FeCl_4$], and P[VATMA][BF_4]. These PILs demonstrate dielectric constants ranging from 3.7 to 5.3 and dielectric loss factors from 0.18 to 0.37, which are comparable to some semiconductors. Regrettably, the RL of these PILs was not calculated in this study, and there have been no further reports regarding the absorption performance of PILs to date. Based on their dielectric constants, these PILs are expected to exhibit standard absorption performance due to their low dielectric losses. Nevertheless, the dielectric and magnetic properties of PILs can be enhanced through various strategies, and it is their potential in terms of stability and flexibility that holds the most intriguing appeal.

8.1.4 Perspectives on Dielectric Liquid Medium Gel EMW-Absorbing Materials

Dielectric liquid medium gels inherit the properties of liquids and encompass additional characteristics, including stretchability, adhesion, self-healing, biocompatibility, and thermal conductivity. They also possess the capacity to respond to environmental stimuli, such as changes in temperature, pH, and electric fields. These attributes showcase the potential of polymer frameworks in emulating the properties of liquid mediums. Despite the significance of this subject, research in this area remains limited, and numerous unexplored questions persist. For instance, (1) there is potential for the development of additional liquid and polymer frameworks to expand the EM database of gels; (2) the exploration of magnetic gels and suitable liquid mediums for lower frequencies is crucial due to the inadequate performance of current gel-based absorbers; (3) comprehensive theoretical studies and physical models are necessary to differentiate the contribution of ionic conductivity to MW absorption from electronic conductivity; (4) the instability of the gel can lead to volatilization, leakage, freezing, and moisture absorption, rendering it unsuitable for high-/low-temperature environments; (5) the diverse applications of gel-based absorbers, such as in smart windows, MW therapy, and intelligent devices, warrant further investigation. We envision an optimal medical absorber in the future that aligns seamlessly with the process of targeted aggregation, microwave-induced heating, controlled drug release, tumor eradication, and biodegradation.

8.2 Dielectric Solid Medium Gel EMW Absorption Materials

Aerogels (dielectric solid medium gel EMW absorption materials) present a host of compelling advantages as MAMs. Their low-density and extensive porosity create a lightweight structure with a substantial surface area, facilitating efficient interaction with microwaves. This attribute, coupled with the capacity to tailor properties like density and composition, renders aerogels versatile for diverse frequency ranges and application requirements. Additionally, their thermal insulation properties contribute to effective heat management, while their mechanical flexibility enables versatile integration into various form factors. We will concentrate on introducing metal-based aerogels and ceramic-based aerogels.

8.2.1 Ceramic-Based Aerogel EMW Absorption Materials

It is crucial to develop next-generation stealth protection materials with both good thermal insulation and EMW-absorbing properties at medium and high temperatures to cope with complex and changing service environments. Ceramic-based absorption materials are widely used in high-temperature environments due to their superior thermal stability and controllable dielectric loss capability. In particular, several stand-alone ceramic-based aerogels such as Al_2O_3, SiC, SiCN, SiOC, and Si_3N_4 aerogels have recently been developed, which offer ultralow density, thermal superinsulation, intentional mechanical strength, good thermal stability, and chemical stability, all of which are critical for applications in harsh environments, especially high-temperature thermally corrosive environments [28–32].

8.2.1.1 Preparation Method of Ceramic-Based Aerogel EMW Absorption Materials

Ceramic-based aerogels are materials with high porosity, low density, and excellent thermal stability and mechanical properties, which are widely used in the fields of thermal insulation materials, acoustic materials, catalyst carriers, and so on. Their main preparation methods include the sol-gel method, supercritical drying method, template method, and so on.

The sol-gel method is one of the most commonly used methods for the preparation of ceramic-based aerogels. The ceramic precursor is first dissolved in a suitable solvent to form a sol-gel. Then, the gelation reaction occurs through heating, stirring, and ultrasonic treatment to form a gel. Finally, ceramic-based aerogel materials are obtained through steps such as drying and heat treatment. Freeze or supercritical drying techniques are commonly used to manufacture these materials from precursor solutions based on polymers [28, 29]. The supercritical drying method is to place the sol directly in a supercritical fluid and make it undergo a drying reaction under high pressure and high temperature to obtain ceramic-based aerogel materials [30]. The template method involves dissolving the ceramic precursor in a suitable solvent in the presence of an organic or inorganic template to form a sol. The sol is then impregnated onto the surface of the template and gelatinized by heat treatment or chemical reaction to ultimately obtain the ceramic-based aerogel material [31, 32].

8.2.1.2 Ceramic-Based Aerogel EMW Absorber Applications and Research Progress

Wang [33] constructed lightweight and strong silicon carbide (SiC) aerogels with a three-dimensional architecture, and in situ grown SiC nanowires are obtained from natural eggplants by direct graphitization and subsequent carbothermal reduction. The SiC aerogels show a very low thermal conductivity of 0.035 W m^{-1} K^{-1}, and the SiC aerogels show a minimum reflection of –43 dB and an effective bandwidth of 4.0 GHz at a 2.0 mm thickness. Subsequently, in order to further enhance their application in high-temperature environments, Wang and coworkers [34] prepared SiC/Si_3N_4 aerogels with multicomponent microstructures by freeze-drying and carbothermal reduction reaction. The porous structure composed of nanowires resulted in an aerogel with a density of only 8.67 mg cm^{-3}, which was thermally stable in an air atmosphere below 1100 °C, exhibiting a minimum reflection loss of −25 dB and an effective absorption band of 4.3 GHz. These results indicate that the developed SiC aerogels are outstanding materials with potential applications in high-performance EM absorption and thermal insulation in the aviation and aerospace fields. Two examples show the disadvantage of single-component ceramic aerogel is that its performance and application range are limited, such as lower compressive strength, insufficient heat resistance, limited adsorption capacity, and other issues. With the development of science and technology, composite ceramic aerogel has gradually become an effective way to solve these problems. Composite ceramic aerogel mixes different kinds of ceramic aerogel materials or surface modifications, which make its performance significantly improved. Through the composite technology, the compressive strength, heat resistance, and EMW absorption capacity of ceramic aerogel have been significantly improved, making it more widely used in complexity. Therefore, the emergence of composite ceramic aerogel provides new ideas and methods to solve the shortcomings of single-component ceramic aerogel and also brings new hope for the development of ceramic aerogel materials.

Fan's group [29] fabricated multifunctional $SiC@SiO_2$ ceramic aerogels ($SiC@SiO_2$ NFA) with a 3D porous crosslinked structure using a simple chemical vapor deposition method and subsequent calcination treatment process (Figure 8.4a). As shown in Figure 8.5b–d,

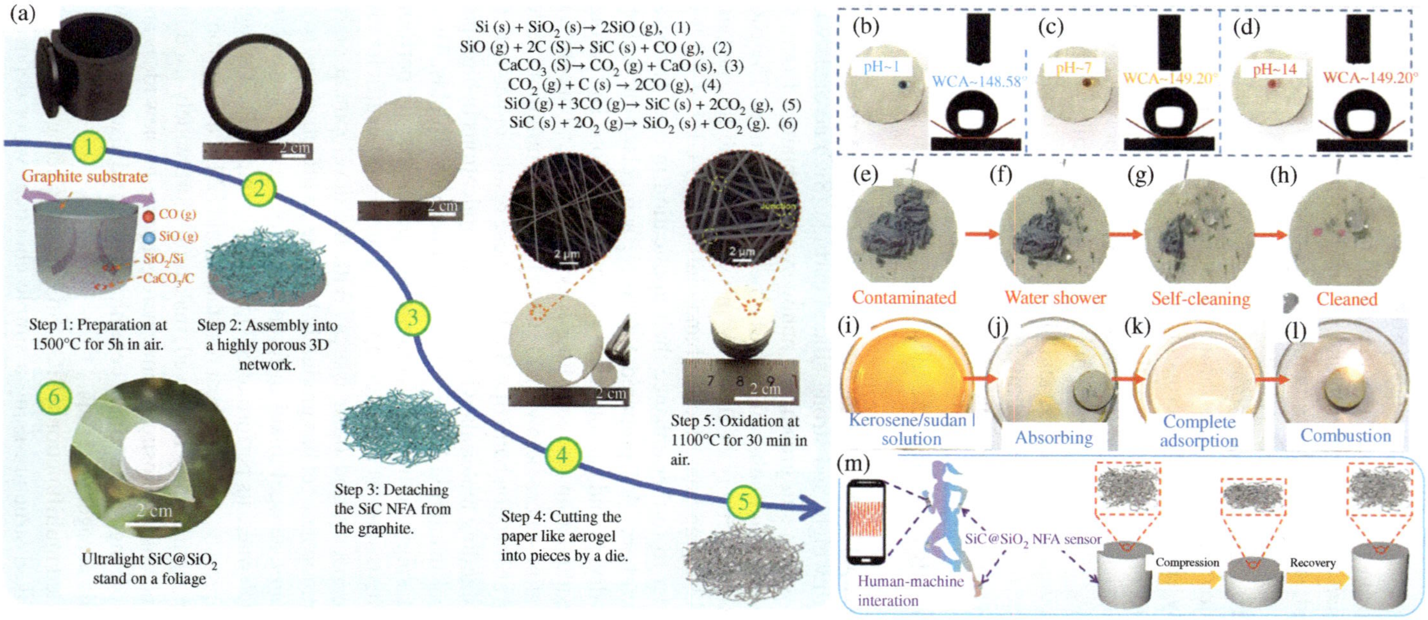

Figure 8.4 (a) Preparation process for the SiC@SiO_2 NFA. High absorption capacities for organic liquids and the self-cleaning property of the oil-modified SiC@SiO_2 NFA. (b–d) Digital photographs of water droplets with pH values of 1, 7, and 14 on an oil-modified SiC@SiO_2 NFA and the corresponding WCA images, respectively. (e–h) Self-cleaning process of the oil-modified SiC@SiO_2 NFA. (i–l) Process of absorption of methyl orange aqueous solution by the oil-modified SiC@SiO_2 NFA and a subsequent combustion test. (m) Applications of the SiC@SiO_2 NFA pressure sensor to detect body activities and tiny pressures. Source: Reproduced with permission Zhu et al. [29]/with permission of Springer Nature.

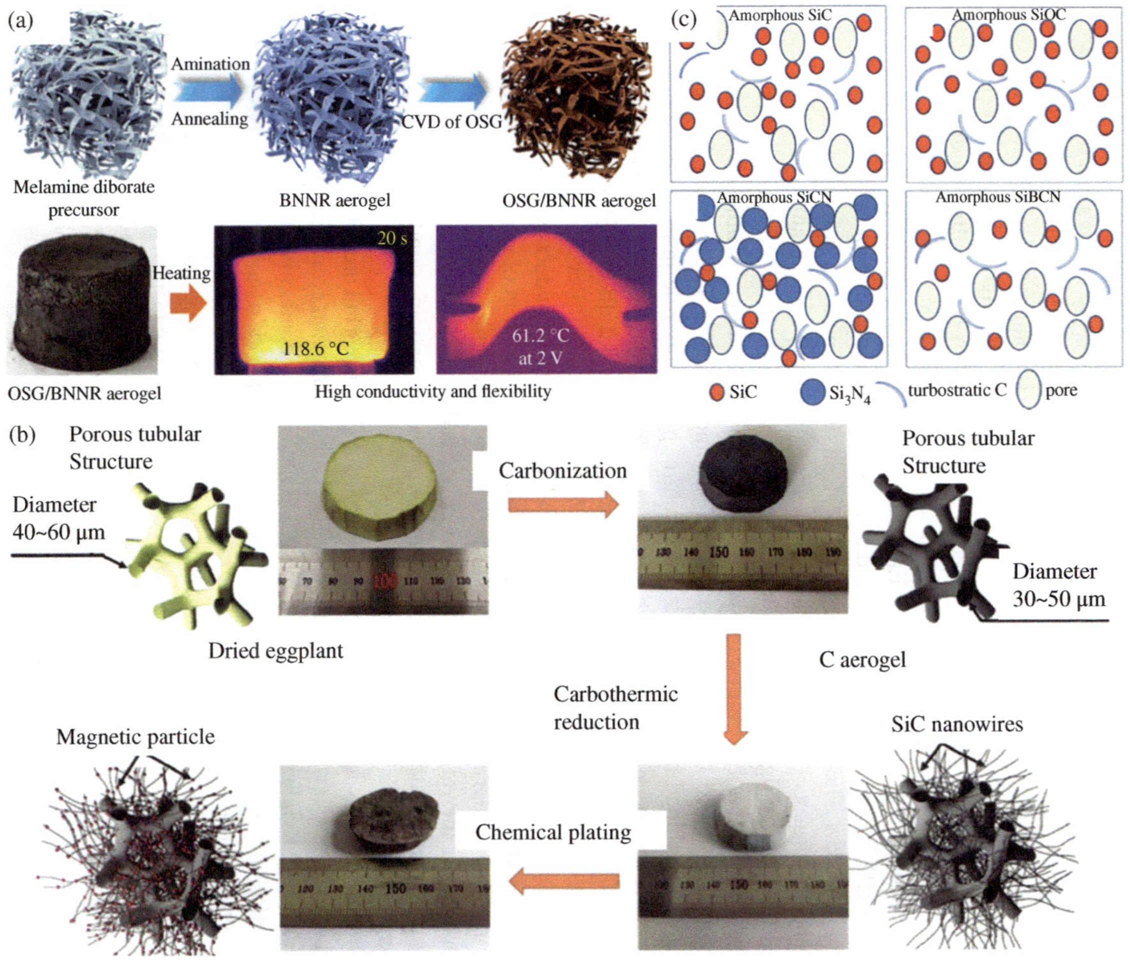

Figure 8.5 (a) Schematic illustration of the fabrication procedure of OSG/BNNR hybrid aerogels and thermal infrared image of 12-OSG/BNNR aerogel upon bending state at 2 V. Source: Feng et al. [35]/with permission of American Chemical Society. (b) Synthesis process of NiCo–SiC aerogel. Source: Zhang et al. [36]/with permission of Elsevier. (c) Microstructure model for ceramic-derived polycarbosilanes, polysiloxane, polysilazane, and polyborosilazane at high temperature. Source: Duan et al. [28]/with permission of Elsevier.

the droplet contact angle tests at different pH values showed excellent hydrophobicity. As displayed in Figure 8.5e,h, a 2.5 mL syringe was used to flush a piece of dusty sample surface with water droplets to confirm the self-cleaning performance of the obtained aerogel. In addition, the excellent hydrophobicity of oil-modified SiC@SiO_2 NFA is the key to its use as a high-performance EMW-absorbing material (Figure 8.4i–l). Figure 8.4m shows the schematic diagram of the SiC@SiO_2 NFA sensor used to detect human motion and the mechanism of piezoresistive sensing performance. In practice, pressure sensors can be used to detect human activity and minute pressures and then transmit this information to a mobile phone. This is achieved by the principle that SiC@SiO_2 NFA is compressed and bent when loaded, which results in the nanofibers coming into contact with neighboring nanofibers and interconnecting with each other. The prepared SiC@SiO_2 NFA has an ultralow density (~11 mg^{-1} cm^{-3}), leading to its remarkable strain-related piezoresistive sensing behavior. In addition, SiC@SiO_2 NFA exhibits excellent EMW absorption performance with a minimum reflection loss (RL_{min}) value of −50.36 dB and a maximum effective absorption bandwidth (EAB_{max}) of 8.6 GHz. The successful preparation of this multifunctional aerogel material offers a promising prospect for the design and fabrication of cutting-edge ceramic materials. Shao et al. [31] prepared lightweight silicon oxynitride of carbon (SiOCN) ceramic aerogels by pyrolysis of bridged poly(sesqui-carbazone) aerogels, which were pre-synthesized using a solgel method followed by vacuum drying. The structural order and content of the free carbon phase, determined by the pyrolysis temperature, play a key role in modulating the impedance matching and attenuation capabilities. The prepared SiOCN aerogel was pyrolysised at 1000 °C and achieved a minimum reflection loss of −64.2 dB at 8.96 GHz and a wide bandwidth of 5.4 GHz at a low thickness of 2.15 mm. The multilevel pore structure, rich heterogeneous amorphous SiOCN/free carbon interface, and conductive free carbon phase facilitate excellent absorption performance through the formation of good impedance matching, multiple scattering, interfacial polarization, and conduction loss. This work provides an insight for the rational design of polymer-derived ceramic (PDC) aerogel-based microwave absorption materials for potential application under extreme conditions such as high temperature and oxidation environments.

Three-dimensional elastic aerogels enable a variety of applications but are often limited by their low thermal and electrical transfer efficiency. Recently, Song and coworkers [35] demonstrated a strategy to fabricate highly thermally and electrically conductive aerogels using hybrid carbon/ceramic structural units made of hexagonal boron nitride nanoribbons (BNNRs) and in situ–grown orthorhombic structured graphene (OSG), as illustrated in Figure 8.5a. The high-aspect-ratio BNNRs were first interconnected into a 3D elastic and thermally conductive backbone, where the horizontal graphene layers of the OSG provided additional super channels for electron and phonon conduction, and the vertical graphene sheets of the OSG greatly improved the surface roughness and charge-polarization capacity of the entire backbone. The obtained OSG/BNNR hybrid aerogels exhibit very high thermal and electrical conductivities (up to 7.84 W m^{-1} K^{-1} and 340 S m^{-1}, respectively) at a low density of 45.8 mg cm^{-3}, which should prove to be a great advantage over the reported carbon and/or ceramic aerogels. In addition, the hybrid aerogels exhibit wide-temperature-invariant super-elasticity (−196 to 600 °C), low-voltage-driven Joule heating (up to 42–134°C at 1–4 V), strong hydrophobicity (contact angles up to 156.1°), and strong broadband electromagnetic interference (EMI) shielding (up to 70.9 dB at a thickness of 2 mm), which is well maintained under repeated mechanical deformation and long-term immersion in strong acid or alkali solutions. The shielding is well maintained even after repeated mechanical deformation and long-term immersion in strong acid or alkali solutions. Using these extraordinary combined properties, the team demonstrated the great potential of OSG/BNNR hybrid aerogels in wearable electronics for regulating body temperature, waterproofing against contamination, de-icing, and protecting human health from EMI.

Furthermore, in order to achieve optimal performance of an EMW absorber, additives are also critical in improving its overall EM response mechanism. To this end, several new additives have been developed. Representative highly conductive carbon nanotubes, graphite, metal particles, alloys, and so on are usually incorporated into composites using various techniques such as insertion, hybridization, interpenetration, lamination, and encapsulation [37–40]. These additions can significantly improve the charge transport and polarization behavior, even with very small additions. As presented in Figure 8.5b, Zhang et al. [36] coated magnetic NiCo layers onto SiC aerogels (NiCo–SiC) by chemical plating. NiCo–SiC aerogels have a very low density (0.194 g cm^{-3}) consisting mainly of porous SiC skeletons, SiC nanowires, and magnetic NiCo alloy nanoparticles distributed on the surface of the SiC skeletons and nanowires. The presence of magnetic NiCo alloy nanoparticles in the aerogel increases the magnetic and dielectric losses due to the formation of heterogeneous interfaces. As a result, the NiCo–SiC aerogel with a thickness of 3 mm has a RL_{min} of −44.5 dB and an effective absorption bandwidth (RL < −10 dB) of 7.6 GHz, demonstrating in lightweight and excellent EMW absorber.

8.2.1.3 Polymer-Derived Ceramics Aerogels: EMW Absorber Applications and Research Progress

With the development of modern aircraft toward high Mach numbers and long endurance, high-temperature EMW absorption has become a cutting-edge challenge for EMW functional materials used in harsh environmental applications, such as stealth aircraft and aerodynamically heated components of aero-engines [41, 42]. The PDCs route is characterized by excellent thermo-mechanical properties, stability, good availability, and processability compared to existing ceramic production technologies. Shown is a schematic illustration of the formation of one-dimensional nanostructures during polymer-to-ceramic conversion with the addition of catalysts at high temperatures [43, 44]. These 1D nanostructures typically exhibit unique and often superior properties to their bulk and/or microscale counterparts. Both SiC and carbon 1D nanostructures exhibit high electrical conductivity and specific surface area (SSA), which promote conductive polarization and interfacial polarization [45, 46]. They favor the improvement of the absorption properties of PDCs.

Li and coworkers [47] prepared SiC-BN@CNT ceramic composites by applying BN coatings to the surface of carbon nanotubes and introducing them into polymer-converted silicon carbide ceramics (PDC–SiC) with a periodic structural design of ceramic composites. It is shown that the effective absorption bandwidth of the material reaches 15.52 GHz at 1273 K, which covers almost the whole S, C, X, and Ku bands. The mass loss of SiC-BN@CNT ceramic composites after undergoing oxidation for two hours at 200, 400, and 600 °C is 0.28, 0.34, and 0.65 wt.%, respectively, which exhibit excellent thermal stability. The formation of a SiO_2 oxide layer on the surface after oxidation at 600 °C prevents further oxidation of the ceramic composites and improves their impedance matching ability. This allows the composite ceramic aerogel to move toward extreme high-temperature applications. As shown in Figure 8.5c, Yin and coworkers [28] gave a microstructural model of various ceramic-derived polymers at high temperatures in which the microstructures of nanodomains were characterized. Excellent microwave-absorbing materials should consist of an electrically insulating matrix (phase) and an electrically lossy phase (phase B). For PDCs, the amorphous phase and pores are the A phase, and both the vortex crystal C and nanocrystalline SiC can be considered as the B phase, which suggests that PDC can be an ideal candidate for radar-absorbing materials.

8.2.2 Metal-Based Aerogel EMW Absorption Materials

Metallic aerogels are an emerging class of nanomaterials that combine the properties of both metals and aerogels, with features that are different from those of other aerogels, such as excellent electrical conductivity, high catalytic activity, plasmonic optical properties, magnetic properties, and so on. These characteristics show great potential for application in the fields of electrocatalysis, energy storage, sensing, and so on. However, most of the metal aerogels reported in the literature nowadays lack sufficient mechanical strength, which leads to the related materials that still have a certain distance from being used in practical applications.

8.2.2.1 Preparation Method of Metal-Based Aerogel Absorption Materials

Metal-based aerogel is a new type of porous material with high porosity and a large SSA, which has excellent adsorption, catalytic, and mechanical properties and is widely used in the fields of catalysts, adsorbents, sensors, energy storage, and so on. The main preparation methods include the sol-gel method, co-precipitation method, and template method.

The sol-gel method is one of the most commonly used methods for the preparation of metal-based aerogels. The metal salt is first dissolved in a suitable solvent to form a sol-gel. Then, the gelation reaction occurs through heating, stirring, and ultrasonic treatment to form a gel. Finally, the metal-based aerogel material is obtained through steps such as drying and heat treatment [48, 49]. The co-precipitation method is to dissolve two or more metal salts and one or more precipitating agents in a suitable solvent, and by mixing the solutions, the metal salts undergo a precipitation reaction, so as to obtain metal-based aerogel [50]. The template method involves dissolving metal salts or metal organics in a suitable solvent in the presence of an organic or inorganic template to form a sol. The sol is then impregnated on the surface of the template and gelled by heat treatment or chemical reaction to ultimately obtain a metal-based aerogel material [2, 51]. In addition to the above methods, there are some other preparation methods, such as microemulsion method and supercritical fluid method. Each of these methods has its own advantages and disadvantages, and the appropriate preparation method can be selected according to the specific needs and actual situation. With the continuous development of science and technology, the preparation methods of metal-based aerogels are also being improved and perfected, providing more possibilities for their application in various fields.

8.2.2.2 Metal Aerogels: EMW Absorber Applications and Research Progress

Yang Cheng's team [48] can prepare highly compressible and air-dryable nickel nanowire aerogels (NNWA) through a facile magnetic field-assisted gelation strategy, as displayed in Figure 8.6a–c. Its unique anisotropic layered structure and abundant cold-welded junctions of entangled nanowires enable excellent electrical conductivity and compression fatigue resistance. This free-standing NNWA with ultralow density (20 mg cm^{-3}) can be directly used as an electrode-conducting backbone for energy storage and electrocatalytic purposes

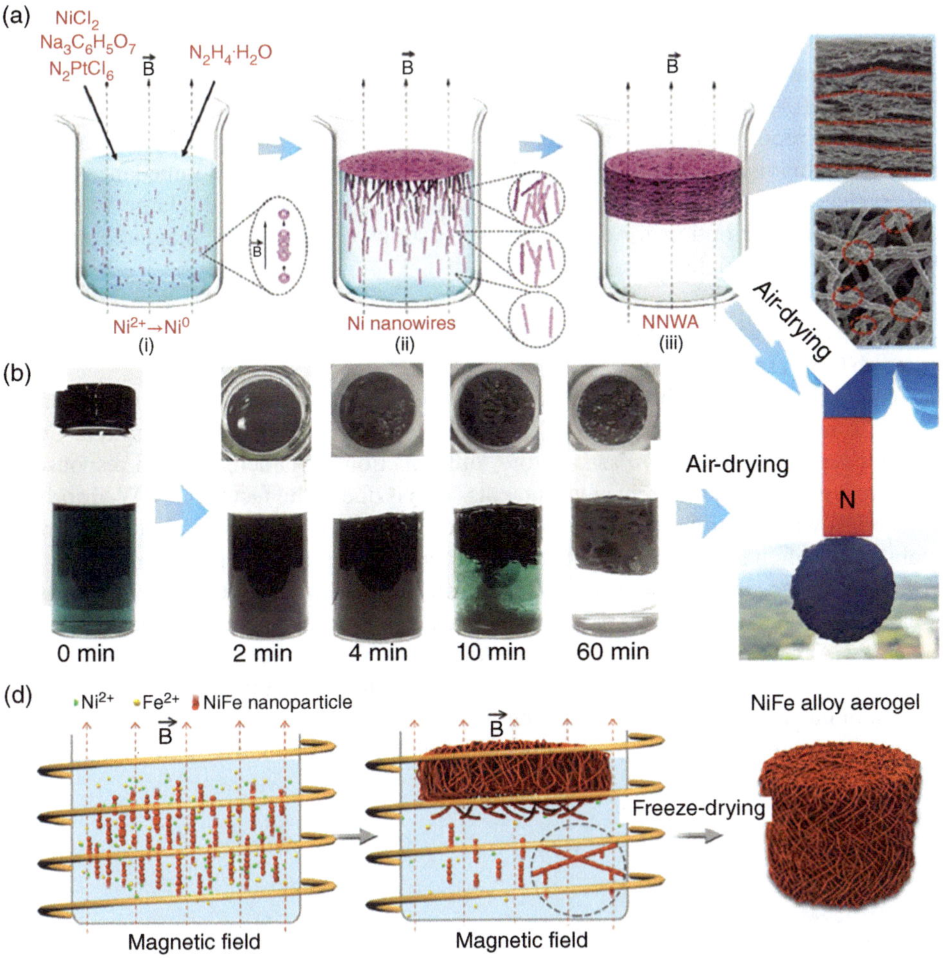

Figure 8.6 NNWAs during different preparation stages. (a) Schematic diagram of the magnetic field-assisted preparation process of the NNWAs. (b) Digital photos of the gel preparation process under a magnetic field. (c) Photograph of an NNWA attracted by a permanent magnet. Source: Pan et al. [48]/with permission of John Wiley & Sons. (d) Preparation and morphology of the NiFe aerogel. Schematic of the synthesis of NiFe aerogels with the presence of a Helmholtz coil and freeze-drying. Source: [49]/with permission of John Wiley & Sons.

even at high mass loading (10 mg cm^{-2}). Second, NiFe aerogels [49] were synthesized via Helmholtz coils ($Ni_{0.80}Fe_{0.20}$ aerogel) by modulating the magnetic field strength and introducing ferric ions, and freeze-drying conditions, as presented in Figure 8.6d. The interconnections and metallic bonding between the NiFe nanowires explain the excellent mechanical strength and electrical conductivity of the $Ni_{0.80}Fe_{0.20}$ aerogels. Its excellent overall properties and mass production also enable it to serve as a versatile platform in many other important fields, such as EMW absorption and energy storage. Recently, our group [50] has successfully synthesized mono-metallic aerogels with a 3D network structure under the synergistic effect of $NaBH_4$ reduction and subsequent release of H_2 in this work and modulated their microstructural and EM parameters by varying the molar ratio of $NaBH_4$. The metal salt was reduced to $NaBH_4$. Notably, the Co aerogel (C-AG-0, Co^{2+}: $NaBH_4$ = 1:3) has a density of 0.16 g cm^{-3} and exhibits the best EMW absorption properties. It has an effective absorption bandwidth (EAB) of 6.24 GHz (10.88–17.12 GHz) and a thickness of 1.7 mm, which even exceeds the reported metal matrix composites. The excellent absorption performance is attributed to the highly porous 3D lattice structure, which provides good impedance matching and conduction loss, as well as interfacial polarization induced by the large SSA. This work provides another option for the facile preparation of lightweight magnetic metallic EMW absorbers with efficient absorption.

8.2.2.3 Composite Metal-Based Aerogels: EMW Absorber Applications and Research Progress

The development of lightweight and efficient magnetic metal microwave-absorbing materials is essential to broaden the applications of metal-based materials in EMW absorption. Composite metal aerogels have shown satisfactory multifunctional characteristics in response to functional requirements in different environments. This is due to the fact that EMW-absorbing EMA materials with multiple functions such as hydrophobicity, thermal insulation, and nonflammability are often required in complex and variable application environments.

Liu and coworkers [51] developed a silicone oil domain-limited gelation method with strong generality and successfully prepared a series of metal aerogel/porous skeleton composites (Pt_xPd_y AGs/CC, $Pt_{50}Pd_{50}$ AGs/CC, $Pd_{50}Ag_{50}$ AGs/CC, $Au_{50}Ag_{50}$ AGs/CC, and $Pt_{50}Ag_{50}$ AGs/CFF), which overcame the critical problem of inherent structural collapse during the preparation of electrodes from metal aerogels, as illustrated in Figure 8.7a–d. The preservation of the pristine multilevel pore structure and continuous conductive nanowire network

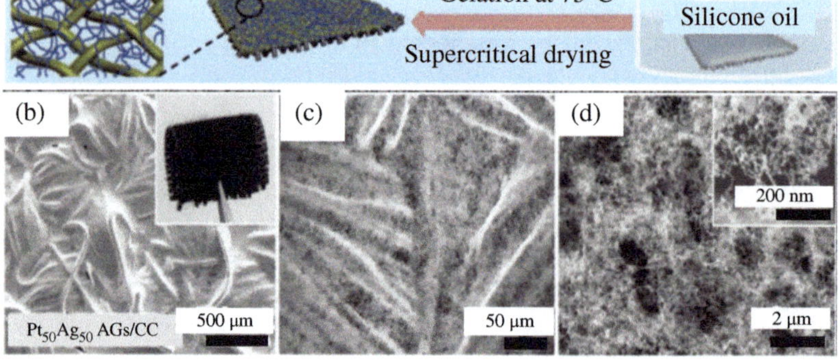

Figure 8.7 (a) Schematic illustration of the preparation of Pt_xAg_y AGs/CC. (b–d) SEM images of the $Pt_{50}Ag_{50}$ AGs/CC. Source: Zheng et al. [51] with permission of John Wiley & Sons.

structure of the metal aerogels can further accelerate the mass transfer and electron transfer rate, which can effectively avoid Oswald ripening and thus enhance their electrocatalytic activity and stability. Similarly, these metal aerogels are uniformly covered on the CC surface like a thin film composed of interconnected nanowires. The diverse selectivity of the composition and support materials is expected to further expand the flexible applications of metal aerogels, such as flexible sensing in the field of artificial intelligence and EMW-absorbing materials with heterogeneous flexible coatings.

Zhang and coworkers [2] have prepared ultralight Ni/MXene/rGO (NiMR-H) composite metal aerogels by directional freezing and hydrazine vapor reduction using MXene/reduced graphene oxide (rGO) aerogels anchored by magnetic nickel nanochains. Additionally, in order to counteract the interference caused by rainy weather, the self-cleaning property was studied using hydrophobicity. The contact angle (β) between water and the sample is used as an indicator of hydrophobicity, with $\beta > 90°$ and $\beta > 150°$ indicating hydrophobicity and super-hydrophobicity, respectively. In the case of NiMR-H, it is found that water droplets tend to form and remain on the surface due to surface tension, making it difficult for the water to penetrate the aerogel. A high level of hydrophobicity with a β of 121.2° is measured, which suggests the remarkable self-cleaning property.

The water contact angles (WCAs) of NiR-H, NiMR-H, and NiMR-A were 118.3°, 125.5°, and 91.1°, respectively, showing effective hydrophobicity. As one of the potential applications, the porosity and hydrophobicity of NiMR-H aerogels give them an interesting oil-water separation capability. The absorption capacity for xylene, methylene chloride, and pump oil reached an impressive 8600, 8200, and 7500 wt.% of their own weight, respectively.

In addition, NiMR-H and NiMR-A aerogels also outperformed commercial expanded polystyrene (EPS) and polyurethane (PU) foams in terms of thermal conductivity, which were 25.7 and 24.2 $mW\ m^{-1}\ K^{-1}$, respectively. Among them, NiMR-H aerogel with a thickness of 1.5 cm can effectively block the high-temperature heat source generated by the alcohol lamp. With the assistance of the aerogel, the intensity of infrared radiation is reduced by a factor of 1.9, demonstrating its effective infrared stealth. This stealth capability can be attributed to the presence of cavities containing well-insulated air at various scales, as well as a high thermal resistance at the interface due to a mismatch of the phonon spectrum among the heterogeneous composition.

There are fewer studies on the integral metal-based aerogel in the field of EMW absorption, and there is a great potential for the future research.

8.3 Prospect of Integral Gel EMW Absorption Materials

Integral-type gel EMW-absorbing materials have attracted much attention in the current scientific research field, and their research outlook and application prospects are very broad.

As an important EMW absorption material with excellent electrical and thermal conductivity, metal-based aerogel can be used for high-frequency EMW absorption, which is especially suitable for radar stealth and EMI fields. In the future, it is expected to further improve the EMW-absorbing properties of metal-based aerogels and expand their applications in aerospace, communication equipment, and electronic devices. Ceramic-based aerogels are widely used in microwave windows, EM shielding, and radomes due to their excellent high-temperature resistance and chemical stability. Future research directions include optimizing the microstructure and composition of ceramic-based aerogels to improve their EMW-absorbing properties and expand their applications in high-temperature environments.

As an environmentally friendly and biocompatible EMW-absorbing material, hydrogel has great potential for applications. Future research will be devoted to improving the EMW-absorbing properties and stability of hydrogels to meet their needs in the medical field, food packaging, and environmental monitoring. Ionic gels and organic gels have broad application prospects in the field of flexible electronic devices and flexible antennas due to their excellent flexibility and plasticity. Future research will focus on improving their EMW-absorbing properties and expanding their applications in wearable devices, smart homes, and flexible communication devices. In summary, the research outlook and application prospects of integral gel EMW-absorbing materials are very broad. Future research will focus on improving the performance of EMW-absorbing materials, developing new types of EMW-absorbing materials, and expanding their applications in aerospace, communications, medical, and smart equipment to meet the demand for EMW absorption materials in different fields.

References

1 Qin, M., Zhang, L., Zhao, X., and Wu, H. (2021). Lightweight Ni foam-based ultra-broadband electromagnetic wave absorber. *Adv. Funct. Mater.* 31: 2103436.

2 Liang, L., Li, Q., Yan, X. et al. (2021). Multifunctional magnetic $Ti_3C_2T_x$ MXene/graphene aerogel with superior electromagnetic wave absorption performance. *ACS Nano* 15: 6622–6632.

3 Li, Y., Liu, X., Nie, X. et al. (2019). Multifunctional organic–inorganic hybrid aerogel for self-cleaning, heat-insulating, and highly efficient microwave absorbing material. *Adv. Funct. Mater.* 29: 1807624.

4 Oh, K.I., Rajesh, K., Stanton, J.F., and Baiz, C.R. (2017). Quantifying hydrogen-bond populations in dimethyl sulfoxide/water mixtures. *Angew. Chem. Int. Ed.* 56: 11375.

5 Li, Y., Chen, X., Wei, Q. et al. (2020). Oxygen-sulfur Co-substitutional Fe@C nanocapsules for improving microwave absorption properties. *Sci. Bull.* 65: 623.

6 Song, W.L., Zhang, Y.J., Zhang, K.L. et al. (2020). Ionic conductive gels for optically manipulatable microwave stealth structures. *Adv. Sci.* 7: 1902162.

7 Cheng, H., Pan, Y., Wang, X. et al. (2022). Ni flower/MXene-melamine foam derived 3D magnetic/conductive networks for ultra-efficient microwave absorption and infrared stealth. *Nano-Micro Lett.* 14: 63.

8 Sun, Z., Chen, Y., Zheng, J. et al. (2022). Temperature-dependent electromagnetic microwave absorbing characteristics of stretchable polyurethane composite foams with Ultrawide Bandwidth. *Adv. Eng. Mater.* 24: 2101489.

9 Hu, Q., Yang, R., Yang, S. et al. (2022). Metal–organic framework-derived core–shell nanospheres anchored on Fe-filled carbon nanotube sponge for strong wideband microwave absorption. *ACS Appl. Mater. Interfaces* 14 (8): 10577–10577.

10 Zhang, X., Zhang, D., Fu, Y. et al. (2020). 3-D printed swastika-shaped ultrabroadband water-based microwave absorber. *IEEE Antennas Wireless Propag. Lett.* 19: 821.

11 Chen, X., Du, L., Jiang, G. et al. (2023). Hydrogel-based optically and mechanically manipulable broadband microwave absorber. *Nano Res.* 16 (7): 10175–10182.

12 Guo, Y., Bae, J., Fang, Z. et al. (2020). Hydrogels and hydrogel-derived materials for energy and water sustainability. *Chem. Rev.* 120: 7642–7707.

13 Tao, K., Yu, J., Zhang, J. et al. (2023). Deep-learning enabled active biomimetic multifunctional hydrogel electronic skin. *ACS Nano* 17 (16): 16160–16173.

14 Huang, W., Ding, Q., Wang, H. et al. (2023). Design of stretchable and self-powered sensing device for portable and remote trace biomarkers detection. *Nat. Commun.* 14: 5221.

15 Lu, Z., Manias, E., Macdonald, D.D., and Lanagan, M. (2009). Dielectric relaxation in dimethyl sulfoxide/water mixtures studied by microwave dielectric relaxation spectroscopy. *J. Phys. Chem. A* 113: 12207.

16 Daguenet, C., Dyson, P.J., Krossing, I. et al. (2006). Dielectric response of imidazolium-based room-temperature ionic liquids. *J. Phys. Chem. B* 110 (25): 12682.

17 Wakai, C., Oleinikova, A., Ott, M., and Weingartner, H. (2005). How polar are ionic liquids? Determination of the static dielectric constant of an imidazolium-based ionic liquid by microwave dielectric spectroscopy. *J. Phys. Chem. B* 109 (36): 17028–17030.

18 Gao, Y., Hao, Y., Zhang, W. et al. (2022). Microwave-triggered ionic liquid-based hydrogel dressing with excellent hyperthermia and transdermal drug delivery performance. *Chem. Eng. J.* 429: 131590.

19 Zhu, Y., Yang, Z., Pan, Z. et al. (2022). Metallo-alginate hydrogel can potentiate microwave tumor ablation for synergistic cancer treatment. *Sci. Adv.* 8: eabo5285.

20 Zhao, Z., Zhang, L., and Wu, H. (2022). Hydro/organo/ionogels: "controllable" electromagnetic wave absorbers. *Adv. Mater.* 34: 2205376.

21 Qian, W., Texter, J., and Yan, F. Frontiers in poly(ionic liquid)s: syntheses and applications. *Chem. Soc. Rev.* 46: 1124.

22 Li, Z., Wang, W., Chen, Y. et al. (2016). Constructing efficient ion nanochannels in alkaline anion exchange membranes by the in situ assembly of a poly(ionic liquid) in metal–organic frameworks. *J. Mater. Chem. A* 4: 2340.

23 Chen, S., Xiang, Y., Banks, M.K. et al. (2018). Polyoxometalate-coupled MXene nanohybrid via poly(ionic liquid) linkers and its electrode for enhanced supercapacitive performance. *Nanoscale* 10: 20043.

24 Bi, W., Tian, M., and Row, K.H. (2012). Facile synthesis and application of poly(ionic liquid)-bonded silica hybrid materials. *Analyst* 137: 2017.

25 Yuan, J., Mecerreyes, D., and Antonietti, M. (2013). Poly(ionic liquid)s: an update. *Prog. Polym. Sci.* 38: 1009.

26 Yi, M., Wang, M., Wang, Y. et al. (2022). Poly(ionic liquid)-armored MXene membrane: interlayer engineering for facilitated water transport. *Angew. Chem.* 61: e202202515.

27 Tang, J., Radosz, M., and Shen, Y. (2008). Poly(ionic liquid)s as optically transparent microEMW-absorbing materials. *Macromolecules* 41: 493.

28 Duan, W., Yin, X., Li, Q. et al. (2016). A review of absorption properties in silicon-based polymer derived ceramics. *J. Eur. Ceram. Soc.* 36: 3681–3689.

29 Song, L., Zhang, F., Chen, Y. et al. (2022). Multifunctional SiC@SiO_2 nanofiber aerogel with ultrabroadband electromagnetic wave absorption. *Nano Micro Lett.* 14: 152.

30 Wang, T., Lu, X., and Wang, A. (2020). A review: 3D printing of microwave absorption ceramics. *Int. J. Appl. Ceram. Technol.* 17: 2477.

31 Shao, G., Ding, C., Yu, G. et al. (2023). Bridged polysilsesquioxane-derived SiOCN ceramic aerogels for microwave absorption. *J. Am. Ceram. Soc.* 106: 2407–2419.

32 Ren, S., Yu, H., Wang, L. et al. (2022). State of the art and prospects in metal-organic framework-derived microwave absorption materials. *Nano Micro Lett.* 14: 68.

33 Liang, C. and Wang, Z. (2019). Eggplant-derived SiC aerogels with high-performance electromagnetic wave absorption and thermal insulation properties. *Chem. Eng. J.* 373: 598–605.

34 Dai, D., Lana, X., Wu, L., and Wang, Z. (2022). Designed fabrication of lightweight SiC/Si_3N_4 aerogels for enhanced electromagnetic wave absorption and thermal insulation. *J. Alloys Compd.* 901: 163651.

35 Feng, L., Wei, P., Song, Q. et al. (2022). Superelastic, highly conductive, superhydrophobic, and powerful electromagnetic shielding hybrid aerogels built from orthogonal graphene and boron nitride nanoribbons. *ACS Nano* 16: 17049–17061.

36 Zhang, Y., He, J., Fua, Z., and Yang, H. (2022). Lightweight porous NiCo-SiC aerogel with synergistically dielectric and magnetic losses to enhance electromagnetic wave absorption performances. *J. Alloys Compd.* 926: 166758.

37 Zhang, C., Wu, Z., Xu, C. et al. (2022). Hierarchical $Ti_3C_2T_x$ MXene/carbon nanotubes hollow microsphere with confined magnetic nanospheres for broadband microwave absorption. *Small* 18: 2104380.

38 Wang, F., Liu, Y., Zhao, H. et al. (2022). Controllable seeding of nitrogen-doped carbon nanotubes on three-dimensional Co/C foam for enhanced dielectric loss and microwave absorption characteristics. *Chem. Eng. J.* 450: 138160.

39 Li, Y., Liao, Y., Ji, L. et al. (2022). Quinary high-entropy-alloy@graphite nanocapsules with tunable interfacial impedance matching for optimizing microwave absorption. *Small* 18: 2107265.

40 Zhao, Z., Kou, K., Zhang, L., and Wu, H. (2022). Optimal particle distribution induced interfacial polarization in bouquet-like hierarchical composites for electromagnetic wave absorption. *Carbon* 186: 323.

41 Vakifahmetoglu, C., Colombo, P., Carturan, S.M. et al. (2010). Growth of one-dimensional nanostructures in porous polymer-derived ceramics by catalyst-assisted pyrolysis. Part II: cobalt catalyst. *J. Am. Ceram. Soc.* 93 (11): 3709–3719.

42 Vakifahmetoglu, C., Pippel, E., Woltersdorf, J., and Colombo, P. (2010). Growth of one-dimensional nanostructures in porous polymer-derived ceramics by catalyst-assisted pyrolysis. Part I: iron catalyst. *J. Am. Ceram. Soc.* 93 (4): 959–968.

43 Berger, A., Pippel, E., Woltersdorf, J. et al. (2005). Nanoprocesses in polymer-derived SiOC ceramics: electron microscopic observations and reaction kinetics. *Phys. Status Solidi (A)* 202 (12): 2277–2286.

44 Vakifahmetoglu, C. (2011). Fabrication and properties of ceramic 1D nanostructures from preceramic polymers: a review. *Adv. Appl. Ceram.* 110 (4): 188–204.

45 Liu, X.M., Yin, X.W., Kong, L. et al. (2014). Fabrication and electromagnetic interference shielding effectiveness of carbon nanotube reinforced carbon fiber/pyrolytic carbon composites. *Carbon* 68: 501–510.

46 Seong, H.K., Choi, H.J., Lee, S.K. et al. (2004). Optical and electrical transport properties in silicon carbide nanowires. *Appl. Phys. Lett.* 85 (7): 1256–1258.

47 Ren, B., Deng, Y., Jia, Y. et al. (2023). Achieving broadband electromagnetic absorption at a wide temperature range up to 1273 K by metamaterial design on polymer-derived SiC-BN@CNT ceramic composites. *Chem. Eng. J.* 478: 147251.

48 Pan, W., Liang, C., Sui, Y. et al. (2022). A highly compressible, elastic, and air-dryable metallic aerogels via magnetic field-assisted synthesis. *Adv. Funct. Mater.* 32: 2204166.

49 Liang, C., Pan, W., Zou, P. et al. (2022). Highly conductive and mechanically robust NiFe alloy aerogels: an exceptionally active and durable water oxidation catalyst. *Small* 18: 2203663.

50 Guo, Y., Chang, Q., Jia, J. et al. (2023). Lightweight magnetic metal nanochain aerogels for realizing broadband and strong absorption of electromagnetic wave. *Mater. Today Commun.* 36: 106792.

51 Zheng, Y., Yang, J., Lu, X. et al. (2021). Boosting both electrocatalytic activity and durability of metal aerogels via intrinsic hierarchical porosity and continuous conductive network backbone preservation. *Adv. Energy Mater.* 11: 2002276.

9

Thin-Film Electromagnetic Wave Absorption Materials

Bin Shi

MOE Key Laboratory of Material Physics and Chemistry Under Extraordinary Conditions, School of Physical Science and Technology, Northwestern Polytechnical University, Xi'an, China

CHAPTER MENU

9.1 Introduction

Electromagnetic wave (EMW) absorption materials with good thermal management and flexibility properties are urgently needed to meet the more complex modern electromagnetic environment, especially in the field of smart wearable electronics. The unusual characteristics of EMW metamaterials primarily depend on their shape, size, and geometric structure, rather than the composition of their constituents. By adjusting their structural parameters appropriately, it is possible to independently control the effective permittivity (ε) and permeability (μ) of EMW-absorbing metamaterials. With the ever-growing EMW pollution and the rising demand for flexible absorbing materials pasted onto the surfaces of different geometric surfaces, it is urgent to develop high-efficiency EMW absorption films with good flexibility and mechanical strength. The EMW absorption films should not only exhibit high absorption performance but also be thin, lightweight, and applicable to various substrates. Traditional processing methods endow the film with superior features that can be adapted to targeted geometric surfaces and complex conditions; however, the durability is not satisfactory. Therefore, it is of great significance to design portable, simple, clippable, and stretchable EMW-absorbing films with efficient EMW absorption for a variety of harsh application scenarios.

9.2 Film Electromagnetic Wave Absorption Materials

9.2.1 Carbon-Based Composite Films

In effect, numerous electromagnetic materials explored in recent years almost completely fulfill the performance demands and variations, especially in EMW absorption utility. However,

Electromagnetic Wave Absorbing Materials: Fundamentals and Applications, First Edition.
Edited by Hongjing Wu, Jun Luo, and Meiyin Yang.

the booming evolution of integrated and miniaturized electronic devices, especially the rise of smart wearable electronics, has put forward some higher demands for EMW-absorbing materials. Both the heat management property and flexibility have also become two indispensable and important features that should be considered simultaneously in designing EMW-absorbing materials. Among many candidates, 2D carbon materials are commonly employed in the manufacturing of EMW-absorbing composites owing to their high electrical conductivities and dielectric loss. Moreover, the unique 2D lamellar structures and high aspect ratio increase the path of EMW reflection and loss and cause multiple interlayer internal reflections and scattering. In addition, owing to ultrahigh intrinsic thermal conductivity, it is also expected to be an ideal filler with ultralightweight and high heat conduction properties.

Graphene is another candidate for use in EMW-absorbing films owing to its high dielectric property and low density. Graphene is well dispersed in solvents in the form of graphene derivatives (e.g. graphene oxide) and can be easily fabricated into films. Wang's group fabricated graphene nanosheets/aramid nanofiber (C-GNS/ANF) carbonizing films with nacre-like structures via the blade-coating/carbonization procedure to balance the relation of electromagnetic performance, thermal management, flexibility, and thickness (Figure 9.1(a)). The ingenious configuration from highly ordered alignment GNS interactively connected by a carbonized ANF network can effectively improve the thermal/electrical conductivity of a C-GNS/ANF film. Moreover, the obtained C-GNS/ANF film can be used as a lightweight microwave absorbent, achieving excellent microwave absorption performance with an RL_{min} of −56.07 dB at a thickness of 1.5 mm and an EAB of 5.28 GHz with an addition of only 5 wt%. The conductive network with finite resistance was formed in the GNS and C-ANF. When the incident EMW propagated through the resistive conduction network, the induced microcurrent rapidly decayed and was converted into thermal energy in the resistive network, resulting in the attenuation of the penetrating electromagnetic energy (Figure 9.1(b)). Different phase boundaries in GNS and C-ANF can accumulate charge, which can cause interface polarization. At the same time, the defects in GNS can serve as dipole centers and bring about dipole polarization. Both polarization processes also contribute to attenuating the incident EMW. Furthermore, the ultralightweight, flexible, and nacre-like C-GNS/ANF film also possesses hydrophobic and flame retardant properties, which are highly significant for its practical application. In conclusion, the efficient preparation method and exceptional comprehensive performance of C-GNS/ANF films provide a broad application prospect for forward-looking thermal conductivities and EMW-absorbing materials with ultrathin and flexible features [1]. Furthermore, Wu's group used different methods to modify graphene oxide (GO) and prepared composite films with polyimide (PI) as matrix, by in situ polymerization and phase-separation techniques. The structure and electromagnetic absorbing properties have been investigated. The RL_{min} of composite films reached −18 dB at the frequency of 18 GHz, and the EAB reached 6.3 GHz (11.7–18 GHz) when STAB-GO (GO has been non-covalently modified with surfactant octadecyltrimethyl ammonium bromide) content was 1%. The EAB of composite film reached 9.4 GHz (8.6–18 GHz) with the content of 0.8%. And then the GO-PANI nanocomposite has been prepared by in suit polymerization. The results indicated that long fibrous PANI sparsely and irregularly adhered to graphene nanosheets; the RL_{min} of composite films reached −28.2 dB at a frequency of 9.6 GHz, and the EAB reached 5.2 GHz (11.7–18 GHz) when GO-PANI content was 8%. The EMW-absorbing performance has improved compared to STAB-GO/PI. In order to improve the EMW-absorbing properties of composite films further, covalent modification of GO has

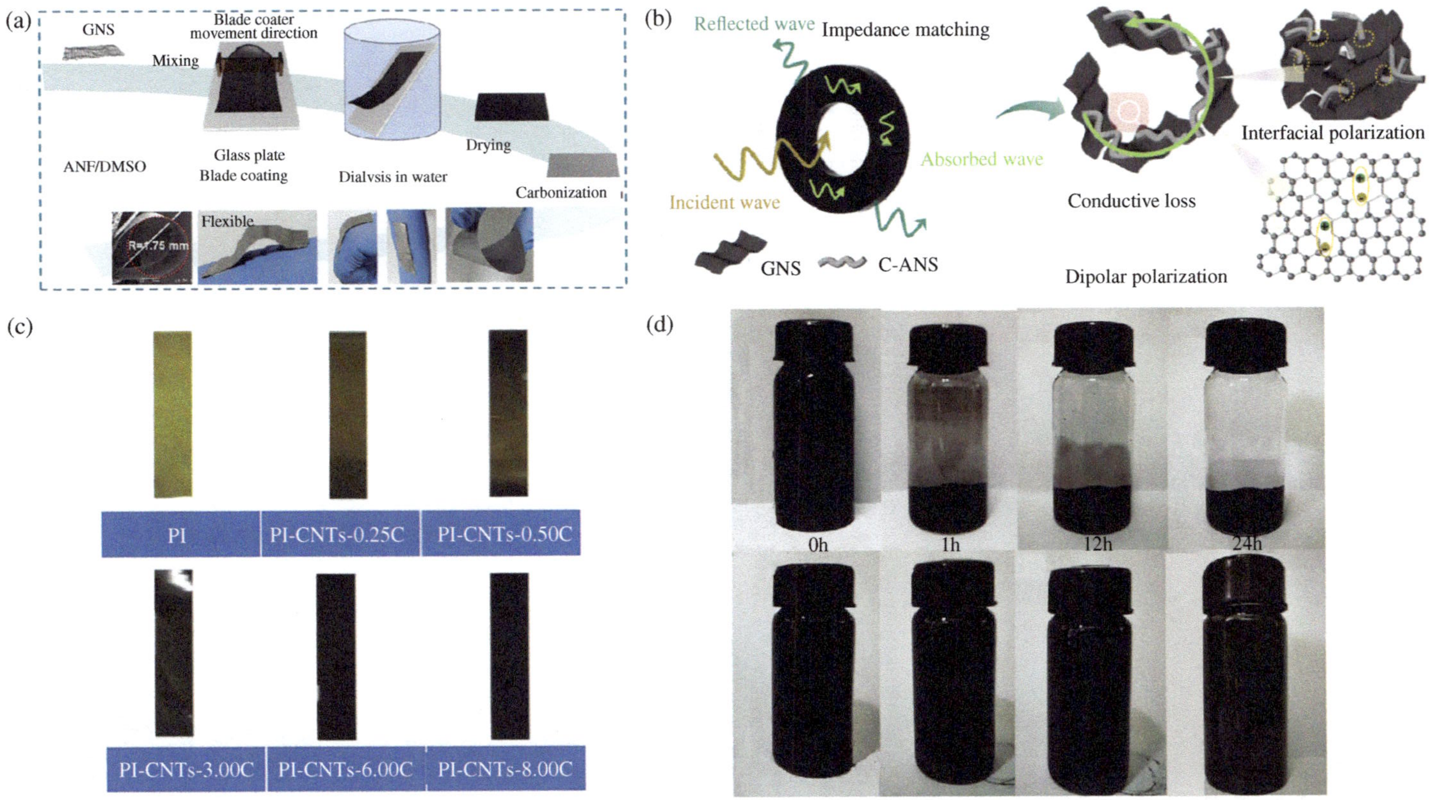

Figure 9.1 (a) Diagram depicting the fabrication and flexibility of the C-GNS/ANF films and (b) schematic diagram of EMW absorption. Source: Li [1]/with permission of American Chemical Society. (c) The images of PI films with different contents of CNTs and (d) comparison of dispersion images of CNTs before and after grinding. Source: Jiang [2]/with permission of John Wiley & Sons.

been carried out. Through covalent modification (CC) to GO, CCGO has been obtained, and PANI has been aggregated in the surface layer of CCGO, and then CCGO-PANI nanocomposite has been prepared. Finally, CCGO-PANI/PI composite films with a thickness of 0.2 mm have been prepared by in suit polymerization and a phase separation process. The composite film had the RL_{min} of −44 dB at 7.7 GHz, and the EAB reached 6 GHz (4.9–10.9 GHz) when the CCGO-PANI content was 8%. Dielectric loss of graphene and PANI was mainly an absorbing mechanism, and the neat arrangement of PANI has played an important role in improving the dielectric loss of composite membranes and multiple reflection loss. With the modification of graphene gravimetric to varying degrees, and dispersion of the nanocomposites in PI substrate have been carried out evenly by using in situ polymerization and composite films had excellent thermal stability and mechanical properties, provided the reference for ideal absorbing composites. Among carbon-based EMW-absorbing films, carbon nanotubes (CNTs) are distinguished by their surface effect, small size effect, and macroscopic quantum tunneling effect. At the same time, the special helical and chiral structures of CNTs make them have special electromagnetic effects. In addition to the above advantages, CNTs also have a high aspect ratio, thermal conductivity, and excellent mechanical properties. Niu's group prepared the polyimide (PI)/CNT films (PI-CNTs) with EMW absorption performance (Figure 9.1c,d). Thanks to the sand mill grinding that promotes the uniform dispersion of CNTs, this method is more convenient and faster compared to the previously reported dispersion methods for nano-carbon materials. PI is employed as an EMS-absorbing polymer matrix, and CNTs are the EMW-absorbing fillers. When the CNT content is 0.5 wt%, and the mechanical properties of the film are the best, the tensile strength increases from 240.2 to 281.3 MPa, and the tensile modulus increases from 4.36 to 4.90 GPa. And the thermal decomposition temperature of all films is between 580 and 601 °C, indicating that they have good heat resistance. The lowest value of the CTE is 8.32 ppm K^{-1}, and Tg of the composite film is higher than that of the pure PI film with the introduction of CNTs. The PICNTs-6.00C composite film achieves the RL_{min} value of −18.62 dB (>90% attenuation) and the EAB of 2.72 GHz with a matching thickness of only 2.0 mm, which makes the composite film an attractive candidate for an EMW absorption material. The result demonstrates a novel and effective way to synthesize PI-CNT composite films. And, more importantly, it shows that PI has a clear competitive edge in the direction of turning it into a highly efficient EMW absorption material [2]. Moreover, Quan's group studied the electromagnetic characteristics and performance optimization of reduced graphene oxide (rGO)/CNTs multifunctional composite EMW-absorbing film (CGP) and the conduction regulation and performance optimization of Ag/Cu intercalated rGO EMW-absorbing film by means of component optimization, conduction regulation, and impedance gradient design. First, the preparation and process optimization of 1D CNTs/2D rGO high-dielectric composite films were realized. GO was used as a thin-film framework, combined with high-dielectric CNTs, through chemical reduction and vacuum filtration self-assembly, to obtain CGP wave-absorbing films. The dielectric constant of the film was controlled by adjusting the ratio of rGO to CNTs, and the EMW absorption performance was further optimized. When the thickness of the CGP2 film is 1.2 mm, the EAB reached 6.5 GHz. On the basis of maintaining excellent performance, the prepared EMW-absorbing films also had high resistance strength, writability, easy processing, and flexible and foldable functions. Moreover, the electronic transmission regulation of rGO 2D materials was realized through the intercalation of nano-metals (Ag, Cu) and studied the correlation model between conduction regulation design and the electromagnetic spectrum to establish the correlation between conduction regulation, electromagnetic spectrum, and performance optimization. By adjusting the binding mode of the metal and rGO film, the conduction regulation of the control film is optimized, so as to further optimize the comprehensive absorbing performance. When the thickness of G60P-Cu film is 1.8 mm, the EAB reached 5.2 GHz. When the thickness of G60P-Ag film is 2.2 mm, the EAB was 6.9 GHz. The prepared nano-metal intercalated rGO films have significantly improved the wave absorption performance compared with CGP films while maintaining the mechanical properties of the films. Finally, an organic combination of computational model design, impedance gradient structure preparation, and broadband absorption performance optimization was implemented to realize the impedance gradient design and preparation of rGO-based thin-film materials, and the prepared multilayer wave-absorbing film showed good broadband absorption performance. According to the predesigned structure, a simple vacuum filtration assembly method, supplemented by heat treatment, prepared three layers of impedance gradient-absorbing film GCo. In contrast, the GO/Hkust-1-derived GCu film without structure design, which is synthesized by embedding the composite particles into the monolayer film, has poor absorbing performance, which also verifies the importance of structure design. Compared with G60P-Ag and G60P-Cu films, the EAB of GCo films with impedance gradient characteristics was significantly improved, and it can reach 13.6 GHz. This lays a theoretical and experimental foundation for the development and application of absorbing films. In order to improve EMW-absorbing properties of graphene-based polymer composites, modify the compatibility of graphene with the polymer [3].

9.2.2 Magnetic Metal Films

Soft magnetic metals (i.e. Fe, Co, Ni, and their alloys) have been studied for EMW-absorbing materials owing to their high magnetic properties. The high saturation magnetization (M_s) values of magnetic metals result in a high Snoek's limit and permeability in a high-frequency region, which leads to a strong magnetic loss ability. Dielectric properties and magnetic properties are the main concerns when designing EMW-absorbing films. Generally, carbon materials and magnetic materials are often utilized as the raw materials, the combination of which can achieve these two dissipations at the same time, and furthermore promote the field response of each other.

Li's group fabricated the FeCo film by the electrodeposition method and cut the film into toroids of the same size then taped them onto the surface of the absorber with molten paraffin (Figure 9.2(a)). By coating a FeCo film on the surface of the absorber, an enhancement of the reflected wave from the air–absorber interface (AAI) was obtained. Furthermore, the RL was greatly improved due to the increase in the reflected wave from AAI. The RL_{min} value reached −45.8 dB at a matching frequency of 2.7 GHz for the FSCI@FeCo film hybrid structure. The result indicated that increasing the reflected wave is an alternative method to obtain excellent microwave-absorbing property in low matching frequency range [3]. However, metal alloys have high densities, which not only increase the weight of EMW-absorbing films but also lead to low dispersion stabilities in solutions for film casting. Kim's group synthesized a free-standing EMW-absorbing film, which consisted of polyurethane and FeCoNi@graphene and based on a modified electroless plating method without any heat treatment to retain the high magnetic properties of FeCoNi and the dielectric properties and dispersion ability of graphene simultaneously. It should be noted that the synthesis of FeCoNi@graphene was performed at a low temperature (<90 °C), and the process is

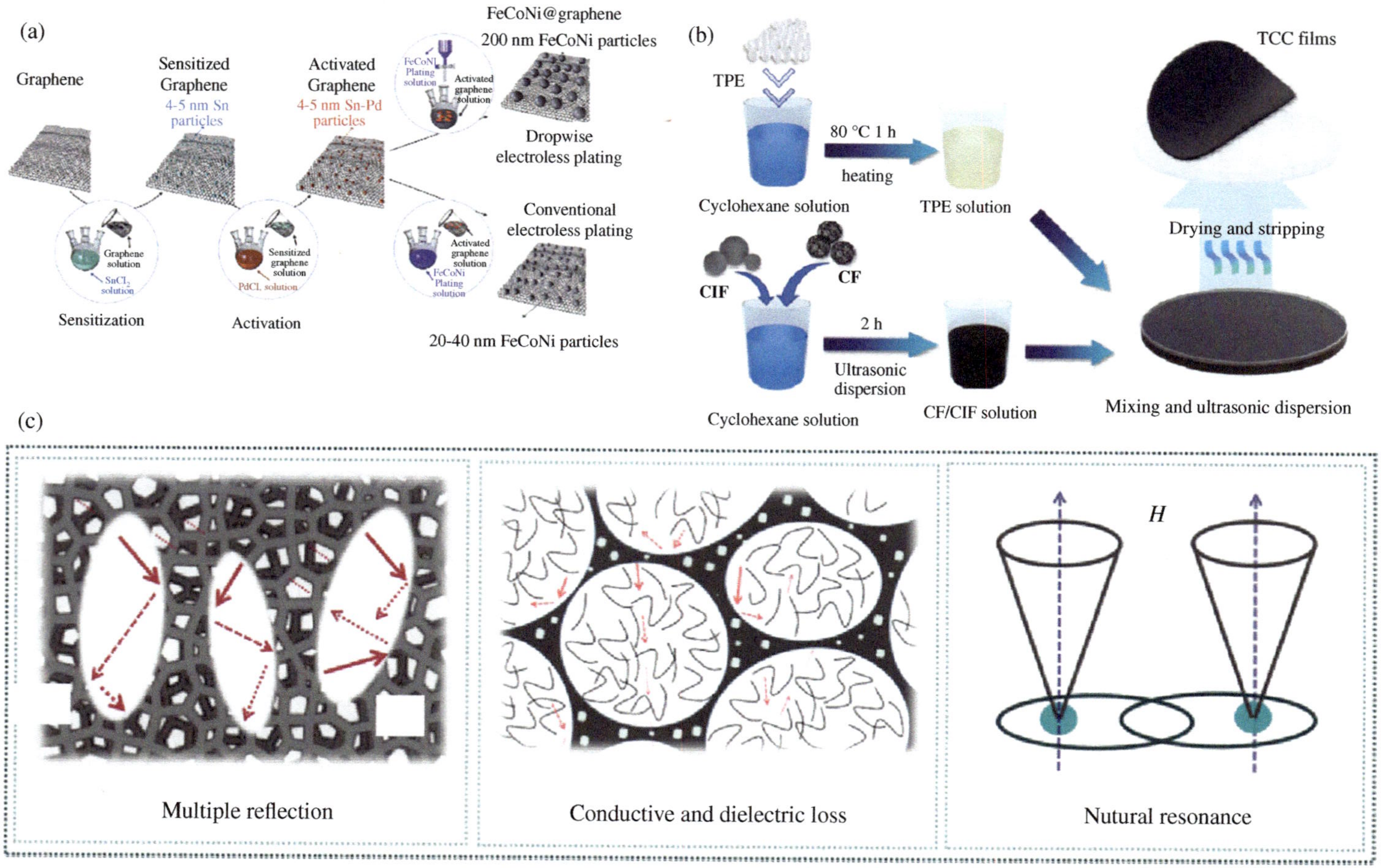

Figure 9.2 (a) A schematic representation of conventional electroless and dropwise electroless plating methods presented. Source: Kim [4]/with permission of Elsevier. (b) A schematic of the preparation processes of TCC film [4]. (c) A schematic illustration of the microwave attenuation mechanisms of porous CFs. Source: Li [5]/with permission of American Chemical Society.

environmentally friendly and scalable. The high permeability of the FeCoNi@graphene film resulted in a high EMW-absorbing performance due to impedance matching and the high magnetic properties of the hybrid material. The new hybrid material based on a FeCoNi@graphene hybrid displayed excellent performance (RL_{min} of −68 dB) and showed a higher saturation magnetization (120.4 emu g^{-1}) than that of a hybrid material prepared by conventional electroless plating (56.8 emu g^{-1}). These results provide an important advancement toward not only the commercialization of EMW-absorbing films for healthcare, electronic reliability, and tactical security applications but also the development of magnetic metal and graphene hybrid materials using a scalable process. Furthermore, the synthesized FeCoNi@graphene hybrid will promote the application of magnetic@graphene hybrids because of its high magnetization saturation (MS) and the functionality of the graphene surface. In addition, the free-standing, flexible, and highly EMW-absorbing film has great potential to advance the development of EMW-absorbing systems for stealth technologies [5]. Moreover, He's group developed the Flaky Fe-based magnetorheological elastomer film (F-Fe/MREF) composite through mixing micron-sized flaky Fe particles into a silicone rubber (SR) matrix by compression molding and vulcanization. The results demonstrated that F-Fe/MREF composites with a filler loading as low as 28 wt% present dual resonant frequency peaks and achieve an RL_{min} value of −53.3 dB and an EAB of 6.0 GHz. Furthermore, anisotropic F-Fe/MREF obtained with the assistance of an external magnetic field demonstrates an effective bandwidth of 12.5 GHz at a particular thickness. Additionally, an MR effect of 39.2% was obtained for anisotropic FFe/MREF due to enhanced interaction between F-Fe particles and the cross-linked matrix. Such remarkable improvements are interpreted as the synergistic effect of dielectric loss and magnetic loss, good impedance matching, along with the promotion of multiple reflections and scatterings. Through particle orientation under curing magnetic fields, flaky Fe particles can be well rearranged in the SR matrix, which substantially enhanced the dielectric/magnetic loss due to the large anisotropy in permittivity/permeability. Overall, this study provides an easy and magnetically controlled approach to synthesize viscoelastic and wide-bandwidth EMW absorbent, along with relatively strong MR effects, which could be a potential candidate for irregular EMW-absorbing applications [6]. The biomass-derived carbon materials have attracted much attention due to their simple processability, low cost, and interesting porous morphology for EMW absorption. Porous carbon blocks can be directly obtained by pyrolysis using the porous biomass as precursors and templates. A 3D conductive network can effectively enhance multiple reflections, resulting in repeated absorption and a highly efficient EMW-absorbing performance. Li's group used a porous polyimide (PI) film with high thermal stability as the precursor and the magnetic metallic salts of $Fe(acac)_3$ and $Ni(acac)_2$ as the magnetic component to design novel porous carbon materials exhibiting a hierarchical 3D conductive/dielectric/magnetic network structure. First, the porous PI composite films containing magnetic metallic salts were prepared via phase separation, and the derived magnetic microporous carbon composite films were fabricated by pyrolysis. Additionally, magnetic nanoparticles can be in situ fabricated by a carbon thermal reaction and uniformly dispersed within the carbon matrix owing to the homogeneous dissolution of magnetic metallic salts. In comparison with CF-20Fe, CF-20Ni shows more effective EMW-absorbing performance with enhanced absorption intensity and absorption bandwidth. When the thickness is 5.1 mm, CF-20Ni can achieve a minimum RL_{min} value of −30.2 dB at 15.0 GHz, and the corresponding effective absorption frequency range is 14.1–16.0 GHz. The EMW-absorbing performance of magnetic porous CFs can be ascribed to the synergistic attenuation effects of dielectric/conductive/magnetic loss and multiple reflections of EMWs within hierarchically 3D porous carbon networks. This work provides a novel and feasible idea for designing multifunctional and hierarchical porous carbon structures for EMW absorption [4]. Moreover, Huang's group prepared the TPE-based (thermoplastic elastomer-based) magnetic/dielectric two-phase mixed EMW-absorbing films (TCC films) by the thermally assisted homogeneous separation technology (Figure 9.2(b)). In detail, EMW response from 4.31 to 18 GHz can be realized by adjusting the thickness, and the absorption peaks move to low frequency when changing the feeding ratio. The EAB reached up to 2.95 GHz at 1.5 mm. Additionally, the abundant loss mechanisms, derived from the intrinsic loss of magnetic/dielectric loss as well as their synergetic effect (conductive pathway of carbon fibers (CF), magnetic response of carbonyl iron spheres (CIS), dielectric-magnetic coupling, and interfacial polarizations of CF-CIS mixing, dielectric polarizations at nano-micro scale), endow the TCC films with fine EMW absorption performance, such as the large effective absorption bandwidth and strong EM response at low thickness (Figure 9.2(c)). Furthermore, benefited from the chain-bridge and adhesion effect via TPE solution-based solid–liquid separation, the as-prepared TCC films exhibit good flexibility, machinability, and mechanical strength. For example, it has a low Young's modulus of 323.8 Pa and a large elongation three times their original length, which allows the film to be easily stretched and shaped as a potential wearable film. In all, this work opens up a new pathway to realize the integration of strong EM response, good flexibility, and fine mechanical properties, and the obtained films exhibit great potential as flexible EMW absorption materials that can be pasted on or peeled off the targeted surfaces at any time [7].

9.2.3 Thin-Film Materials Composite with Metal Oxides

Nickel-zinc ferrite ($Ni_xZn_{(1-x)}Fe_2O_4$) has magnetic properties and is a kind of magnetic loss absorbing agent with excellent properties, but it has some problems such as high density and poor high-frequency absorbing effect. Polyaniline (PANI) is a dielectric loss absorbent, which can be used in combination with nickel-zinc ferrite and other magnetic loss absorbers to broaden the effective wave absorption frequency band. PI has attracted much attention and research because of its excellent thermal stability and mechanical properties and can be used as a matrix resin for high-performance absorbing materials. Taking the advantages of the above materials into account, Liu's group prepared nickel-zinc ferrite/polyaniline/polyimide (NPPI) ternary porous composite films using nickel-zinc ferrite/polyimide composite filler as an absorbing agent and polyimide as the matrix by in situ polymerization and thermal phase separation technology. The addition of composite fillers has no obvious effect on the formation of imide and the crystallinity of polyimide, and the prepared NPPI has a porous structure, and a network channel is formed inside. Various performance tests and analyses show that NPPI porous composite films have good thermal stability and tensile properties. With the increase of composite filler content, the conductivity of NPPI increases, and NPPI has good dielectric and magnetic loss characteristics. When the composite filler content is 20 wt%, the NPPI porous composite film has good thermal and tensile properties. The absolute reflection loss of NPPI can reach 30.01 dB at a thickness of 4 mm, which has good absorbing performance. Because the polyimide film has a certain wave permeability, when the external EMWs enters the surface of the polyimide film, a small part of the EMW is reflected back onto the surface of the film, while most of it enters the film, and a small part of the EMW is transmitted through the film to the outside world. Due to the network channel effect inside the polyimide film, EMWs can be reflected several times inside the film, thus increasing the relaxation time, that is, increasing the residence time of the EMW inside the film. With the increase of mobile carriers inside the film, the double effect of absorbing agent and network channel can effectively improve the reflection loss of EMWs, so that the total absorption performance of the sample increases. In general, the effective reflection loss is mainly attributed to the impedance matching between the absorbing agent (nickel-zinc ferrite-polyaniline), the double loss mechanism of the absorbing agent, the impedance matching between the absorbing agent and the PI matrix, and the conductive network channel formed by the porous structure [8].

Meanwhile, in order to enhance impedance matching, which can minimize the reflection of the incident EM wave, and absorption of both electric and magnetic fields, the improved decoration and blending of magnetic metal or metal oxide particles have become effective approaches for researchers. Wang's group reported an Fe_3O_4/CNT composite film with a cratered surface structure. This film is prepared by continuously shrinking and winding a cylindrical CNT assembly, and the cratered structure formed by inducing a reaction between the introduced ferric acetylacetone and the Al substrate of the film (Figure 9.3(a)). While the pristine CNT film and the Fe_3O_4/CNT composite film with a smooth and flat surface show good microwave absorption only at large thicknesses (4 mm and above), the composite film with a cratered surface structure starts to show effective absorption over wide C and X radar bands at a thickness as small as 0.4 mm. The RL reached a minimum of −34.97 dB at 6.51 GHz, and the effective absorption bandwidth covered 5.82 GHz at a thickness as small as 400 μm. When the EMW propagates onto the surface of the film, a little EMW is directly reflected, and most of it enters the film due to its appropriate values of permittivity and permeability. Scattering is another important mechanism for the attenuation of EMWs. At the macroscopic scale, the chance of scattering EMWs is increased by the surface with a crater structure. And at the microscopic scale, multiple scattering occurs inside or between the components of the film, such as CNTs, amorphous carbon, and iron-containing particles. This is the main reason why the absorbing performance of the cratered film surpasses that of the flat film (measured by different methods). Ultimately, a minority of the EMWs passes through the film and reflects back off the standard aluminum plate. However, little electromagnetic signal can be detected by the receiving antenna. This study provides a cratered surface strategy for developing microwave-absorbing materials with thin thickness, lightweight, and strong wideband absorption [9].

High reflection loss, wide bandwidth, and low density are desirable attributes of EMW absorption materials. Among the many varieties of magnetic particles, Fe_3O_4 is unique in its large constant magnetic moment and negligible conductivity, which give rise to both magnetic loss and dielectric loss. Here, Chou's group synthesized an ultrathin MnO_2/Fe_3O_4/CNT film in two steps using solvothermal deposition and electrochemical deposition. The Fe_3O_4 nanoparticles and the MnO_2 sheets form a nano-scale core–shell structure on the surface of the CNT film. A series of this hierarchical composite film with different shell structures and MnO_2 weight ratios was successfully synthesized. Their unique microstructure greatly enhances the EMW absorption capability of the composites. Intrinsic ferromagnetism is confirmed by the magnetic hysteresis loops. The strong couplings among dielectric loss, hysteresis loss, and multiple scattering resulted in the reflection loss up to −42.2 dB with the bandwidth of 7.1 GHz at 10% MnO_2 weight ratio for a flexible composite film of 1 mm thickness. The MS values confirm that the MnO_2/Fe_3O_4/CNT composite films with different MnO_2 weight ratios exhibit ferromagnetism. The diamagnetic MnO_2 reduces the ferromagnetism. However, the composite of a 10% weight ratio of MnO_2 with the special core–shell

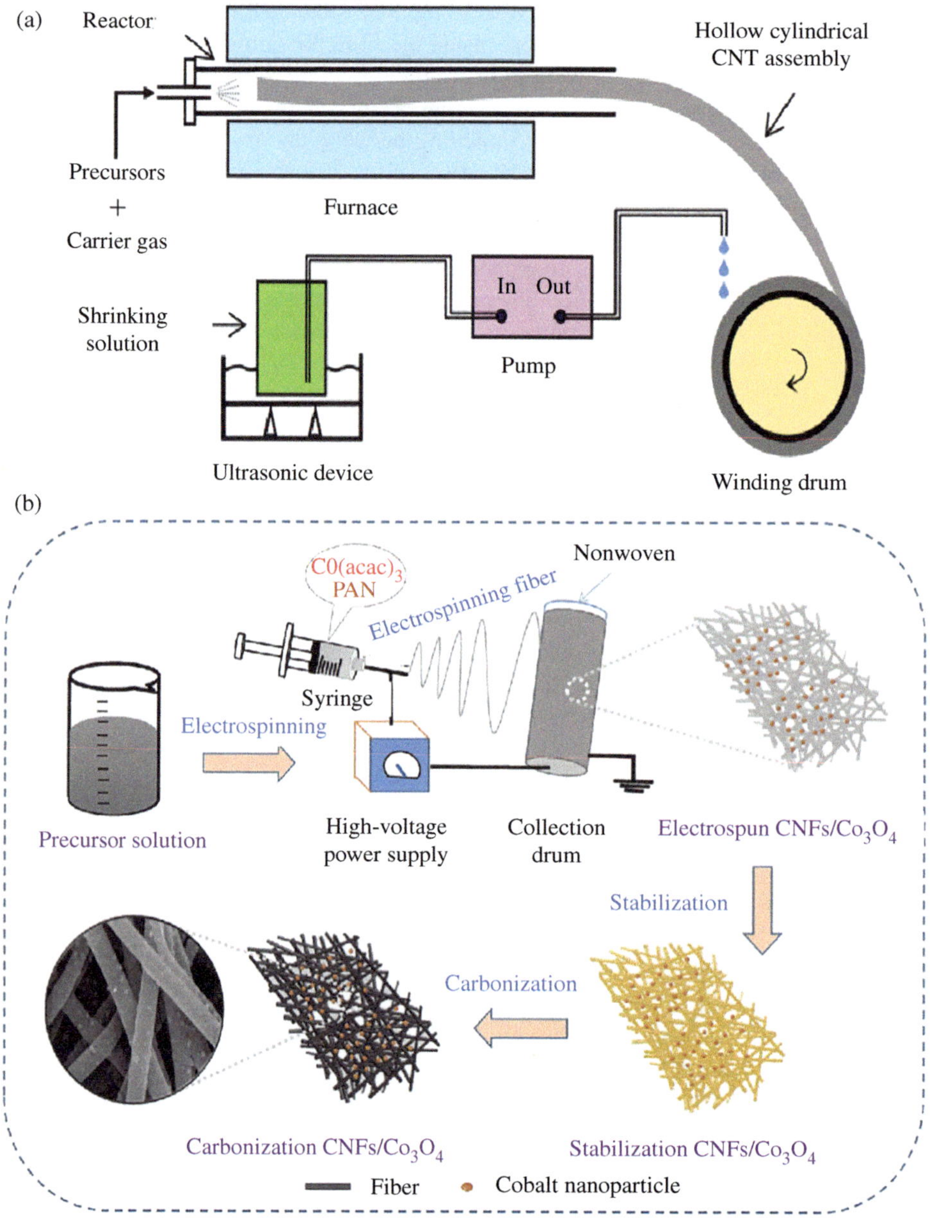

Figure 9.3 (a) Schematic illustration of the preparation process of the microwave-absorbing film from a cylindrical CNT assembly. Source: Wu [9]/with permission of Springer Nature. (b) Schematic diagram showing the synthetic process of CNFs/Co_3O_4 composite samples. Source: Abdalla [10]/Springer Nature/CC BY 4.0.

structure enhances the reflection loss. Additionally, the self-assembled ultrathin composite film with the unique microstructure increased scattering reflection and dissipated EM wave energy. Moreover, the broad bandwidth is due to the combination of dielectric loss and magnetic loss, as well as the matching impedance in the frequency range of 2–18 GHz. The CNT film, MnO_2 nanosheets, and the interfacial polarization of the hierarchical structure enhanced dielectric loss; the Fe_3O_4 nanoparticles and the Fe_3O_4/MnO_2 core–shell structure enhanced the magnetic loss. In addition, the unique cavernous structure of MnO_2 and the loosely stacked core–shell particles formed a 3D porous structure, which is beneficial for multiple scattering. Therefore, the novel design of the ultrathin and flexible MnO_2/Fe_3O_4/ CNT film may be promising for application in EM wave-absorbing composite systems [11]. Theoretically, the structural designability advantage of multilayer composites is more prominent in absorbing applications than in shielding applications. Cai's group assembled polycarbosilane nanofiber-ferroferric oxide (PCS_{nf}-Fe_3O_4) films on both sides of CNT-polyaniline buckypaper (CNT-PANI BP) by electrospinning to obtain PCS_{nf}-Fe_3O_4/CNT-PANI BP/PCS_{nf} Fe_3O_4 sandwich porous composites, with a diameter of 110–120 mm and a thickness of 150–200 μm. The Fe_3O_4 nanoparticles adhered to the PCS nanofibers formed a bamboo-like or plum-like structure. The PANI nanoparticles were uniformly embedded in the CNT network. Both PCS_{nf}-Fe_3O_4 film skin layers were well bonded to the CNT-PANI BP core layer, with two main forms of the interlayer interface of rattan-like and root-like entanglement. The sandwich composite film (SCF) with 15.1 wt% Fe_3O_4 in the skin layers had a density of 0.97 g cm^{-3}, a porosity of 61% and a total pore area of 30.1 m^2 g^{-1}, with a multimodal pore size distribution. The SCF with 11.8 wt% Fe_3O_4 in its skin layers achieved optimal dielectric and magnetic loss matching, with a minimum reflection loss of −43.54 dB and a maximum effective absorption bandwidth of 6.97 GHz. In the SCFs, a variety of absorption and loss mechanisms worked together, including repeated reflection, scattering, and absorption of EMWs at the interlayer interface; repeated scattering, absorption, and transmission of EMWs in the PCS nanofiber and CNT porous networks; polarization loss of interlayer interface and interphase interface; dipole polarization loss; the conduction loss of CNT and PANI nanoparticles; and the resonance loss of Fe_3O_4 nanoparticles. Therefore, in this study, an interlayer interface with rattan- or root-like entanglement characteristics was constructed to realize the integrated assembly of sandwich composite materials, and a well-bonded and ultra-flexible interlayer interface was obtained. The super-flexible sandwich composite absorbent has good application potential in wearable electronic devices [12].

Chou's group used electrophoresis to assemble geometry-controllable 3Dreduced graphene oxide (rGO) network layers on CNT-Fe_3O_4 films. The fabricated microstructure consists of an interconnected network of graphene and CNTs with magnetic Fe_3O_4 particles as the fast transport channel of charge carriers for high electrical conductivity. The minimum reflection loss of the CNT film-Fe_3O_4-rGO-polydimethylsiloxane (PDMS) composite with four layers is −50.5 dB, and the absorption bandwidth at −10 dB is 5.7 GHz, respectively; the thickness of this material is only 1.42 mm. Long-term cycling tests in bending and twisting deformations show >93% retention of EMW absorption after 2000 cycles. For the multilayer CNT film-Fe_3O_4-rGO composite, the rGO network on the CNT film-Fe_3O_4 composite provided not only high-dielectric loss by polarization but also high magnetic loss by eddy current, as well as the multiple reflections because of a large surface area at the layered interfaces, which facilitated the EMW absorption and broadened the absorption bandwidth. Furthermore, the CNT film was used as both substrate and dielectric absorber, and the Fe_3O_4 nanoparticles provided not only magnetic loss but also the level of impedance matching.

This research may pave the way for developing novel flexible EMW-absorbing materials and potential applications in portable electronics and wearable devices [13].

Aissa's group reported on the electromagnetic properties in the high-frequency (HF) domain of multiferroic Bi_2FeCrO_6 (BFCO) thin films. The films were epitaxially grown on $SrTiO_3$ substrates by pulsed laser ablation. Typical 50 nm-thick BFCO films having both (111) and (001) orientations were investigated. The films exhibit systematically deep EM absorption narrow bands, localized in the HF X-band domain, with an attenuation reaching as high as −24 dB. The magnitude and the shape of EM absorption depend on the crystal orientation of the film. The improved microwave absorption properties observed in BFCO (001) can be attributed to the enhancement of magnetization and leakage current density in the films. Domain walls could behave as a pathway through which electrical conduction can occur and thereby contribute to the EM attenuation. BFCO thin films show a promising potential for microwave applications as notch filters in specific frequency bands [14].

Xue's group assembled an excellent photothermal-thermoelectric polydimethylsiloxane/single-walled CNT@Fe_3O_4 film for the first time, owning superior EMW attenuation, self-healing, and low-temperature resistance by synergistically manipulating multiple factors, such as compositions, structures, and morphologies. It is found that the optimized composite film delivered a strong absorption of −60.8 dB with an effective absorption band of 3 GHz (8.2–11.2 GHz) and a small thickness (2.79 mm) in X-band, while single-walled CNTs with an extremely low content of 0.08 wt% and prepared nanocage-like Fe_3O_4 are employed to compound with polydimethylsiloxane matrix. And noticeable photothermal effect and thermoelectric effect with p/n-type are simultaneously acquired via adjusting the contents of single-walled CNTs, along with flexible hot-press/photo/electro-driven self-healing and low-temperature resistance of −19.7 °C, making it the only indirect solar power generation EMW absorption material by photothermal and thermoelectric effect with extreme environmental tolerance. The excellent impedance matching, attributing to the balance of dielectric and magnetic properties, the special nanocage-like microstructure of Fe_3O_4 and layered morphology, would encourage incident EMW to enter the composites without obviously reflecting on the surface of PSFN. Subsequently, the unique microstructure of nanocage-like Fe_3O_4 and layered structure of the PSFN would lead to multiple internal reflection of the incident EMW. More importantly, the multiple heterogeneous interfaces of Fe_3O_4 nanocage/SWCNT/polymer matrix can generate the aggregation and nonuniform dispersion of space charges at the heterointerfaces, forming a capacitor-like structure that contributes to interfacial polarization and attenuation of the incident wave's power by aligning polar bonds or charges under the alternating electromagnetic field. On the other hand, dipole polarization originated from functionalized SWCNT, and functional groups from other components would occur as well. Additionally, the 3D conductive net constructed by SFN would contribute to conduction loss via migrating electrons and hopping electrons. Finally, the existence of the ferromagnetic Fe_3O_4 endows the magnetic loss of as-prepared PSFN film in the form of eddy current loss and magnetic resonance, contributing to a remarkably improving the microwave attenuation capacity [15].

Li's group report a scalable strategy to create Co_3O_4/carbon composite nanofibrous membrane by electrospinning technique followed by stabilization and carbonization processes (Figure 9.3(b)). The electromagnetic absorbing behavior of the nanofibrous membranes was studied as EM wave absorbers in the frequency range of 2–18 GHz. The obtained samples exhibit improved absorbing performance toward higher frequencies with an increase in the cobalt nanoparticle concentration. Carbon nanofibers have reduced the magnetization and coercivity of cobalt nanoparticles, which can finally cause a decrease in magnetic permeability in composite membranes. Co_3O_4/carbon nanofibrous material, mixed with 20 wt% of paraffin, affords superior absorption performance compared with pure carbon nanofibrous material with the same content of paraffin. The optimal RL is 36.27 dB with a coating thickness of 2 mm at 13.76 GHz [10].

9.2.4 MXene Films

In 2011, the scientists at Drexel University discovered a new family of 2D materials named as MXenes. The MXenes family consists of transition metal carbides, nitrides, and carbonitrides. The general formula can be written as $M_{n+1}X_nT_x$, where M represents the early transition metal, X represents the C or N element, and T_x represents a surface-terminating group (such as —F and —OH/═O). $Ti_3C_2T_x$ MXene is a typical representative of the MXenes family with a multilayer structure, high specific surface area, hydrophilic surface, abundant surface functional groups, and adjustable electrical and chemical properties. Generally, $Ti_3C_2T_x$ MXenes are fabricated by the selectively removing metal layers from their parent Ti_3AlC_2 MAX ceramics. In addition, changes in the etching process conditions (such as etching time and etching concentration) lead to the formation of multilayered or few-layered MXene, while the conductivity of few-layered MXene is much higher than that of multilayered MXene. Therefore, few-layered MXene is confirmed to be an effective and potential material for the enhancement of MA performance. Therefore, $Ti_3C_2T_x$ MXene has been proven to be a material of high application potential in the field of electromagnetic protection.

Fan's group reported the synthesis of lightweight $Ti_3C_2T_x$ MXene/polymer (TEP, DOP, GI, and PVB) films (MPFs) and a multilayer impedance gradient absorbent by tape casting and laminated modeling methods, and the developed method is also suitable for large-scale fabrication. The $Ti_3C_2T_x$ MXene films with varying MXene content, as well as a multilayer absorber with S2.5–S17.5 films arranged in the impedance gradient order, were designed and prepared, and the electromagnetic and mechanical properties of the materials were subsequently characterized. For MPFs, the polymer phase with low permittivity and dielectric loss ability coats, wraps, and bonds the conductive $Ti_3C_2T_x$ MXene phase to form a conductive network, with a plenty of heterogeneous interfaces and multistage hole structures, thus enhancing and adjusting the electromagnetic and mechanical properties. Compared with cast films, the multilayer absorber exhibits a superior EMW absorption efficiency. For example, the ANSYS Electronics Desktop 2019 (HFSS) calculation reveals that the effective absorption bandwidth is capable of covering the whole X-band (8.2–12.4 GHz) with only a micro-adjustment of the thickness, with the minimum reflection coefficient (RC_{min}) reaching up to –26.10 dB. On one hand, the multilayer composites with excellent performance have been successfully developed in this study by laminating the cast films. On the other hand, the optimization of impedance matching has been confirmed as an efficient method to improve the absorption capacity of the materials. Therefore, by adjusting the content of the absorbing agent, the electromagnetic parameters of MPFs can be controlled, and the multilayer impedance gradient design improves the EMW absorption capacity and broadens the absorption bandwidth. In addition, this study not only proposes the effective design of a novel, multilayer impedance gradient absorber but also provides a new fundamental basis for combining the simulation technology and material design [16].

Che's group reported an electrostatic self-assembly strategy to fabricate a magnetized MXene-rGO/CoNi film. rGO with a similar structure is selected as the intercalation modifier, which destroys the continuity of the original pure MXene film and reduces the excessively high conductivity to make it more in line with the design requirements of microwave absorption materials (Figure 9.4(a)). By changing the amount of intercalation of rGO, precise control of its electromagnetic parameters can be achieved. In addition, a large number of highly dispersed magnetic alloy particles are firmly attached to the surface of the graphene so that the entire film exhibits magnetic characteristics and greatly improves the magnetic loss capability. The magnetic coupling behavior is increased from the original nanometer scale to the centimeter scale, so that the MXene-rGO/CoNi film exhibits excellent microwave absorption property with a maximum reflection loss value of −54.1 dB at 13.28 GHz [17].

However, the ultrahigh electrical conductivity of MXene results in poor impedance matching with free space, leading to the reflection behavior of EMWs on its surface due to the skin effect. To address this issue, designing vertically stacked heterostructures of MXene-based microwave absorbents was proposed. Zhou's group developed an MXene/rGO film with a novel vertical heterostructure, delivering different impedance gradients for the first time. The synthesized MXene/rGO films that are highly time-efficient and cost-saving show excellent EMW-absorbing performance. In particular, with an optimized gradient, the MXene/rGO film with 80 wt% rGO content exhibits microwave absorption with an RL_{min} value of −44.3 dB at 7.06 GHz (>99.99% of the EMW is absorbed) and an EAB of 4.84 GHz (12.78–17.62 GHz) at a thickness of 1.5 mm. This performance is markedly better than that of the best-known thin-film absorbent, and the enhanced dielectric loss capacity was attributed to vertically stacked heterostructures, providing new electron transport and leap channels along the vertical direction of the films (Figure 9.4(b)). Moreover, visualized power loss density simulations were first used to investigate the evolution of the loss densities induced by the vertical heterostructures, further confirming the synergistic interaction between MXene and rGO can substantially increase microwave absorption. The impedance matching and attenuation capability of the films can be controlled by simply changing the mass ratio and stacking order of MXene and rGO to achieve frequency-selective absorption of the films. The laminar stacking structure with a large number of heterogeneous interfaces and defect sites leads to strong polarization behavior and conductive loss capability. This finding provides an alternative approach for developing efficient, customizable, simple, and low-cost 2D material-based absorbers by controlling the formation of vertical heterostructures [18].

Moreover, Che's group prepared a flexible MXene/FeCo film by the electrostatic interaction between negatively charged MXene and positively charged FeCo alloy and via vacuum-assisted filtration. The morphology of the FeCo alloy can be controlled by the facile hydrothermal synthesis method by simply changing the molar ratios of Fe^{2+} to Co^{2+}. Benefiting from the introduction of magnetic nanoparticles, both MXene/FeCo films showed the enhanced EMW absorption performance, while the morphological distinctions led to the total differences in dielectric/magnetic loss and further microwave attenuation capability. The MXene-FeCo nanoflower-like film demonstrated better EMW absorption performance, with an RL_{min} value of −43.7 dB at 3.76 GHz, while the RL_{min} reached −36.2 dB at 10.16 GHz for the MXene-FeCo nanoplate-like film. The continuous layered structure composed of single-layered MXene stacking was destroyed due to the intercalation of the magnetic nanoparticles, resulting in the reduced conductivity and proper impedance matching. Due to the anisotropy associated with different morphologies, the rough surface of nanoflower-like FeCo alloy has better interfacial polarization and magnetic loss capabilities

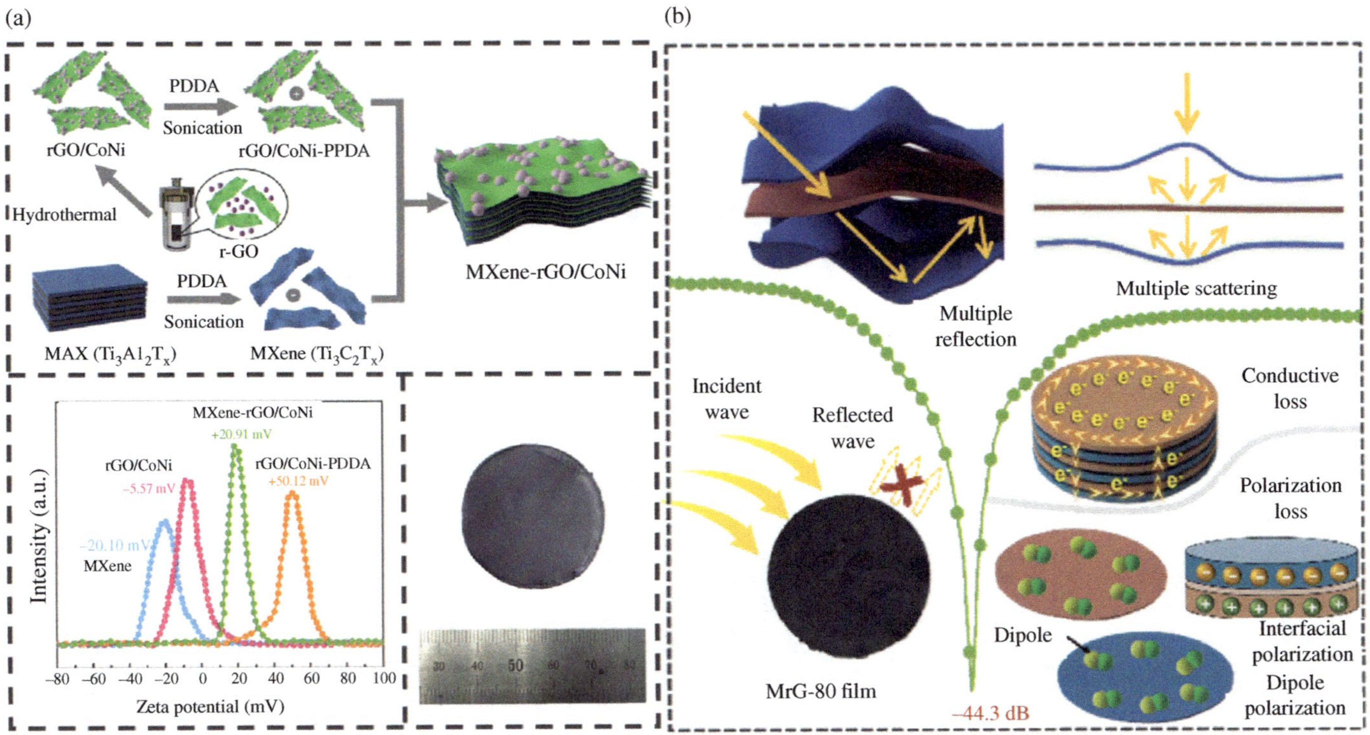

Figure 9.4 (a) Schematic illustration for the synthesis of the MXene-rGO/CoNi hybrids. Source: Li [17]/Springer Nature/CC BY 4.0. (b) Schematic diagram of the microwave absorption mechanism of MrG-80 film. Source: Li [18]/with permission of Elsevier.

induced by the more active response sites. Thus, this effective frequency adjustment also expands the practical applications of absorbers to deal with different frequencies of the outer microwave changes. It is expected that regulating morphology might shed new light on the absorbers designing with tunable MA performance. The flexible film is composed of alternating layers of magnetic FeCo alloys and few-layered MXene nanosheets, which surpass many traditional magnetoelectric composite methods [19].

9.2.5 Thin-Film Material Composite with Sulfide

Among the various responsive films, high-efficiency EMW absorption films have become more important due to less secondary pollution and better stealth performance. However, their design remains a challenge owing to impedance mismatch, insufficient attenuation, and instability at small thicknesses. In extreme conditions—such as acid/alkali and ocean environments, as well as at low/high temperatures—the multifunctionality of EMW absorption films has not yet been fully achieved, including excellent mechanical flexibility, thermal stability, and long-term stability. As such, the components and microstructures of the films must be simultaneously fine-tuned to achieve maximum absorption performance while maintaining their indispensable physical and chemical properties. Wu's group proposed a series of PI reinforced composite films with periodic copper-based electroconductive units that possess tunable electrical properties for specific applications. These periodic units optimize the electric field distribution of the films in an alternating electromagnetic field, resulting in stronger power loss. Furthermore, the hierarchical structure and abundant heterogeneous interfaces of each unit contribute to multiple scattering and polarization loss. The obtained copper mesh@ $Cu_{2-x}S$@PI (CCP) film shows excellent EM shielding performance (63.7 dB), while the Air@$Cu_{2-x}S$@PI (ACP) film delivers a superior specific absorbency value of 391.43% mm^{-1} in the X-band. The incident EMW in the vertical direction penetrated the PI shell and $Cu_{2-x}S$ layer with preliminary attenuation. Afterward, the EMW exhibited different transmission behaviors in the CCP and ACP films. For the CCP film, owing to the high conductivity of the copper core, the penetrated EMW was mainly reflected, resulting in excellent shielding performance (Figure 9.5(a)). By contrast, the free space in the ACP film incorporated into the $Cu_{2-x}S$ layer formed multiple interfaces, which maximized the propagation paths of the penetrated EMW and increased EMW attenuation inside the film. Furthermore, owing to their mechanical properties, thermal conduction/dissipation performance, and excellent environmental stability characteristics, these films offer significant potential as high-performance EMW response materials with resistance to harsh conditions [20]. In addition, hierarchical structural design has been proved to be a new strategy for fabricating high-performance microwave absorbents, so some researchers use low-dimensional components to construct continuous networks for better performance. Huang's group constructed a hierarchical core–sheath composite with ZIF-67 as an intermediate layer and MoS_2 as the outer layer to realize the EMW absorption. In detail, they make use of an 1D carbon fibers as the inner core due to their fine mechanical properties, so that ZIF-67 particles can be firmly dispersed and form uniform films. This novel strategy by adjusting the load of ZIF-67 and constructing a hierarchical structure can obtain excellent absorption performance across a wide frequency range. CPZM-1 possessed an RL_{min} value of −72.7 dB at 10.0 GHz when the thickness was 2.6 mm. Relaxation loss was mainly attributed to dipole polarization and interfacial polarization (Figure 9.5(b)). First, dipole polarization arose from defects and pores in the metal–organic framework and MoS_2 layers. Moreover, interfacial polarization was derived

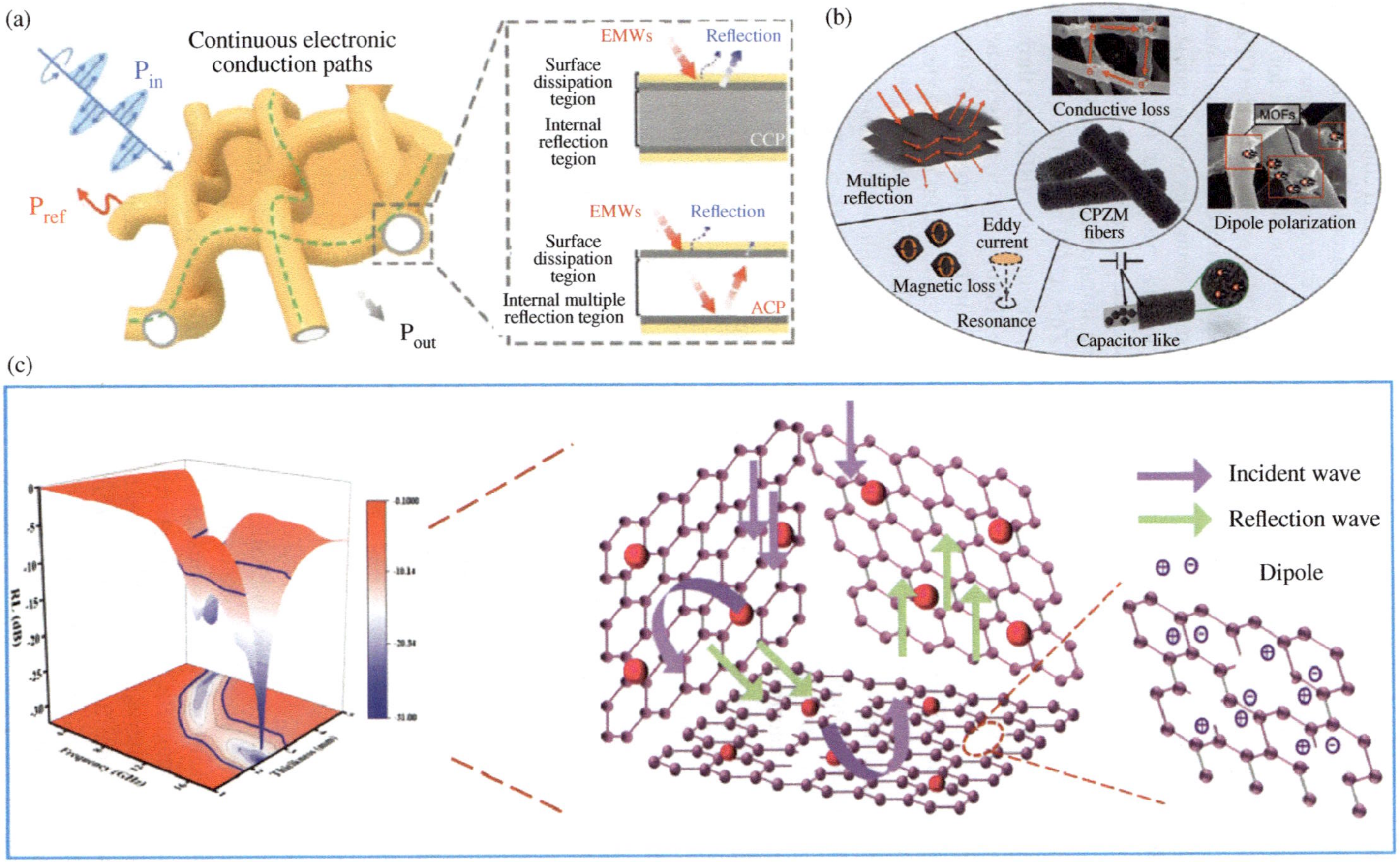

Figure 9.5 (a) Schematic diagram of incident EMW transmission paths. Source: Zhao [20]/with permission of John Wiley & Sons. (b) Schematic illustration of the improved EMW absorption mechanism. Source: Han [21]/with permission of Elsevier. (c) The schematic illustration of the possible microwave absorption mechanism for the sample. Source: Huang [22]/with permission of Elsevier.

from abundant interfaces between carbon fibers, MOF and MoS_2 layers. These interfaces between different components caused a barrier capacitance effect, which contributed to the improving absorption performance. Meanwhile, magnetic Co_3O_4 particles induced magnetic loss properties, which were mainly ascribed to natural resonance, exchange resonance, and eddy current loss. In addition, the continuous film provided more scattering sites for EM, thus increasing the propagation distance and achieving more EMW attenuation. The CPZM films as absorbers meet wearable and flexible requirements in electronics and artificial intelligence fields [21]. In addition, CuS and ZnS are p-type semiconductors which can store electromagnetic energy and have appropriate dielectric constants. Moreover, the band gap energy between S and Cu and Zn is relatively small, and surface plasmon resonance occurs in the near-infrared region, which is beneficial in reducing infrared emissivity. Ji's group used sodium alginate (SA) as the crosslinking agent, CuS/ZnS (CZS) as the precursor, and then mixed it with GO to prepare rGO-coated CuS/ZnS composite films (RGCZS) through a simple method. Different absorption characteristics can be obtained by changing the addition ratio (v) of rGO to the precursor. Compared with rGO, the infrared emissivity of the two composites decreased to 0.577 and 0.625 at 8–14 μm. The composite film has an extensive effective absorption bandwidth (f_e), up to 10.14 GHz under the filling ratio of 12%, covering the X-band and Ku-band. More importantly, the absorption bandwidth could reach 6.38 GHz with the an ultralow filling rate of 6%. The lamellar structure of RGO has a large specific surface area, which can enhance the reflection and scattering of incident waves. RGCZS is rich in dipoles and interfacial polarization, and the existence of CZS particles blocks the propagation of micro-currents, which will consume some energy when EMWs pass through CZS nanoparticles (Figure 9.5(c)). Additionally, when the content of GO is increased, the conductivity of the composite can be improved, which makes it have better dielectric loss characteristics and electron transfer ability, which is conducive to the transition of free electrons and can effectively attenuate the incident wave. Besides, the simulation results showed that the plate covered by RGCZS-2-6% has the maximum RCS reduction value of 26.3 dB m^2, indicating good radar stealth characteristic. These results are of great significance for designing and synthesizing high-performance materials for microwave-infrared compatible pollution protection. This work meets the requirements of lightweight and broadband and is helpful in exploring the relationship between transition metal sulfides and EMWs absorption characteristics [22].

9.3 The Conclusion and Prospect

This paper summarizes and discusses the research progress of flexible composite wave-absorbing thin-film materials. For the carbon-based flexible composite thin-film materials, they are introduced at five levels: composites of different carbon materials, composite of carbon materials and magnetic metals, composites of carbon materials and metal oxides, composites of carbon materials and MXene, and composites of carbon materials and sulfides. It can be seen that the flexible wave-absorbing thin-film materials are developing in the direction of multifunctional composites.

In conclusion, wave-absorbing materials are not only widely used in the military but also used increasingly in daily production and life. With the further development of new wearable devices, flexible robots, foldable devices, and flexible wave-absorbing materials are moving toward the direction of composites, multi-spectrum compatibility, intelligence, flexibility, and so on. If these flexible composite wave-absorbing materials can provide biocompatibility for wearable devices and also have a certain sensing function, they can be used for human activity or health detection. This will broaden the application of flexible composite wave-absorbing materials in new fields. In the future, the discovery of new technologies and materials will continue to promote the flexible composite wave-absorbing materials to achieve greater progress and development. The application of these materials will become more and more extensive.

References

1 Li, L. (2023). Flexible and ultrathin graphene/aramid nanofiber carbonizing films with nacre-like structures for heat-conducting electromagnetic wave shielding/absorption. *ACS Appl Mater. Interfaces* 15 (12): 15872–15883.

2 Jiang, M. (2021). Preparation and properties of polyimide/carbon nanotube composite films with electromagnetic wave absorption performance. *Polym. Eng. Sci.* 61 (10): 2691–2700.

3 Han, R. (2014). An enhancement of reflection loss by controlling the reflected electromagnetic wave at air-absorber interface. *Appl. Phys. A* 119: 201–204.

4 Kim, T. (2019). Magnetic and dispersible FeCoNi-graphene film produced without heat treatment for electromagnetic wave absorption. *Chem. Eng. J.* 361 (1): 1182–1189.

5 Li, J. (2022). In situ fabrication of magnetic and hierarchically porous carbon films for efficient electromagnetic wave shielding and absorption. *ACS Appl Mater. Interfaces* 14 (29): 33675–33685.

6 Zheng, J. (2020). Viscoelastic and magnetically aligned flaky Fe-based magnetorheological elastomer film for wide-bandwidth electromagnetic wave absorption. *Ind. Eng. Chem. Res.* 25 (8): 3425–3437.

7 Quan, B. (2023). Synergistically enhanced flexibility, mechanical strength and microwave absorption performances of TPE-based hybrid films via thermally assisted homogeneous separation technology. *Chem. Eng. J.* 206 (25): 392–401.

8 Zhang, C. (2022). Rational design and fabrication of lightweight porous polyimide composites containing polyaniline modified graphene oxide and multiwalled carbon nanotube hybrid fillers for heat-resistant electromagnetic interference shielding. *Polymer* 224: 1123742.

9 Wu, G. (2020). A novel Fe_3O_4/carbon nanotube composite film with a cratered surface structure for effective microwave absorption. *J. Mater. Sci. Mater. Electron.* 31: 11508–11519.

10 Abdalla, I. (2018). Co_3O_4/carbon composite nanofibrous membrane enabled high-efficiency electromagnetic wave absorption. *Sci. Rep.* 8: 12402.

11 Shao, Y. (2018). Flexible ultra-thin Fe_3O_4/MnO_2 core-shell decorated CNT composite with enhanced electromagnetic wave absorption performance. *Compos. Part B Eng.* 144 (1): 111–117.

12 Cai, Y. (2024). Superflexible porous PCS_{nf}-Fe_3O_4/CNT-PANI BP/PCS_{nf}-Fe_3O_4 sandwich composite films with excellent electromagnetic wave absorption performance based on integral assembly by electrospinning. *J. Alloy. Compd.* 976 (5): 173080.

13 Li, J. (2018). Flexible electromagnetic wave absorbing composite based on 3D rGO-CNT-Fe_3O_4 ternary films. *Carbon* 129: 76–84.

14 Aissa, B. (2011). High-frequency electromagnetic properties of epitaxial Bi_2FeCrO_6 thin films grown by pulsed laser deposition. *Appl. Phys. Lett.* 99 (18): 183505.

15 Wu, S. (2022). Photothermal-thermoelectric composite film with excellent self-healing and low temperature resistance properties for electromagnetic wave shielding and absorption. *Carbon* 196 (30): 163–175.

16 Ji, B. (2021). Microwave absorption properties of multilayer impedance gradient absorber consisting of $Ti_3C_2T_x$ MXene/polymer films. *Carbon* 181 (30): 130–142.

17 Li, X. (2022). Self-assembly MXene-rGO/CoNi film with massive continuous heterointerfaces and enhanced magnetic coupling for superior microwave absorber. *Nano-Micro Lett.* 14: 73.

18 Li, X. (2023). Vertically stacked heterostructures of MXene/rGO films with enhanced gradient impedance for high-performance microwave absorption. *Carbon* 208: 374–383.

19 Li, X. (2021). MXene/FeCo films with distinct and tunable electromagnetic wave absorption by morphology control and magnetic anisotropy. *Carbon* 175 (30): 509–518.

20 Zhao, Z. (2022). A flexible, mechanically strong, and anti-corrosion electromagnetic wave absorption composite film with periodic electroconductive patterns. *Adv. Funct. Mater.* 32 (15): 2111045.

21 Han, X. (2022). The flexible carbon fibers@ZIF-67 decorated with MoS_2 layers towards tunable and high-performance electromagnetic wave absorption. *Compos. Part B: Eng.* 239 (15): 109965.

22 Huang, Q. (2022). A graphene-based compatible flexible film with ultra-wideband microwave absorption and low infrared emissivity. *Comp. Commun.* 35: 101349.

Index

Electromagnetic Wave Absorbing Materials: Fundamentals and Applications, First Edition.
Edited by Hongjing Wu, Jun Luo, and Meiyin Yang.

d

e

i

l

m

Printed in the USA/Agawam, MA
September 16, 2026

910661.184